W9-BAP-301

THE DELUGE
AND THE ARK

BOOKS BY DALE PETERSON

A Mad People's History of Madness

Big Things from Little Computers

Genesis II

Intelligent Schoolhouse

COCO Logo
(with Don Inman and Ramon Zamora)

The Deluge and the Ark

THE DELUGE AND THE ARK

A Journey into Primate Worlds

DALE
PETERSON

Houghton Mifflin Company
Boston 1989

Library of Congress Cataloging-in-Publication Data

Peterson, Dale.
The deluge and the Ark : a journey into primate worlds / Dale Peterson.
p. cm.
Bibliography: p.
Includes index.
ISBN 0-395-51039-2
1. Primates — Popular works. 2. Endangered species — Popular works. I. Title.
QL737.P9P422 1989
599.8 — dc20 89-33912 CIP

Printed in the United States of America

V 10 9 8 7 6 5 4 3 2 1

For Britt and Bayne

Contents

Foreword

BY JANE GOODALL

When I was a child I was fascinated by monkeys and apes, our closest living relatives. Partly, perhaps, because I was given a large toy chimpanzee when I was two years old. But mainly, I think, because I loved to dream about the remote forests and bushlands where they lived.

Places like the great rain forests of the Amazon and the Belgian Congo (as it was then) held for me a special magic. Wild, impenetrable, utterly remote. How could we have guessed then that fifty years later so much of the world's great rain forests would have been destroyed? And so many places that were wild and untamed when I was young have changed forever. Wilderness has disappeared before the relentless advance of humans, with their needs and their greed and their frightening lack of understanding and concern for nature.

This shameless disregard for the natural world is casting an ever-lengthening shadow over the animals with whom we share, or should share, this planet. There are about 180 species of primates — monkeys, apes, and their little cousins the prosimians — in existence today: over half of these species are now threatened with extinction, and others soon will be. If we could tally up the number of individual primates who have died as a direct result of human exploitation during the half century of my life, the figure would be so huge as to be meaningless to most of us. All because one species of primate, *Homo sapiens,* the human primate, far from declining in numbers, is increasing relentlessly, year by year. And so the nonhuman primates are inexorably crowded out. If, when their homelands have been turned into wasteland, they try to share the crops grown by men as a substitute for the natural food sources that have been destroyed, the primates are shot. In some places they are hunted for food. And often adults are shot so that the young ones can be dragged from their dead or dying mothers, taken to unscrupulous animal dealers, and shipped to the far ends of the earth for entertainment or for medical research.

The Deluge and the Ark is an accurate, highly readable, and well-written story. It tells of the Deluge perpetrated by human primates as they kill and exploit the world's nonhuman primates and destroy their environment. And it tells of the Ark, the sanctuaries, reserves, and national parks where some pieces of wilderness are preserved unspoiled and where some nonhuman primates and other animals can live out their lives as nature intended.

There are two messages for us in this book. The first is a warning: if things go on as they have, then before long most nonhuman primates will be gone, along with countless other animal species. The second message carries a faint ray of hope. The reader learns about the captive breeding programs, about rehabilitation efforts and, above all, about the growing number of people who are at last becoming aware of the true situation and who want to do something to help. We learn of the successes of particular conservation projects — projects that have succeeded because of the dedicated efforts of individuals who not only understand the problems, but care enough to try to do something about them.

Only when we understand will we care, only when we care will we help. Without our help our fellow primates will be gone, and the world will be the poorer. Dale Peterson is a wonderful example of someone who cared enough to throw himself into the battle. He is to be congratulated: This book will add to people's understanding of the primates' worlds, and more people will begin to care and begin to help.

Preface

I first became interested in primates one day a few years ago, when my wife showed me a newspaper article.

The article told about a monkey in southeastern Brazil, the largest monkey of South and Central America. The monkey was sometimes called the woolly spider monkey and sometimes the muriqui, but its name was becoming irrelevant. According to the article, this once-common monkey species had lately been reduced to about 350 individuals. Reading that, I was amazed to think that an entire species might be collapsed to such a small number. Three hundred fifty individuals was roughly the size of my high school.

The article went on to describe an old man in Brazil, a coffee grower, who preserved on his coffee plantation a forest containing one of the largest remaining groups of muriquis, almost fifty animals. He had preserved that forest for most of his lifetime, but he was growing old. His sons, who would inherit the land, according to this article were considering cutting down the trees to sell as timber. But the old man was offering to sell the land to conservationists. For a down payment of a hundred and fifty thousand dollars, anyone could gain title to that land in Brazil and perhaps save the muriqui, the largest monkey of South and Central America.

The article continued with a description of the monkey and a general statement of ecological problems in southeastern Brazil, but by then my eyes had skipped off the page, and I was reading things in my mind. An entire species. Not just any species, but a beautiful and dramatic one. The largest monkey of South America. A population shrunk to the size of my high school. Nearly extinct. But someone could do something about it. For the price of a house, we might save the muriqui.

I didn't have the price of a house, but I did live in a society where people spend billions, even trillions, of dollars for objects that are supposed to fly into the air and blow up. So it seemed to me entirely possible to raise a hundred and a half thousand dollars. I located the source of that newspaper article, telephoned her, and said I would help raise the money. I said I would write a brochure about the muriqui and that forest.

After a while, I decided the brochure ought to be a book.

After several weeks of reading and studying and some writing, however, I learned that the situation of the muriqui is not at all unique. I learned that of the approximately 180 species of nonhuman primates in the world, roughly a third to a half are threatened and declining toward extinction with various degrees of rapidity, including not only the largest of South and Central America, the muriqui, but the largest in the world, the gorilla; the smartest, the pygmy chimpanzee; and the most beautiful, the golden lion tamarin. The decline of the primates, I learned, is a global, not a local, event, so I determined to write a global book. It seemed to me that the story of an isolated species such as the muriqui or any of several other highly endangered primates might make a compelling and coherent and dramatic story. But I wanted to lead both myself and my reader into a solid understanding of the entire story: why the primates as a group are dying and what we can do to save them. I wanted enlightenment. The decline of a single species is detail without pattern. I wanted to move through detail into pattern.

And so I began writing *The Deluge and the Ark.* I acquired an entrance ticket to Harvard University's great library at the Museum of Comparative Zoology and began what would turn out to be three years of research. I learned about the Deluge, which consists partly of subsistence hunting of primates in the tropics and partly of the international trade in live primates, but is largely driven by our century's unprecedented destruction of tropical forests. And I learned about the Ark, which includes national parks and other sorts of forest protection, zoos, and ultimately the earth itself.

After some study and consideration, I decided it would be most strategic to concentrate on the stories of individual species to represent the larger group of primates. So I selected twelve species (in some cases, subspecies) to represent the primate order in terms of geography, type, problem, and solution. I selected, in other words, species from each major geographical location, species to represent the major types of primates — monkeys, apes, and prosimians — and species that would illustrate particular aspects of the Deluge and the Ark.

And one day I had a manuscript, based largely on my research at the Harvard Museum of Comparative Zoology library. I still hadn't seen most of those primates, however. I had tried to see them in my mind. I had read everything I could read, looked at pictures, seen movies, seen stuffed animals in glass cases, gone to zoos. But I realized then that I actually had to go out and see my primates. I don't particularly like to travel, but one day I realized I had to travel around the world and find my twelve representative species.

This book is based upon that journey around the world, into the Deluge and onto the Ark, in search of the world's endangered primates. It was an exciting journey for me, one of the great adventures of my life, and I have styled this book in part as a travel narrative to communicate some of that excitement and adventure. But travel without knowledge is a fleeting movement through surface reality, providing a patina of present impressions. To know, really know, the present, one has to know the past and have some structure for speculating about the future. Thus I have also included in this book material from another journey, my journey of research and understanding. Why are the primates declining? What can we do about it?

We live in a world of disasters, most of them growing from a single root: our century's explosion of human populations with expanding technological powers, an explosion that is continuing in the third world. We count the disasters and potential disasters — the greenhouse effect, the ozone hole, pollution of lakes and rivers and oceans, acid rain, the threat of nuclear holocaust — and add to that list the destruction of tropical forests and the concomitant decline of the primates. Some people prefer to ignore or deny such problems and thereby feel better, but for those of us who choose to look at the present and think of the future, the vision of these emerging global problems is disturbing indeed — sobering, numbing, or depressing. And the story of the Deluge may also be depressing. But what has elevated my spirits during the journeys of this book is the story and concept of the Ark. Primates *are* being saved; people *can* act with wisdom and compassion.

I think of the whales, that other major group of large-brained mammals. Two or three decades ago several whale species were rapidly moving toward extinction as whalers from several major nations pursued them with great and destructive technologies. But the sheer concern of ordinary people, their identification with whales and their desire to affect a powerful historical trend, has altered that situation. Today only Japan, Norway, and Iceland continue to massacre whales, and some of these great and beautiful species have been reprieved, drawn back from the edge of extinction. They may survive.

I believe that great and organized public concern can move in similar ways to save the primates. And I hereby declare this last decade of our millennium as the first Decade of the Primate. We can save the primates, and in doing so we can begin to save ourselves, the primate most greatly burdened by its capacity to destroy and to create.

Writing is a lonely obsession, by and large, but this book has been written with the occasional companionship and regular guidance of many people who have generously shown me the way into the Deluge and onto the Ark. I wish particularly to thank Anbalagon of Tirunelveli (India), Dr. Anthony Anderson of Museo Paraense Emilio Goeldi (Brazil), Professor Lynne Ausman of Tufts University and the New England Regional Primate Center, Dr. José Marcio Ayres of Museo Paraense Emilio Goeldi (Brazil), "Baiano" (Ubaldino Vieira Lima) of Base 4 at Tucuruí Reservoir (Brazil), Dr. Andrew Baker of the Poço das Antas Biological Reserve (Brazil), Maurice Ratsizakanana of Andasibe (Madagascar), Dr. Kurt Benirschke of the San Diego Zoo, Dr. Richard Bierregaard of the Minimum Critical Size Project in Amazonia (Brazil), José Eugenio Binder of Centro de Proteçao Ambiental in Electronorte's Permanent Village (Brazil), Roger Birkel of the St. Louis Zoo, Dr. Diane Brockman of the San Diego Zoo, Admiral Ibsen de Gusmão Câmara of the Brazilian Conservation Association (Brazil), Dr. Suzanne Chevalier-Skolnikoff of the University of California at Berkeley, Dr. Aldemar Coimbra-Filho of the Rio de Janeiro Primate Center (Brazil), Demondiny of Maroansetra (Madagascar), Dr. Betsy Dresser of the Cincinnati Zoo, Abdala Miguel Feliciano of Fazenda Montes Claros (Brazil), Dr. Lester Fisher of the Lincoln Park Zoo, Dr. Nathan R. Flesness of the Minnesota Zoo, Gale Foley of the San Diego Zoo, Dr. Thomas J. Foose of the American Association of Zoological Parks and Aquariums, Laurence Gledhill of the Woodland Park Zoo, Professor Steven Green of the University of Miami, Dr. Ronald Hunt of the New England Regional Primate Center, Dr. Andrew Johns of Danum Valley Field Studies Center (Malaysia), Dr. Lorna Johnson of the New England Regional Primate Center, Professor Alison Jolly of Rockefeller University, Marvin Jones of the San Diego Zoo, Dr. Michael Kavanagh of the World Wildlife Fund–Malaysia, Dr. Devra G. Kleiman of the National Zoo in Washington, Dr. Duane C. Kraemer of Texas A&M University, Professor Lois Lippold of San Diego State University, Dr. Shirley McGreal of the International Primate Protection League, Jack McHenry of the International School of Kenya, Rita Mesquita of the Minimum Critical Size Project in Amazonia (Brazil), Marion Meyer of Centro de Proteçao Ambiental at Electronorte's Permanent Village (Brazil), Minah of the Tiwai Island Wildlife Sanctuary (Sierra Leone), Dr. Russell Mittermeier of the World Wildlife Fund, Dr. José A.P.C. Muniz of the Centro Nacional de Primatas in Belém (Brazil), Dr. Norman Myers (England), Professor John Oates of Hunter College, Agustus Nadeak of Padang (Indonesia), Antonius Napitupulu of Muara Siberut in the Mentawai Islands (Indonesia), Dr. Francine Patterson of the Gorilla Foundation, Dr. Andrew Petto of the New England Regional Primate Center, Pedro Pimentel of Base 4 at Tucuruí Reservoir (Brazil), Dr. Laurenz Pindar of the Poço das Antas Biological Reserve (Brazil), Dr. Earle Pope of the Cook Springs Primate Facility, Karlo Saddeau of Rok-Dok village in Siberut (Indonesia), Professor Charles Snowden of the University of Wisconsin, Expedito Pereira Sobrinho of Base 4 at Tucuruí Reservoir (Brazil), Professor Robert Sussman of Washington University, Dr. Suzette Tardif of Oak Ridge Associated Universities, Dr. Linda Taylor of the Duke University Primate Center, Professor Richard Tenaza of the University of the Pacific, Anne Todd of the Tiwai Island Wildlife Sanctuary (Sierra Leone), Professor Celio Valle of Universidade Fédérale de Minais Gerais (Brazil), Eduardo Marcelino Veado of the Caratinga Biological Station at Montes Claros (Brazil), Dr. Amy Vedder of the University of Wisconsin, Warden S. Viswanathan of the Kalakad Sanctuary (India), John Waugh of the U.S. Peace Corps (Sierra Leone), Professor Bruce Wilcox of the Center for Conservation Biology at Stanford University, Dr. Marina Wong of the Minimum Critical Size Project in Amazonia (Brazil), and Dr. Pat Wright of the Duke University Primate Center. I must doubly thank four people who were kind enough to read and criticize significant parts of my manuscript: Dr. Norman Myers, Professor John Oates, Professor Richard Tenaza, and Dr. Amy Vedder.

I should also acknowledge my special reliance on four significant "guidebooks": Michael Kavanagh, *A Complete Guide to Monkeys, Apes and Other Primates* (1983); Ramona Morris and Desmond Morris, *Men and Apes* (1966); Norman Myers, *The Primary Source* (1984); and Jaclyn Wolfheim, *Primates of the World* (1983). My guide into the jungle of North American publishing, Peter Matson, deserves particular gratitude, as do my guides at Houghton Mifflin, Harry Foster and Peg Anderson.

I am grateful to the Arlington Arts Council of Arlington, Massachusetts, for providing encouragement and actual assistance in the form of a Massachusetts Arts Lottery Grant. I am also grateful for an Arlington Literary Competition Award, sponsored by the Arlington Arts Council. I am most indebted, figuratively far more than literally, to two individuals who so very generously lent me enough money to travel around the world: my brother Lt. Col. Dwight Peterson and my friend Susan Avila. Dwight and Susan were my biggest and most trusting backers at a crucial time in the writing.

My wife, Wyn Kelley, and my children have, as always, provided steady delight, pleasure, and consolation. Dr. Kelley doubles as my favorite critic and fiercest defender, and to her I am grateful beyond telling. To my children, I dedicate the book.

Make yourself an ark of gopher wood; make rooms in the ark, and cover it inside and out with pitch. This is how you are to make it: the length of the ark three hundred cubits, its breadth fifty cubits, and its height thirty cubits. Make a roof for the ark, and finish it to a cubit above; and set the door of the ark in its side; make it with lower, second, and third decks.

GENESIS 6:14–16

And the waters prevailed so mightily upon the earth that all the high mountains under the whole heaven were covered; the waters prevailed above the mountains, covering them fifteen cubits deep. And all flesh died that moved upon the earth, birds, cattle, beasts, all swarming creatures that swarm upon the earth, and every man; everything on the dry land in whose nostrils was the breath of life died.

GENESIS 7:19–22

THE DELUGE AND THE ARK

One

WORLDS

ONE JANUARY DAY I left my home in Arlington, Massachusetts, to find twelve of the world's most endangered primate species. I kissed my wife and two children, then walked down a hill, punching holes in the new snow with my new leather boots, balancing an old military surplus pack on my back. Snow the consistency of wool settled into my hair, eyebrows, beard, and a sharp wind penetrated my thin windbreaker.

I sat on an airplane, which roared and tilted and burst into the sky. I read, listened to music, watched a movie, slept. Time passed in and out of dreams until I woke to see below me thin morning light washing a hornet's nest, a puffy gray and layered sky, into which a hole was scratched, through which I descended on a cotton rope, entering at last a hot world where crowds of people vibrated impatiently and babbled impenetrably. Thunk: my visa was stamped.

Then I was breathing steam, crushed in a crowded bus, rolling across fields of concrete amid shuttling steel, through slums, past factories and fences, cranes and ships, until the bus deposited me on a sidewalk somewhere, and I began to ask directions, in Spanish, for the Hotel Aeroporto. No one understood Spanish, but eventually I determined that the hotel must be over there, across that overpass, on the other side of that highway. I walked across the overpass, past seedy bums and sinister loiterers, along a broken sidewalk, and at last penetrated the hotel entrance. It was a cheap hotel, but it looked tolerable, and the clerk spoke a little English. I handed him a large-denomination cruzado note. "For me?" he said with a big smile and a look of gosh. "No, it's not for you. I would like some change, please."

A porter carried my bag for me. I was hot and tired. We took the old elevator up to the fifth floor, came at last to my door, my room. Inside, it stank and steamed. The single window provided a view of brick. The porter turned on the rattling air conditioner and the television. I tipped him, and he left.

Samba music began to emerge from the television, and on the screen a snowy image in black and white slowly assembled itself. A dancer in G-string and pasties waved her crotch at the camera. The camera flashed to ankle level and moved up between the dancer's legs, closer, closer. A fat man's leering face and porous proboscis appeared on the screen, and he cackled and chattered in Portuguese. Then more dancers in G-strings. I switched channels to find more samba dancers, a soccer game, and two channels of crazily violent cartoons. I turned it off.

I looked at myself in the mirror and asked the standard questions, such as *What am I doing here?* I washed, rearranged my possessions, put my money and passport in a belt, and left the hotel for a walk. I crossed the highway again and eventually made my way over to the sea. I stood on the edge of garbage heaps and rubble, and began noticing all the shacks and people on the garbage and rubble. A black man wearing only cotton underpants stood next to rusted metal and an old bedspring; near him sat a woman tending stew bubbling in a tin can over a wood splinter fire. The man smiled and motioned me over. I smiled back but declined the invitation and continued walking. I looked out beyond the shacks and rubble and saw, across water, a famous mountain shaped like a breast, Sugarloaf.

I continued along the water until I came to the beaches and walked next to waves green and clear that poured and hissed onto sand. I was still wearing my leather boots — I hadn't thought to bring lighter footwear on this trip. I also hadn't thought to bring shorts. I was wearing long pants, as well as an undershirt and a long-sleeved shirt. I felt hot and obvious. Obvious: I looked up and down the beach and saw no one — no one! — wearing so many clothes or looking so uncomfortable. The handkerchief in my pocket I could have used for a bathing suit on this beach. The shirt on my back, enough cloth to provide for three dozen bathers here. I was drenched with sweat, swollen with heat. A woman walked by wearing a loose bathing suit, arms up, carrying a wooden beach chair on her head. One of her breasts drooped out of its loose cloth cup. She noticed, and casually pushed it back in. Other bathers lounged on the beach in suits constructed from dental floss and chewing gum. Some of them lifted barbells or comic books, but most simply sprawled under the sun. They baked and browned, displayed and mingled, kissed and fondled.

I returned to my hotel room, turned the air conditioner on high, and tried to sleep. Hours later I left my room again and entered the elevator. It was evening then, but still unbearably hot, even hot for Brazilians, it seemed, since the two couples in the elevator with me were dramatically waving their hands in front of their faces. One of the men said to me, "Calor!" which made me very excited because it was the first word of Portuguese I understood. "Calor!" I repeated and waved my hand in front of my face.

By the time the elevator reached the bottom floor, the two couples had invited me — entirely through pantomime — to have dinner with them. I accepted, and the five of us left the hotel together. This part of town was deserted, and we walked along dark and desolate streets for some time before catching a bus that careened many miles into the Copacabana district. We waded through great crowds of people and finally found a restaurant with an empty outdoor table where we ordered beer and Coke and things to eat.

The table was just across a street from the beach. The sky was black, the sea black, the beach dark, but crowds of people swarmed over the sand. A wind combed the high fronds of palms along the beach. But the air was hot and so moist I felt underwater.

We ate and drank and talked. The two couples with me talked in Portuguese and laughed, and I talked to them in Spanish. I've just come from Boston, just yesterday, I told them, where it was snowing. Snow falling down. I made snowing motions with my fingers. I'm here in Brazil to look at endangered primates — monkeys. You understand? But they understood nothing. And even though I listened so intently, like a dog before steak, to every word spoken, I understood very little. They were in Rio on vacation from somewhere else in Brazil. One of the men was an auto mechanic; one of the women was a domestic. Meanwhile, sidewalk merchants and beggars approached us. People tried to sell us ships in bottles, spinning light makers, blankets, castanets, magazines, peanuts. A woman with a baby asked for money for her baby. A little girl offered flowers for sale and asked if she could finish a half-drunk Coke.

On the bus back to the hotel, one of my companions offered me a stick of chewing gum. I accepted and began chewing it. Then I noticed they were all laughing at me. One of the men stuck his tongue out and pointed at it, laughed, and handed me a small mirror. I took the mirror and looked at hairy face punctuated by black tongue. I laughed: practical joke chewing gum.

The primate world I had come to see began about 65 million years ago — but let's look back a little further for a moment. Around 250 million years ago, according to fossil hieroglyphics, the earliest mammalian forms appeared, having evolved from reptiles. Those early mammals were furtive little creatures that hid and scurried about, but for a long time never developed into the larger mammals we recognize today, possibly because 225 million years ago one group of reptiles, the dinosaurs, came to dominate most environments. Some dinosaurs became extraordinarily successful and plentiful. During their 160 million years of dominance they evolved into a multitude of manifestations, then suddenly, 65 million years ago, they disappeared, leaving only fossilized footprints, bones, teeth, eggs, dung, and several small reptilian descendants by which to be remembered.

Upon the demise of the dinosaurs, the trees presented an open and attractive environment, a relatively unoccupied niche for mammals. Offering food — fruits, nuts, leaves, and insects — and safety from ground-dwelling predators, the trees became an open theater for primate adaptation.

The concept of adaptation to the trees is simple. Simple enough, actually, to illustrate the general process of adaptation and evolution described by Charles Darwin more than a century ago. Darwin's theory begins by recognizing two facts of life: that parents tend to produce an excess of offspring and that the offspring are diverse — they differ in strength, build, aggression, intelligence, and so on. The excess ensures that a group of plants or animals will survive over time, even though some members of the group die before reproducing. Diversity means that some members of the group are better suited to their environment, tend to survive longer, and therefore tend to reproduce more. In other words, those individuals best suited to the environment tend to pass on their distinctive genetic packages to succeeding generations. Over time, then, over hundreds, thousands, or millions of generations, the collective genetic definition of a group alters according to the pressures of the environment.

In the case of animals inhabiting the trees, one particular environmental pressure is terribly obvious. *You can fall.* Those animals that by virtue of even very minor differences are less likely to fall are more likely to stay alive, to reproduce, and to pass on their genetic proclivities to succeeding generations: over great time, an adaptation in bodily form to the problem of gravity in high places. Most primate species alive now still live in the trees. Yet even those that have left the trees — the ground dwellers, including most baboon species of Africa, many macaque species of Africa and Asia, gorillas, and humans — still carry, in body and behavior, certain qualities that recall the history of that long adaptation. What are those qualities?

First, the grasping hand. The primates retained an existing mammalian structure, the paw or hand with five digits, but the digits became longer and more flexible. In most primates the claws at the ends of the digits have flattened and become nails that protect the sensitive fingertips. A newborn tree dweller needs protection from falling; and in the case of the primates, newborns possess already well developed hands and a remarkably strong grasp. Infants of most primate species can grasp strongly enough that the mother is free to move through the trees with the newborn clinging to the fur of her stomach and chest or to her back.

Second, precise and three-dimensional vision. The eyes of most mammals are located on either side of the head, providing a wide sweep of vision with little or no wasteful overlap in visual field from one eye to the other. In primates, however, the eyes moved (over millions of years) closer together toward the front of the face. The closeness means much visual overlap; one eye sees much of what the other sees. But while the overlap narrows the primate's full visual field, a disadvantage, it also provides a major advantage for tree dwellers: heightened depth perception. The crucially important eyes are protected by ridges of bone around the sockets. And that portion of the primate brain associated with vision is larger than in other mammals.

Third, a minimal sense of smell. Smells linger on the ground, which retains dampness, but smells above the ground quickly dissipate. Thus most, but not all, primate species have smaller and less sensitive noses (with fewer olfactory neurons) than other mammalian types. That portion of the brain associated with smell has also decreased in primates.

Fourth, a large brain and behavioral flexibility. Perhaps at first the expanded brain provided better hand-eye coordination, crucial for survival in the trees. But a larger brain also allowed for behavioral flexibility, which meant both an increased ability to survive in the complicated, three-dimensional world of the trees and an ability to react to changes in the environment, such as seasonally changing food sources. The unusually long period of dependence in the young, a time when the adults earn and the young learn, also promotes flexibility. This extended dependency, in turn, encourages complex, familylike social groups.

The grasping hand, three-dimensional vision, minimized smell, and large brain are primate characteristics that have been obvious for some time, and can easily be measured. Behavioral flexibility, on the other hand, is a quality we have only begun to appreciate in the last few decades, since scientists have begun to study primate species in the wild.

The nonhuman primate world of our time endures mostly in the forests of the tropics, which encircle the middle of the planet as a green and fragmented belt. Zoological classifiers divide the primates into approximately 180 species, of which about 50 live in the New World, the tropical forests of Central and South America.

South America is nearly an island, connected to its northern sibling by a very narrow isthmus, and separated by a vast ocean from the Old World primate region, an expanse of connected and almost connected land from western Africa into Asia and the islands of Southeast Asia. The wide separation of the New World from the Old World accounts for one notable fact about primate distribution. Although all the primates are descended from a single ancestral line that began roughly 65 million years ago, the New World primates have evolved separately from their Old World relatives for around 40 million years.

No one is quite sure how the first ancestors of today's New World primates arrived. A few fossils of very primitive primates have been discovered as far north as North America; perhaps today's Central and South American primates descended from these North Americans (themselves emigrants from Africa), which over millennia may have rafted from island to island across the body of water that at the time separated North and South America. Or the New World primates may have come more directly from western Africa, slowly drifting or storm-driven on rough rafts of flotsam across the ocean, relatively narrow then, separating Africa from South America. In any event, fossils indicate that the first obvious ancestors of today's New World primates arrived in South America about 40 million years ago. The New World primates then evolved to fill the wide spectrum of forest environments on the South American continent and diverged to become the approximately 50 distinct New World species — all monkeys — that we know today.

While the New World monkeys living today are the result of 40 million years of evolution in their hemisphere, the last forty years, the last millionth of their evolutionary history, have brought disastrous change. On average, the human populations of Central American nations are now doubling every twenty-seven years, and those of tropical South American nations every thirty-one years. Nearly everywhere in the hemisphere's tropics, the primary forests are transected by roads, invaded by slash-and-burn farmers, and cut down, often wastefully, to accommodate the needs of uncontrolled human expansion. Primates in many areas are threatened by unprecedented habitat loss, and in some places they are hunted for food. Fully half of the New World monkey species are declining to some degree and ultimately threatened with extinction.

I LEFT Rio de Janeiro, flew north a few hundred miles to the city of Belo Horizonte, took an all-night, standing-room-only bus ride farther north to the town of Caratinga. I rested there on a cot at the ABC Hotel for a few hours the next morning, nearly electrocuted myself in the hotel's shower, then rode east along a rough dirt road in the back of a pickup truck, and arrived that afternoon at a coffee plantation and its small forest in the Brazilian state of Minas Gerais.

That small forest is an Ark, less than four square miles' worth, in the middle of Fazenda Montes Claros, a coffee plantation owned by a wonderful old man, Abdala Miguel Feliciano. I had read a good deal about Senhor Feliciano and Montes Claros and its endangered monkeys, had even sat through a movie on the subject (*The Cry of the Muriqui*, filmed by Andrew Young), so when I arrived at the farm and looked up to see Senhor Feliciano leaning out the big open window of his farmhouse, he and his farm seemed as familiar as an old dream.

I walked up the wooden steps to the house. The patriarch (bald, tall, barrel-chested) greeted me with a crushing handshake. I entered the farmhouse and looked around: bare, cracked plaster walls; a calendar and a few tattered pictures tacked here and there; a cobwebby, roof-bottom ceiling; a single bare light bulb hanging from a wire. Like most of the rural houses I saw in that part of Brazil, the windows were square openings in walls. No glass or screens obstructed the motion of air and insects.

We passed into the kitchen. A chicken ran out the back door, and Feliciano's housekeeper came in. A pot of water was simmering over a wood fire on the stone-block stove. We went out a side door onto an open porch, and I was invited to sit at a wooden table. Feliciano pulled two saucers and two cups out of a screened cupboard and placed them on the table. The cups and saucers were covered with swarms of tiny ants, which he at first didn't notice and then did, and so gave them to the housekeeper to wash. Eventually she returned with tea in the cups.

Feliciano knew I had come to see the monkeys. We sat at the table with our tea, facing each other, he speaking Portuguese, I speaking English and then Spanish, until we got tired of not understanding each other, and just sat and smiled.

Next day I walked through the forest. It was the start of the rainy season in that part of Brazil, but the day was extraordinarily beautiful. Where I could see the sky through the trees, it was a deep blue, brushed only by a few wispy cirrus clouds and some soft cumulus:

fishbones and dreams. My guide in the forest was a young Brazilian scientist, Dida Mendes, who led me up and down small hills and occasionally paused to hack with his machete a passage through obstructing vines and ground-level plants. Trees supported hanging gardens of vegetation: dense clusters of vines, thick lianas hanging straight down like ropes or twisting around like snakes, quiet green explosions of sword-leafed bromeliads nestled in tree crotches, and, all around us, small, bright, flowered surprises.

The forest seemed a baffle for sound. We kept passing through pockets of sound — whistle of bird, song of tree frog, rainfall percussion of cicada. In places the forest opened into the groan and roar of howler monkeys, then closed again, the sound fading quickly. The birds and frogs and insects were mostly invisible. Once a toucan appeared and disappeared. Once we walked past a few brown howler monkeys sitting high in the trees, hunchbacked and heavy-chested, with long, full brown tails hanging down and deadpan faces.

Late in the morning Dida turned to say something to me, when suddenly he glanced to his left and pointed. Looking to where he had pointed, I saw at a distance, in the trees, the motion of a furry arm or leg. Dida and I made our way through the forest until we were almost underneath her: a female muriqui, the largest and one of the rarest primates of South America, moving high through the leaves, her fur beige (the color of the tree trunks around us), her face bare and brown. Over her shoulder appeared the tiny face of a baby clinging to her back.

Almost at the moment we saw her fully above us, two other muriquis burst into view through the leaves. They were bigger than I had expected. But they were quiet, and so we listened only to the splash and rustle of moving leaves.

I watched a muriqui reach from the branch and greenery of one tree to the long, hanging branch of another. In a single easy motion, he transferred his weight to the branch — holding on with an arm, two legs, and his tail — and like a long pendulum, like a big playground swing, the entire branch swayed across a gap and carried him over to another hanging branch. He transferred his weight to the next branch and climbed up it hand over hand, foot over foot, while his tail, freed from grasping, curled up into a tight spiral over his back. Up in the sunlight now, he looped that long, strong tail around an overhanging branch, held on with both feet, hung down almost upside down, and reached out with both hands to draw a yellow, vase-shaped flower to his mouth. He ate the flower.

We saw others, then, in the trees around us. Quiet, not at all provoked by our presence, they moved at a leisurely pace. I watched them grasp free branches, pull them back, then climb on and swing forward on leafy pendulums. They grasped with hands and feet, and hung upside down with the help of a hand or foot and that long, thick, remarkable tail. (When the tail was not engaged, it curled back tightly into a watchspring.) I saw them climb up thick, solid tree limbs on all fours; pause to look down at us and consider; reach out to eat a flower or a leaf; climb from branch to branch; gracefully cross from one tree to the next. They were entirely at home up there, unhurried and graceful, while Dida and I clumsily fought clutching vines and pricking branches, stumbling on the rough ground below, positioning ourselves to get a better view.

One of the monkeys bent over above me — an ungainly posture from my perspective — and I saw a large scrotum and small penis. Another, in profile against the sky, seemed pregnant. I mentioned this to Dida, who told me they all have potbellies. Indeed, then, I noticed that the others also had rounded bellies. Still, they were anything but graceless. These were among the most sure and elegant creatures I'd ever seen in my life.

As I watched them, and as they once in a while looked down and watched me, I began to imagine I had entered another world. The muriquis were a community of animals, a society, a sphere of existence; they were, moreover, obviously alert and intelligent, an intelligence in the trees, as aware of me as I was of them. And aside from tail and fur, they looked, as many primates do, uncannily humanlike. For a moment I let flutter the fantasy of having just dropped in from another planet, seeing for the first time an alternative, intelligent life form.

We saw other muriquis at a distance and moved over to watch them more closely. But it was hard to move on the ground, and while we were tripping through the underbrush, they suddenly disappeared. All we saw at last were splashes in the leaves and an occasional tan arm or leg passing into distant green. Dida and I stumbled around this way and that, trying to figure out where they had gone, but we had no idea. A minute earlier we had seen several, perhaps a dozen, but now we saw nothing, no moving branches or leaves. When we stopped to listen, we heard nothing except the background music of birds and insects and a whisper of leaves in the treetops, wavering in a slow breeze above the forest. They were gone. At last we gave up and sat down, slightly disoriented, and laughed.

When I remember that experience now, I think of laughter. The time seems suffused with bright laughter, an emotion of sunlight.

And it seems to me now, as I reflect on it, that the animals we watched for those few minutes were *having fun*. Their fluid progression through the trees reminds me now of the most basic of human pleasures: motion — my own children swinging in a playground swing; an expert gymnast at once performing and playing; a superb basketball player leaping and passing and, even in the sweating chaos of intense competition, feeling excellent and graceful. Animals, especially young ones, do play. These animals were not playing. But their ordinary efforts in seeking food and moving through the forest seemed so extraordinarily attuned to their abilities that I imagine their emotional experience to have been one of rightness, a modality where intelligence, effort, and event merge to produce steady pleasure.

The muriquis I saw that January day in Brazil are members of a species with a double distinction. Weighing up to thirty-five pounds, measuring almost five feet from head to tail tip, the muriqui is the largest nonhuman primate of the New World, as well as Brazil's largest mammal. It is also probably the most critically endangered primate in the New World.

Muriquis live in relatively large social groups. In the 1940s, groups of thirty to forty were frequently seen. But by the 1970s, given the disastrous decline of the full species, their social groups were much smaller, around six to twenty. Although a muriqui social group usually sleeps, forages, and moves together, scientific field studies reveal that the adults tend to form sexually segregated subgroups, which may stay more than 160 feet — even as much as 3,300 feet — away from each other. At other times all members of the social group gather tightly together, feeding in a dense area of leaves, for instance, or in an attractive concentration of fruit. The subgroups forage and travel along parallel paths, with the adult females (carrying their infants piggyback) and juveniles leading the way and the adult males following, keeping track of each other with loud vocalizations. The muriquis I saw were quiet, intent on moving and feeding. But they are said to use at least eight distinguishable vocalizations, including loud horselike neighs before and during travel, puffs, whines, alarm screams, grunts, and belches of satisfaction.

During a typical day the full group will feed, then rest, then travel rapidly through the trees to new feeding and resting sites. In late afternoon the group gathers for the night, males frequently embracing one another in gestures of reassurance — they are great huggers — before eventually bedding down high in one or two trees, huddled together like great woolly pillows for warmth and comfort.

Not very long ago muriquis lived in abundance throughout southeastern Brazil's Atlantic forests. Before 1500 the species' total population may have been nearly half a million. But intense hunting and massive habitat destruction reduced that number to an estimated 3,000 individuals by 1971. In 1972 the estimate was revised to 2,000. Today only 350 to 400 muriquis are known to exist, with perhaps a few others surviving in the coastal mountain forests of São Paulo State.

The remaining muriquis are carried on ten small Arks: isolated pockets of forest in southeastern Brazil. The most important Arks happen to be on privately owned farms. Senhor Feliciano's Fazenda Montes Claros protects what is probably the second largest concentration of muriquis, more than fifty individuals living in two intact social groups. Montes Claros also harbors significant groups of two other threatened primate types, the southern brown howler and the buffy-headed marmoset. Only the 13-square-mile forest at Fazenda Barreiro Rico in São Paulo State holds more muriquis — some eighty-five to one hundred — while most of the other Arks contain populations so small and isolated that they are probably doomed.

For the past forty years Feliciano has protected his forest and the endangered monkeys living there. But he is old now, approaching his eightieth year, and it is not entirely clear what will happen to Montes Claros after he dies. He once offered to sell the forest to various conservation groups, but later withdrew the offer for unclear reasons. For now, he generously allows scientists to enter that Ark and study its passengers.

It was my good fortune that Russell Mittermeier, a prominent scientist and conservationist, director of the World Wildlife Fund's Primate Division, happened to be visiting Montes Claros when I was there. I had never met Mittermeier, although I had several times spoken to him via telephone and had read several articles by and one about him. The article about him somewhat frivolously described him as "Russell of the Apes." And when I first met him, on a dirt road in the Montes Claros forest, he seemed to my impressionable self almost indeed a fictional character of the jungle. He was wearing a fancy photographer's camouflage cloth vest with half a dozen handy pockets, and shorts and knee socks. A machete was strapped to his belt. His dark hair spread like a round helmet over his head and terminated in a fringe of premature white across his forehead. His eyes were clear blue, his manner open and easy.

Actually, Mittermeier's childhood hero was Tarzan. He read all seventy-six of Edgar Rice Burroughs's books, some several times, including the two dozen of the Tarzan series. "Even when I was a very small kid," he later told me, "I knew what I wanted to be: a jungle explorer." And now he is one of those rare people whose adult life thoroughly meshes with his oldest and deepest fantasies. He has traveled across South America and explored some very obscure places in the Amazon. When he was just out of college, he helped organize an expedition into the Peruvian Amazon that resulted in the rediscovery of Peru's largest monkey, the yellow-tailed woolly monkey, which hadn't been seen for half a century. As director of the World Wildlife Fund's Primate Division, Mittermeier regularly visits critical primate habitats around the world. He is fluent in six languages, a well-respected herpetologist and primatologist and, perhaps most important in his work with the WWF, an excellent politician. I was delighted to meet Mittermeier and, later, pleased to travel with him in a rented VW van from Montes Claros several hundred miles south to Rio de Janeiro.

Riding in a crowded public bus and the back of a truck to Montes Claros, and then with Mittermeier in the VW van to Rio, I spent many hours gazing across what seemed an endless ocean of rolling, soft, emerald-green hills. The country houses along the way were weathered white adobe and brick, with pale blue wooden doors and window frames, and earth-red tile roofs. But the predominant colors were green and earth colors. Almost no road signs, no advertising or billboards, and very little urban litter broke the meditative vision of green hills, which were sometimes gently contoured with cow trails, occasionally patched with small coffee plantations and farms, dotted here and there by solitary trees, or topped with the briefest crowns of forest.

It was beautiful. But I was looking, across hundreds of miles, at the Deluge. In southeastern Brazil the Deluge is so complete now that it requires imagination and some historical knowledge to recognize what has happened. These rolling, grassy hills, now mostly infertile, used to be forested. The small, dark crowns of forest on the hilltops are the tiniest remembrance of a great, luxuriant and diverse tropical forest, Brazil's Atlantic forest, which once extended 2,000 miles along the Atlantic coast and inland, from the state of Rio Grande do Norte in the northeast down to the southernmost state of Rio Grande do Sul. The forest covered about 400,000 square miles, roughly the area of all the Atlantic coastal states in the United States from Maine to Florida. It included coastal forests along the narrow eastern coastal plain and higher forests into the foothills and lower slopes of the Serra do Mar mountains, which generally parallel the coast.

Early European explorers believed the Atlantic forest to be one of the world's tallest. Rich and complex, abundant with orchids, bromeliads, and diverse wildlife, the forest was a unique ecosystem (what remains of it still is), entirely distinct from the great Amazonian forests to the north. Over half of the tree species, for instance, are found only there. But the area was the first to be colonized in Brazil and is now the most populated portion of a large and rapidly growing nation. The region includes Rio de Janeiro and São Paulo, two of South America's three largest cities. São Paulo, in fact, is one of the largest cities in the world, with more than 10 million inhabitants.

The destruction of the Atlantic forest, begun with the first colonization of Brazil 450 years ago, greatly accelerated by industrial and economic expansion in the last two decades, is nearly complete. About 2 percent of the original forest endures in small, scattered oases. Although many of the remaining forest fragments are privately or publicly protected, in fact the land remains vulnerable to illegal use. Altogether, Brazil's Atlantic forest region rivals the island of Madagascar as the most devastated major primate habitat in the world. It is, in the words of one expert, "a dying forest formation." The animals inhabiting the forest are dying too. Some twenty-one primate species and subspecies still live there; sixteen are found nowhere else. Including the muriqui, half a dozen of those primate types are almost extinct, while a total of about fourteen are endangered.

Hunting is an additional pressure in the southeast. South America's smallest primates, the marmosets and tamarins, were and are rarely hunted for food, because the economics of hunting makes larger animals more attractive. (Other factors being equal, the price of a pound of meat — as measured in time and cartridges, darts, or arrows — diminishes as the size of the target increases.) The New World's largest primate, the muriqui, has obviously been a prime target. When much of its habitat was intact and muriquis were still plentiful, the dynamics of reproduction probably compensated for the drain from hunting. In the early nineteenth century the German naturalist-explorer Prince Maximilian du Zeid found muriquis abundant enough to provide a steady source of meat for his Brazilian expedition. But in this century the luxury of that abundance is gone. Brazil started legally protecting all its primates in 1967, but hunters seeking food and sport continue to threaten remaining individuals of this now nearly extinct species. And the uncounted few that may still inhabit unprotected forests of coastal São Paulo State are particularly vulnerable to hunting.

I had come to Brazil already aware of the devastation in the southeast, knowing that historically the nation has regarded its forests as green obstacles to progress and imagining that Brazilians were indifferent, if not antipathetic, to conservation and the dying primates. Mittermeier helped me revise that notion. "In ten years," he told me, as we drove south in the VW van, "Brazilian conservation could be among the best in the world, a model of conservation in the Third World. There's such a level of awareness and responsiveness here. The press is always eager and willing to print conservation news, unlike in the United States, where the press still thinks conservationists are old ladies in tennis shoes. That eagerness combines with the fact that there's still so much in the north of Brazil, in Amazonia, that's still intact. Plus, the Atlantic forest gives Brazilians an example of a forest that's already all screwed up."

We stopped at a gas station in the small city of Manhuaçu and greeted the gas station manager, a young man named Eduardo Pinheiro. Short and stocky, Pinheiro wore a white T-shirt, and his eyes were obscured by aviator sunglasses. But pinned to his T-shirt was a World Wildlife Fund panda. Around his neck on a gold chain hung two medallions, a gold Christian cross and a small gold gorilla.

We went into the gas station office and talked with Pinheiro for a while. Against one wall hung a poster of a motorcyclist driving through flames on a motorcycle that had flames painted on it. A saint in a box hung in one corner of the office above a stack of tires, and elsewhere four scrawny birds were hopping around inside four wooden cages.

Four years earlier, Pinheiro told me, a film about the muriqui was shown in town. After seeing the film, he went up to the woman who had shown it and said, "I know a place around here that has those monkeys." The place was a small forest, several miles outside of town, where he had often gone fishing as a boy. Pinheiro contacted the director of a nearby national park, and together they went out to the Mata do Sossêgo, the Forest of Tranquility, where they counted about twenty muriquis. The park director proceeded with legal applications to have the site declared a reserve while Pinheiro began an interim campaign to protect it. Pinheiro went to all six farmers who own pieces of the forest and persuaded them to protect it and to agree, in principle, to sell it. The selling price changes with the price of coffee, but all told the land may cost up to a quarter of a million dollars. The World Wildlife Fund has given a token donation, but the campaign to raise the money is almost entirely a grassroots campaign, enthusiastically supported by a local community, organized by a gas station manager, run out of a gas station office.

We began noticing, on the windows of cars pulling into the gas station, stickers with a picture of the muriqui and a slogan that translated into "The survival of the muriqui depends on us."

FERDINAND MAGELLAN, the Portuguese explorer who demonstrated the sphericity of the earth by sailing around it, paused off the coast of Brazil before wintering in Patagonia in 1519. We know from the words of Magellan's chronicler that members of the expedition sighted a number of small monkeys in Brazil's coastal forests that were "beautiful simian-like cats similar to small lions." We recognize those "small lions" today as golden lion tamarins.

While the muriqui is South America's largest monkey, the golden lion tamarin is the largest of the smallest (the marmoset and tamarin group). It is also among the most beautiful primates in the world, covered (except for the face, which is bare and dark brown) with shimmering, golden-red fur dominated by a long, swept-back golden mane. The adults weigh little more than a pound and stretch nineteen inches from nose to tail tip, with half that length as tail.

I went to see golden lion tamarins in their most significant Ark, the Poço das Antas (Pool of the Tapirs) reserve in Rio de Janeiro State. Poço das Antas is easy to get to, only two or three hours by car north of Rio, but hard to enter. The public is not invited, and to get into the reserve one normally needs written permission from the Brazilian Forestry Development Institute. Once in the reserve, it is nearly impossible to find the animals without assistance from the Brazilian and North American scientists working there, who require permission from *their* sponsor, the National Zoo of Washington, D.C. In the United States representatives of the National Zoo had provided me with only mild encouragement and no promises, and, because of the vagaries of mail strikes and vacations in Brazil, I was never able to contact anyone at the Brazilian Forestry Development Institute. I had begun to visualize hitchhiking, fence climbing, and bush beating, but happily I was driven to Poço das Antas, carried through the gates, and deposited at the driveway of the research project headquarters by two people who don't require the usual permissions, Russell Mittermeier and Admiral Ibsen de Gusmão Câmara, formerly second in command of the Brazilian navy, now president of the Brazilian Conservation Association.

At the research headquarters we met several scientists, including Andy Baker, an American who was studying the lion tamarins as part of his doctoral research. Baker agreed to help us find some of his research group, so we drove down a chalky white dirt road into the middle of the reserve and disembarked. Baker pulled out a small radio receiver — a box and dials, encased in plastic, with an aluminum antenna looking like a small television antenna, and earphones. He suspended the box across his shoulder with a strap, placed the earphones over his ears, and held the antenna in his hand. Standing in the middle of the dirt road, he pointed the antenna out in front of him and, like a priest blessing the world with a cross, slowly turned. Baker's group had been collared with transmitters, and he was listening for their beep. Eventually he gave the nod, and we left the road and entered the forest, which was actually a swamp forest.

The tall forest at Montes Claros had been the first of many superbly beautiful tropical forests I was to explore in my primate trip around the world, but I have to say that the low, dense swamp forest at Poço das Antas was the most oppressive. The day was hot to begin with, and the forest itself was very humid. Almost immediately my clothes were sopping with sweat. Mosquitoes rose in swarms. The vegetation was a terrible tangle of spiny palms and nettles and brush and vines. We pushed through vegetation, jumped over small rivulets and pools, stepped into ankle- and calf-deep mud and, whenever possible, precariously navigated byzantine bridges of roots. As we followed the radio trail into the swamp, I tried my best to kill the mosquitoes that were landing on my face, hands, neck, and nearly everywhere else. Once I looked down to see five of them on my arm, lined up like pigs at a trough. I slapped: five in one blow. "Twenty-four-hour-a-day mosquitoes here," someone commented.

After many minutes of this tormented journey, however, a small monkey appeared. Bright reddish gold, it sat on a tree limb about as high as my chest, faced us, and let out a small cry, displaying a dark, open mouth framed by four small, sharp canines. That was the "lead mother" of the group, Baker explained. She turned, leapt to another branch, scurried up higher. Soon after, we sighted the other members of the group.

The golden lion tamarins looked like the miraculous offspring of a squirrel and a bird: smaller than squirrels, slightly larger than most birds; as quick and agile as squirrels, scrambling here and there on branches, then suddenly casting off in superb leaps from one branch to another, in near-flight, their tails twisting one way and another as rudders and balance weights. They cried out with birdlike, chittering calls and occasionally turned in my direction to display mouthfuls of little sharp teeth. Most distinctive was that bright reddish golden fur, covering everything except their bare, brown monkey faces.

The group consisted of one adult male, two adult females, two maturing adolescents, and twin toddlers. The twins, around two months old, were mature enough not to be carried around very much by the adults, but still unsteady and cautious. Their fur was lighter in color and finer than the others', a blond-yellow wool, and their tails were long and thin. The adults moved steadily through the vegetation, scurrying and leaping, but the toddlers followed hesitantly. They would fall behind, call out with tiny cries of alarm, asking to be rescued, then stop calling and bravely move on after the receding adults. Before leaping they would pause, nervously frozen for a moment, judging the precipice, then risk a mighty and often awkward leap that, so it seemed, might indeed not carry them far enough.

Golden lion tamarins eat fruits and insects and the occasional tree frog and lizard; this group was feeding as it moved. We watched one climb a palm tree and pause to eat a few dark, grape-sized palm fruits. One of the fruits fell to the ground, and I picked it up and tasted it. It tasted like wine. We watched another lion tamarin dig into the leafen inner cup of a large bromeliad and pluck out a leggy insect. It placed the insect in its mouth and chewed and swallowed; waving legs were the last to disappear.

We watched all seven monkeys climb two fruit trees to feed: walking along a limb, hanging upside down, or sitting on a branch with tail hanging straight down, fruit clutched in a hand, or resting for a moment on a branch with all four legs hanging down. Several times I watched two run headlong at each other from opposite ends of a branch. Just as they were about to collide head to head, with absolutely no hesitation, one leapt over and the other scurried under, and they kept on going.

Lion tamarins prefer the middle layer of their forest, about ten to thirty feet above the ground: a dense area of vines and interlacing branches that protects them from flying predators above and walking or crawling predators below. At night they sleep in a den fashioned from the hollow of an older tree trunk. And, since lion tamarin families often inhabit a single favorite tree hollow for years, the old home becomes insulated and snug as a layer of shed fur accumulates. In younger or secondary forests, where tree hollows are uncommon, lion tamarins probably fashion similar homes within dense vine tangles.

The usual printed wisdom says that lion tamarins are strictly monogamous, that they mate for life, and that the nuclear family, the mated male and female plus offspring, forms their important social group, usually three or four animals altogether. Andy Baker's research at Poço das Antas, though, has begun to indicate a more fluid and complex social structure, with occasional divorces, remarriages, and bigamous liaisons.

Like many other species of the marmoset and tamarin group, lion tamarins typically bear twins. During the first week after birth, the infants cling to the mother's fur. Thereafter the father takes over transportation until the young, at four months, are able to travel independently. At about one year the animals are adult-sized, and at about two years they reach sexual maturity. At that point they are driven out of the family group by their parents, Oedipus-style, and strike out on their own.

On my way to Poço das Antas, I had passed a spot of some historical interest. Along the road, in front of a small supermarket in the town of Silva Jardim, a man known as Joachim Sagué, or Tamarin Joe, used to sit and sell captured golden lion tamarins from among the last of the species to anyone willing to stop and bargain.

The trade has a long history. Early colonists in Brazil found the golden lion tamarin beautiful, charmingly delicate, and easily tamed, so the animals were often exported to Europe, where they were kept as pets by the nobility. Madame de Pompadour, girlfriend of Louis XV, owned one. In this century golden lion tamarins have served as pets and zoo animals and were sometimes used in medical research. Between 1960 and 1965, even though they were already quite rare, some two to three hundred were exported yearly from Brazil. The two hundred wild golden lion tamarins left today would obviously sustain only a few months of such traffic.

The International Union of Directors of Zoological Gardens prohibited member zoos from importing golden lion tamarins in 1967. At the same time Brazil declared illegal the hunting, capture, and export of all primates. Tamarin Joe was forced to close shop.

But the live animal trade is only a secondary aspect of the Deluge for these primates. Golden lion tamarins require lowland forest; they never move into territory higher than about 1,000 feet, so their range has always been delimited by the ocean on one side and the foothills of the Serra do Mar mountains on the other, north of Rio de Janeiro. That original habitat began to shrink in the seventeenth century, with extensive clear-cutting for sugarcane and coffee plantations. Deforestation for banana plantations, housing projects, and bridge and highway construction augmented the loss, until by 1968 this monkey's original range of about 8,000 square miles had been reduced by more than 99 percent, to less than 80 square miles. By 1971 its habitat consisted of only isolated fragments of degraded forest, which were still being cleared for farmland and charcoal production.

Altogether, the Ark that carries the dying plants and animals of Brazil's nearly dead Atlantic forest consists of ten national parks, five nationally administered biological reserves, and several small state and private reserves. Unfortunately, most of these sanctuaries are moderately or poorly protected, or entirely unprotected. They are small: the total area of federal lands is less than that of the smallest of Brazil's Amazonian national parks. And they are scattered: islands of forest typically separated by many miles of open land, so that the plant and animal types within are subject to inevitable decline because of physical and genetic isolation.

I have been told that some 175 wild golden lion tamarins now live in the twenty-square-mile Poço das Antas reserve, which may be close to saturation for this particular primate species. Moreover, Poço das Antas is the only viable habitat left. One other habitat area, a nine-mile strip of coastal forest near the mouth of the São João River, is now set for destruction. Although a small part is owned and protected by the Brazilian navy, the major portion has already been surveyed and divided into lots for beachfront housing.

Will the Ark of Poço das Antas adequately carry golden lion tamarins into the next century? When the area was first chosen as a reserve site in 1971, more than 70 percent of the land was densely forested. By the time it was actually protected, though, large areas had been burned, logged, and replaced by pasture grass and eucalyptus. Squatters lived within the reserve, and cattle from nearby ranches sometimes grazed on the reserve area. Only a third of the reserve was forested, and only a third of that included mature forest. A railroad and a road cross the land, dividing it into four sections. A recently completed dam has flooded a good deal of the reserve, and in building the dam, construction workers leveled an entire forested hill on reserve land.

In short, conservationists have had good reason to be concerned about the future seaworthiness of this Ark. I went to Poço das Antas expecting to review the bad news. Instead, I was pleasurably surprised to hear and see the good news. The squatters are gone from the reserve, and cattle no longer graze there. A tree-planting project combined with normal regrowth has covered several previously deforested parts of the reserve with substantial secondary growth. A road and railroad do indeed cut through the reserve, but traffic on the road is strictly limited, and only one slow train a day creaks along the tracks. Perhaps most significantly, the reservoir created by the dam, instead of destroying large areas of habitat, may actually have increased the total habitat by creating new swamps and swamp forest. The reserve might yet be expanded: an additional 120 acres of forested land on the edge of the reserve could be added to the protected area, should the Brazilian government choose to do so.

Golden lion tamarins have long been favored by zoos, but until recently they seldom lasted long in a cage. Zoo keepers once considered them to be entirely fruit eaters. Thus, they were deprived of the animal protein they need. They were commonly given dietary supplements of vitamin D_2, but South American primates need vitamin D_3. Adult golden lion tamarins in captivity suffered a very high rate of mortality from viral and bacterial infection. In the abnormal conditions of captivity, they did not always care for their young appropriately, and many young animals died from neglect or were killed by their parents. After Brazil stopped exporting the animals in 1967, the world captive population — permanently locked in its own cage — gradually dwindled, until by 1972 only about seventy golden lion tamarins survived in a dozen zoos.

In Brazil at about that time, Adelmar Coimbra-Filho and Alçeo Magnanini established a breeding station known as the Tijuca Bank, located in Tijuca National Park within the city limits of Rio de Janeiro, to preserve and breed lion tamarins and other endangered primates of southeastern Brazil. Animals from this station were later transferred to what is now South America's most important captive breeding project for endangered primates, the Rio de Janeiro Primate Center, located just northwest of the city in a steep canyon within still-forested foothills of the Serra do Mar. By the time I visited it, the Primate Center was providing a caged Ark for almost 250 monkeys, representing several of the highly endangered southeastern species, including roughly three dozen golden lion tamarins.*

Of course, such small numbers of captive animals are subject to the

*At the Primate Center I walked past a very large empty pen prepared for muriquis in the expectation that individuals from doomed populations in some of the very small remaining forest pockets will soon be taken there. Coimbra-Filho has planted near the pen several trees that should produce the right food for muriquis. However, captive breeding is most appropriate for the smallest monkeys of South America, the marmosets and tamarins, which commonly produce twins yearly and survive in captivity quite well. Muriquis produce a single offspring once every few years and will undoubtedly prove sensitive to the constraints of captivity. Captive breeding will never become an important conservation measure for this large monkey.

effects of inbreeding. Arranging for a genetic interchange — interbreeding — between these groups and their wild counterparts could mitigate that problem. Also, with all the breeding animals kept at a single location, an outbreak of disease could quickly eradicate the full captive population. Satellite breeding colonies will probably be established at other locations in the near future.

The Primate Center is a model captive breeding facility, supported entirely by donated funds from private industry in Brazil, managed superbly by Brazil's father of primate conservation, Adelmar Coimbra-Filho. Yet the most extensive breeding effort with golden lion tamarins has taken place not in Brazil but in North America. In 1972, when the species appeared to be approaching extinction both in the wild and in captivity, three major North American zoos sponsored the Saving the Lion Marmoset Conference at the National Zoo in Washington. This Conference and later work by concerned experts resulted in a serious attempt to breed the golden lion tamarin in the United States. The effort included the creation of an international studbook, a cooperative breeding program to encourage loans among zoo collections, research on the animals' sexual and social behavior, and studies to improve general zoo management of the captive animals. Eventually, all the important holders of golden lion tamarins agreed upon a series of management standards for captive breeding.

The results were impressive. Infant and juvenile mortality for the first year of life in North American zoos among golden lion tamarins declined from 60 percent between 1970 and 1975 to about 45 percent in 1982. And the individual reproduction rate increased tremendously, from 1.8 in an average litter in 1975 to 2.1 by 1980. By the early 1980s a full 20 percent of all litters were triplets, and in 1981 the first quadruplet litter was born in the Los Angeles Zoo. In addition, females were frequently producing two and sometimes three litters a year. By 1980 the North American zoo population was double that of 1975. By 1984 there were almost 400 captive golden lion tamarins in North America — double the entire wild population in South America — and the number was increasing by about 50 every year.

In 1981 Devra Kleiman of the National Zoo met with Coimbra-Filho to discuss removing a few "genetically surplus" captive animals from North American zoos and sending them back to Brazil. There the animals could have been placed with the breeding group at the Rio de Janeiro Primate Center, but Kleiman and Coimbra-Filho decided to try reintroducing them into the major Ark for wild golden lion tamarins, Poço das Antas.

Fifteen animals were selected for this experimental project. First they were trained to forage for food in scattered places, under objects, and in crevices. Then, in November 1983, they were shipped to Brazil and placed in outdoor cages to habituate them to the new climate and diet. From this group, ten were put in cages within the reserve, where they continued the habituation process. They were divided into appropriate social groups, including already existing family groups and new mating pairs, some captive born, others matching a captive-born animal with a wild-born animal. At last, the lion tamarins were fitted with radio transmitters for tracking and released.

By July 1984, Kleiman was able to report favorably that the ten released animals "are finding food and ranging much farther than we expected. They are behaving much like their wild counterparts and are doing extremely well." By the time I visited Poço das Antas, however, of the total of twenty-six North American zoo animals that had by then been reintroduced into the reserve, only three were still alive. The rest may have succumbed to the stresses of a radically new environment, including competing wild members of their own species and new diseases.

LOOKING UP from a field in southeastern Brazil one night in January, the night before I first saw muriquis at Montes Claros, I saw a sky bursting with stars. Yet the whole sky seemed new and chaotic to me. Nothing, it seemed, was the same as my home sky back in Massachusetts, and the strangeness of the sky increased my own sense of remoteness.

I noticed a constellation I had only heard about before, the Southern Cross: four bright stars in a symmetrical diamond or cross — forever invisible, below the southern horizon, for northern-hemisphere stargazers.

On the night after I visited golden lion tamarins at Poço das Antas, walking back from a late dinner along the deserted and dim main street of a small Brazilian town, I was delighted to recognize the constellation Orion — far enough north in the Brazilian sky to be visible in the southern sky in Massachusetts. I was tired, reluctantly returning to my hotbox room in a fleabag hotel, and the sight of familiar Orion evoked a brief chord of reassurance. One of my dinner companions, Admiral Ibsen, helped me locate Orion's red star and then mentioned that he sometimes used to navigate by the stars. I imagined a more personal navigation: my two children stepping out the back door of my own house, standing on our hill in Arlington,

Massachusetts, considering some of the same stars I considered in Brazil.

I had left a world, a New England winter, and very suddenly entered another world, a Brazilian summer. Brazil itself seemed a new world. I had mailed myself south in an aluminum envelope, dropped out of the sky, and then found myself engaged and provoked, dismayed and excited, by a new language and culture. Then I had entered forest worlds, serene and separate from the surrounding urban and rural worlds of everyday Brazilian life. And I had peered into primate worlds, the private spheres of muriquis and golden lion tamarins.

Now, as I looked up at a whole skyful of stars, I considered starworlds. According to people who are paid to think about it, the known universe contains more or less 100 billion billion stars, suns similar to our own sun. It is not hard to imagine that some of those stars, perhaps merely a billion billion, are right now being orbited by spheres of rock, and that some of those spheres of rock, perhaps merely a single billion altogether, contain the necessary conditions for the emergence of life. Yet if there is life elsewhere in the universe, we can be certain that it is completely different from ours. Nothing up there will ever replace what is down here in our own starworld. The only living primates in the universe are living here and now on this planet. So it seemed to me that to save these animals was a worthwhile cause. I thought, a small human effort right now can preserve the muriqui and the golden lion tamarin and the many other fading primates of southeastern Brazil and carry them through the Deluge, into another and perhaps better time, where my children and others can still know and see them, or just know of their existence and love their beauty and wildness.

Two

FORESTS

ONE EVENING, a couple of weeks and 3,000 miles later, naked and floating on my back in the warm waters of the Negro River a few miles above its confluence with the Amazon, I considered Orion more serenely.

The stars of Orion outline the body of the Hunter, who is forever threatening Taurus the Bull with a gigantic club. Most obvious are the three bright stars, equally spaced, that form Orion's belt. Those three stars, bright, aligned just so, are immediately apparent in the sky at the year's end and its beginning. My eyes fix on them first, then move to define the rest of the constellation.

Strewn out from the three bright stars of Orion's belt are four other bright stars that define his body. The upper two mark his shoulders, the lower two define his feet — or knees, depending upon how you visualize the hunter's stance and shape.

No matter what direction you're facing, you can always tell Orion's top from his bottom by finding the softly winking ruby star named Betelgeuse that marks Orion's right shoulder. Diagonally opposite Betelgeuse, below the belt, burns the constellation's brightest star, a white diamond known as Rigel: Orion's left foot. A triangle of stars marks his head. His left arm holds a lion's skin of stars. His right arm wields a club of stars. A whole stream of stars, some bright, some faint, descend from his belt to form a sword.

Over the next few months, I sought Orion from several perspectives on the globe, both north and south of the equator and almost on top of the equator. Though his position in the sky varied according to time, place, and season, during much of my trip Orion the Hunter was visible on a clear night, serenely drifting, seeming peacefully, eternally the same. If the strewn jewels of Orion gave me familiarity and orientation, a geography of space, though, they also provided a geography of time.

The patterns of the constellations appear absolutely fixed and certain year after year, as if — as ancient and medieval astrologers believed — the stars were forever fixed within a crystal sphere rotating around the earth. In fact, of course, the stars are hurtling through space, just as our own sun hurtles through space. Thus, over great time the relative motions of the stars should become apparent, as their relative positions from an earth-bound perspective slowly shift. Edmond Halley, discoverer of the comet named for him, was first to notice that some stars in the sky had shifted even since Ptolemy had catalogued their positions a mere fourteen centuries earlier. Most of the stars we see are shifting their relative positions much more slowly, but in any case, by the year 200,000 the Big Dipper will no longer hold water: its cup is sliding open. And 75,000 years ago (possibly around the time humans began using language), the three bright stars of Orion's belt were not equally spaced from an earth-bound perspective: the middle star of the belt has been shifting to the right relative to its two outer companions.

Orion provides a life history of stars. That faint and shimmering star in the middle of his sword isn't a star at all. Even through a weak telescope, you'll see it as a mist, a cloud. It is, in fact, a very large cloud of attenuated matter, shrinking in response to its own gravity, becoming gradually denser. Someday it will achieve the density required to begin its own fusion reaction — the steady conversion of hydrogen into helium that powers all stars. Look at the middle of Orion's sword and see the birth of a star.

The faded red star, Betelgeuse, marking Orion's right shoulder, shows the other extreme of stellar existence. Betelgeuse is a dying red giant, a star that has expended most of its fuel. The force of gravitation has weakened, and the star now fades thin, expanded, and comparatively cool — red, and big enough that if the sun were a basketball, Betelgeuse would be the basketball court.

Most of the other bright stars of Orion display the middle phase of stellar life: white-hot adulthood. And — unusual for any constellation — most of them lie about the same distance away from earth, very roughly 1,000 light years away. Orion is unusual for another reason: most of its stars are comparatively young. Betelgeuse may be roughly 75 million years old, but most of the white stars are a good deal younger. Bright Rigel, for instance, probably appeared in the sky only two or three dozen million years ago.

Three dozen million years may seem a respectable amount of time, but as I floated on the Negro River that night, gazing up at Orion, it occurred to me that the river I was floating on was older than the stars I was watching. Indeed, the Amazon rain forest lying at the edge of the river, dark and quiet, was older than those stars. In fact, if we were to use the median lifespan of Orion's stars as a yardstick of time, Amazonia and the other great tropical forests of the world might measure two or three Orions.

MY PRACTICAL EDUCATION in tropical forests began in the geographical middle of the world's biggest rain forest, the Amazon. I started at the river port city of Manaus and was driven in a jeep with some other people north along a main highway, a slash of red soil through green forest, and then onto rougher roads and through several barbed-wire gates, onto and across ranchland.

The road became double red grooves, rutted, muddy, slippery, in places eroded along the edges into gulleys and miniature canyons. The ranchland seemed interminable and looked unlike any I had seen before: miles of small, bleak hills studded with the broken flotsam and jetsam of a cleared forest — stumps, cast-off tree trunks, and a few still standing giants, denuded and dead. Once in a while we approached small herds of cattle, a zebu-crossed stock, I believe, slow and bony, humped, with flapping dewlaps and crescent horns. One time we stopped and waited while Brazilian cowboys on horses herded cattle past us on the road.

We drove along high walls of forest, skidded up and down small steep hills, and eventually stopped at the edge of the forest. Five people climbed out of the jeep. We tossed out our supplies and said good-bye to the driver. We climbed a fence, ducked under the bridge of a fallen tree, and entered the green world. The wind ceased. The air was suddenly still and the humidity so thick I could see it, a fine haze.

Packs on our backs, additional supplies gathered in our hands and slung across our shoulders, we walked some distance through the forest along a rough path, clambered over logs across streams and swampy mud, then climbed one final steep hill to a cleared hole in the forest, which was our campsite. The ground there was flattened sand, protection from snakes and crawling insects; the camp itself consisted of two long lean-tos constructed of vertical and horizontal wood poles and topped with sheets of corrugated fiberglass. We unloaded our packs and supplies, and strung up our hammocks from the cross-poles inside one lean-to. The other lean-to was our kitchen, already supplied with black pots and dirty dishes.

We set up camp there: me, two Brazilian scientists, and two field assistants and cooks. My guide in the forest was a young ornithologist named Rita Mesquita. The other scientist was an entomologist, there to study termites.

We spent the rest of the day in camp. We straightened and cleaned up, ate dinner, and then spent a long, lazy time sitting in aluminum chairs, looking through binoculars, scanning the high treetops around our campsite, spying on toucans and macaws. I began to notice, and after dark I couldn't avoid noticing, how noisy the forest was, the noisiest I had ever been in. Even now, after visiting several tropical forests around the world, I remember that piece of the Amazon as the noisiest: the point and counterpoint of mostly unseen frogs and birds and insects, whistles, croaks, cries, chirps, hums, and buzzes, a knocking, a loud and regular creaking like the opening of a tight wooden door, and once in a while the rough breathing-groaning of howler monkeys at a distance. Sometimes the sounds moved rapidly, and I knew they were the calls of birds as they flew. At night, though, the sounds were mostly stationary. It rained in the middle of the first night, and the downpour crashed across our corrugated roof in loud waves, obliterating all other sounds.

The next day the entomologist went out to check on his termites, while I followed Rita Mesquita and her assistant as they carried out her research. Mesquita is a small, attractive, very energetic woman, with dense, almost furry, arched eyebrows. In Portuguese, she spoke in a dramatic style, often laughing, and expressing quick sentences that regularly cracked into hoarse, histrionic falsetto. But, turning to explain things to me, she would speak English in a sober and steady, if sometimes ungrammatical, manner. Wearing a T-shirt and loose, black pants, a small pack on her back, her black hair tied back with a piece of lavender twine, she walked through the dripping forest as briskly as a New Yorker at rush hour, and I struggled to keep up with her and her assistant as we climbed up and down the many small, steep hills in high heat and humidity.

Our ignorance of tropical forests is immense: Rita Mesquita had come to the Amazon to study the feeding behavior of a particular bird, but before she could study the bird she had to learn about the fruiting cycle of its favorite food source, *Clusia grandiflora*. Thus I followed her through trails in the forest as she conducted one of her regular surveys of this fruit-bearing tree. She was studying thirty-two *Clusia grandiflora* in the forest, surveying them once every fifteen days and noting their stage in the fruiting cycle. We would walk for a long time, following her pencil-drawn map, then stop and look up at one of her thirty-two trees. She would look up through her binoculars — I was regularly amazed at how she threw her head back, bent her neck way back, and looked almost straight up through the glasses — and then make a check on her clipboard checklist. Next she would examine the ground animatedly, looking for fallen fruit; some she picked up as specimens and placed in plastic bags. When the fruit ripened, it cracked open and spread out like a starfish, with seedrows inside. From the older, already decayed fruits, she would scoop out a mucky jam of seeds and place it in a bottle.

The forest was a little frightening to me, but not because of the wildlife. Actually, we saw little mobile wildlife. We encountered no snakes. Once we saw moving leaves and the brown flash of a monkey high above us. We saw some birds and lizards, many strange insects, and a few slow-moving, egg-eyed chameleons. But *we* were the biggest and most dangerous animals in the forest. (The second biggest, the jaguar, avoids humans whenever possible. The second most dangerous may be the twelve-foot-long bushmaster viper, but more likely it is one of the smallest, perhaps the insect-borne protozoan that causes the deadly Chagas' disease.) Still, I had been told never to go out alone in the forest, and as we tromped along, I began to realize why. It would be very easy to get very lost. We were at the bottom of the forest, and the sun was somewhere above the top, but I could never see it or even discern its direction. I saw the sky only intermittently as a bright speckle between leaves. And so I never had even the mildest sense of direction. Occasionally we left the trail for a few yards, and I would look around and realize I had forgotten where the trail was.

For all that, the Amazon I saw was not a jungle. Some people casually use the word *jungle* to describe a tropical forest, but I've grown uncomfortable with the term, for two reasons. First, jungle implies wasteland — lush, perhaps, but wasteland all the same. Yet tropical forests are the richest ecosystems in the world, packed with a wealth of flora and fauna that are beautiful and worthy in themselves and often valuable to humans. Second, jungle implies tangle, an impenetrable mess of vegetation. Secondary forests can be nearly impenetrable. But primary tropical forests, including the Amazon I walked through, are actually quite navigable. Mostly we walked among crowds of small and medium-sized tree trunks, branchless at our level, and approached every few or few dozen yards a giant tree with massive trunk and thick, snaking roots. Here and there flowers growing on the ground or in trees would intrude brightly, looking like birds' beaks or fingers or mammalian sexual organs of various sorts. Vines and lianas hung down in places, sometimes straight, sometimes coiled or kinked. We did occasionally come across rococo tangles of ground-level vegetation, but it was primarily above us that the forest expanded, spread out in a dense and complicated maze.

At one point in our perambulations, I paused to watch a chameleon on a tree trunk. Suddenly I felt dizzy and fatigued. Sweating and very hot, I heard the wonderful sounds of a breeze above and looked up to see the flickering of canopy leaves, swaying, rippling high above me. I thought to myself, *Ah!* I was thirsty for a taste of that breeze. But where I stood, it was hot and still and hazy with humidity. I felt as if I were an insect at the bottom of a bottle or inside a warm terrarium, and I wanted to crawl out.

After a few hours of this exhausting activity, we finally emerged into the clearing of the campsite. We dropped our bundles and settled down for lunch. The campsite was a hole cut in the forest, but it was a big enough hole that immediately upon entering it I felt the entire microclimate change. A pleasant breeze penetrated the campsite, evaporating our sweat and cooling us down; the humidity drifted away; and I was left with the unforgettable idea that the forest, self-enclosed, creates some of its own climate.

"So, what is your first impression of the Amazon rain forest?" Rita asked me over lunch, which was a piece of our dinner from the night before, reheated. My answer was not very original: "Hot. Wet."

Hot and wet. Not very original, but the basis for understanding any tropical forest.

The earth spins, and day passes into night and back into day. As it spins, energy from the sun strikes the earth's surface — most directly in its middle zone, the tropical belt, and at more of an angle across the temperate areas. Thus sunlight falling on the tropics is most concentrated and least diffused by the atmosphere, yielding high levels of solar energy. Sunlight spreading over the temperate areas is less concentrated and further filtered by its longer, angled journey through the atmosphere. Result: a latitudinal circle roughly intersecting Vienna, Paris, Vancouver, and Ulan Bator in Mongolia receives about half as much solar energy as the great circle of the equator.

The earth also revolves around the sun, completing a full revolution in about 365 daily spins. The axis of spin is nearly perpendicular to the plane of revolution around the sun, but not quite. And because the axis is tilted slightly, the daily bath of sun in temperate areas is less direct for part of the year and then more direct: the seasons. The tropical belt is also affected by that tilt, but mildly enough that for all practical purposes it exults in a single season, eternal summer.

Steady sun by itself is not particularly conducive to biological productivity. But steady sun combined with rain is, and those parts of the tropics regularly bathed in both sun and rain are the planet's great garden hothouse, the tropical forests. So: tropical forests are forests benefiting from the sun and rain of the wet tropics. They lie between, occasionally slightly beyond, the two lines that someone long ago drew on the world map called the Tropic of Cancer and the Tropic of Capricorn. There intense solar energy powers their great photosynthesis factories.

In spite of their location within that single bright band on the globe, however, the world's tropical forests are highly varied, their variety determined by such factors as soil, elevation, and local events of biological history — but perhaps most importantly by water, by the intensity and regularity of rainfall. Take New York City as a baseline. New York receives about 40 inches of rain a year. Where, in the tropics, less than half that annual rainfall occurs, only brush and thorny woodland may appear. Where between half and twice that rainfall occurs (20 to 80 inches), broad-leafed, seasonal forests may grow. Where more than twice New York's rain falls yearly (over 80 inches), tropical moist forests may grow that are green the year round. And where more than four times New York's rain falls (over 160 inches), there may emerge the fullest, most abundant, and varied of tropical formations, the true rain forests.

The amount of rainfall, however, is not as important as its regularity. We might finally say that tropical forests emerge in areas free of seasonal frost, where the mean annual temperature is at least 75 degrees Fahrenheit, with a minimum of 80 inches of rain a year distributed so that at least 4 inches falls every month for two out of three years. True rain forests emerge under the same conditions of steady heat, where at least 160 inches (up to 320 inches) of rain falls during the year, with a minimum of 8 inches per month for at least ten out of twelve months. The world's major tropical rain forests grow in lowland Amazonia, the Zaire Basin, Sumatra, Borneo, and Papua New Guinea. The world's major tropical moist and seasonal forests exist generally a little north and south of the true rain forests, in parts of Central and South America, parts of West Africa, Burma and Indochina, some of the Indonesian islands, and northeastern Australia.

Steady sun and steady rain actually have combined with a third force to produce today's tropical forests: time. The tropical forests are by far the world's oldest major terrestrial ecosystems, products of an unbroken continuity reaching back to the era of the dinosaurs, over 65 million years ago. At that time, at least in certain areas of Southeast Asia, tropical forest formations were already thriving and had been for at least 5 million, possibly up to 30 million years. It is true that major environmental changes have affected the forests since then. Quite recently (within the last tens of thousands of years), expansion of the polar ice caps dried out large portions of the forests of South America and Africa and lowered seas enough that great island chains in Southeast Asia were temporarily reconnected to the Asian continent. But those changes, profound though they were, rattled but did not break the tropical forests' chain of evolutionary continuity.

If we think of sun and rain as the primary foods for green life, then we can recognize that life in the temperate zones is less well nourished than in the tropics and in addition is perpetually restrained by the yearly oscillations between feast and famine. Whereas life in the wet tropics has been steadily feasting, unrestrained, for between 50 and 100 million years.*

If tropical forests begin in sun and rain over great time, they end in a living expression, around the world, that can be distinguished from temperate-zone forests in several important ways. Structurally they are layered. Physically they are plenitudinous. Biologically they are extremely diverse, although that diversity is highly dispersed. They are also exceedingly complex: plants and animals depend upon each other in remarkable chains of mutualism. And finally they are amazingly fragile.

Multilayered plenitude. In the great green feast, plants compete with plants, literally climbing on top of each other in a quiet and endless scramble toward the sun. Trees shoot up on tall, often limbless stalks, and then spread out branches and leaves at their tops, forming a high, sun-drenched canopy. Vines snake up, around, and between trees for hundreds of feet, holding on with tendrils and small roots, looping around branches and trunks, ultimately to spread out their own large leaves and burst open their own flowers in the sunlit canopy — as do the very thick, often complexly twisting woody vines called lianas. Multitudinous ferns and other epiphytes (a Greek-based coinage meaning *air plants,* because their nourishment doesn't come from the

*For the sake of expediency, I'll henceforth describe the tropical forests as 75 million years old.

soil), such as orchids and bromeliads, anchor themselves wherever a convenient niche appears — on tree branches and trunks, shrubs, even rocks — and gather sustenance mainly from rain and settled detritus. Some epiphytes prefer sun, others shade, so that a single tree may carry hundreds of smaller, epiphytic plants, as many as thirty species, some flowering brightly. Of the 20,000 to 35,000 orchid species found worldwide, the many tropical types are primarily epiphytic — with long, greenish or white roots that tenaciously grip surfaces, allowing the flowering plants to make their home in unlikely places and positions. The 1,600 tropical bromeliad species, members of the pineapple family, hold water and organic sediment within circular clusters of leaves. These small and not so small ponds (some hold up to twelve gallons of water and detritus) give steady nutrition to the plant holding them; they also provide microenvironments for a host of more mobile forms, for insects, worms, snakes, salamanders, frogs.

The high canopy, thoroughly closed by the intermingling of treetop branches and leaves, additionally interconnected with vines and lianas, in some forests festooned with brilliantly flowering epiphytes, is a world in itself. It harvests sunlight so thoroughly that the ground below is dimly cast in perpetual twilight. It breaks rain so completely that even a tropical downpour will reach the forest floor as mist, drip, and drizzle. It deflects the forces of motion enough that gale winds above will stir only a gentle whisper in the air below. Thus the canopy actually insulates the forest, thereby contributing to the system's hothouselike environmental stability. Perhaps most spectacularly in Southeast Asia, but true in all tropical forests, the canopy possesses its own specialized faunal world: insects and birds that live only there, primates and other mammals that seldom or never descend to the ground, squirrels and lizards and snakes and frogs that, with the assistance of assorted aerodynamic skin flaps, leap and glide great distances horizontally.

High above the continuous canopy, a few towering trees emerge and form their own higher, discontinuous forest. The canopy itself may reach up to 160 feet, while beneath the canopy an understory of many smaller and younger trees, perhaps 100 feet high, stretches toward whatever sunlight is available. (The canopy trees themselves seldom waste energy by sprouting leaves and branches within their own shade.) Beneath the canopy and understory, a third dense stratum of trees may grow in a wide middle zone, and a fourth zone of high shrubs may occupy the space six to ten feet above the floor.

Yet another world thrives on the forest floor. Since only a small proportion of the sunlight spreading across the canopy actually winnows down to the forest floor, the vegetation there is comparatively modest — not the impenetrable jungle of pulp fiction, but a sparse growth of tree seedlings, herbs, ferns, grasses, mosses, and so on. The bare earth may be layered thinly with leaf litter and other decaying vegetation, interrupted here and there by the snaky emergence of roots. Since most tropical forests stand on poor soils, tree roots typically spread wide rather than deep. A tropical forest's dense root world may reach down only a foot into the soil. Yet the tall trees are physically supported, as well as nourished, by immense networks of root spreading from the trunk across the forest floor as much as a football field's distance in all directions. Some tree species are also propped up by dozens of root stilts that anchor overhanging branches to the ground. Other trees are supported by three or four or a half dozen radiating buttresses — thin, hard triangles of exposed root that may reach twice a man's height or more off the ground and extend far outward at their base.

The full spectacle is grand, even by the crudest measures. In the great rain forests of Borneo one finds trees second in height only to a few redwoods and giant sequoias of California. In a forest of 160 to 200 trees per acre in Borneo, the massive, unencumbered columns of some 20 trees will spread into a canopy 160 feet above the ground — the height of Niagara Falls — with occasional four- to six-century-old emergent giants towering into their own broken forest at perhaps twice that height. (If they grew in Manhattan, some of these gargantuan plants would dapple sunlight a quarter of the way up the side of the Empire State Building.) If all the vegetation from one acre of tropical forest were placed on a scale, it would weigh (dry weight) 120 to 200 tons, or even up to 400 tons (with a record of about 480 tons per acre measured in Panama), which is several times the biomass of a temperate forest acre. Measure the growth of a tropical forest, and you'll discover it produces 14 to 24 tons (dry weight) of new plant life per acre per year, which is among the world's highest rates of vegetative growth and about two to four times the productivity of temperate forests.

Tropical forests are layered and plenitudinous. They are unusually productive and very big. Even the most casual walk inside a tropical forest gives the impression of powerful growth and size. In fact, both flora and fauna sometimes take gigantic form. Examine individual species and you'll find not only some of the smallest plants and animals anywhere, but several of the largest: yard-wide flowers (the root-parasite "stinking corpse lilies" of the *Rafflesia* genus), six-foot leaves (of the *Monstera gigantea* vine), grasses up to sixty feet tall (bamboos), seven- to eight-inch-long insects (the Hercules beetle of Amazonia), butterflies and moths with foot-wide wingspans (the Queen Alexandra birdwing butterfly and the owlet moth), eleven-inch millipedes, seeds the size of hens' eggs, and seed carriers (including such nonpredatory birds as macaws) large enough to carry them.

Diversity. Yet height and weight, productivity and gigantism, tell the least about tropical forest plenitude and are measures mainly relevant to loggers. Most relevant to human prosperity at large is the tropical forests' plenitudinous diversity.

The overriding pattern is this: tropical forests cover only 7 percent of the earth's land surface but contain 40 percent or more of the total species. In other words, of the earth's five to ten million different species of plant and animal, some two to four million live in the tropical forests.*

Among the details are these: an acre of temperate forest will have approximately four species of tree, whereas in an acre of tropical forest one may find twenty to eighty-six tree species. In the United States the entire Appalachian forest formation supports a total of only about two dozen tree species, whereas a 25-acre piece of northwest Borneo contains 760 tree species. The entire state of New Hampshire grows about a dozen tree species, whereas a backyard's worth of forest in the Choco region of Colombia contains more than sixteen times that. The whole of the United States sustains about 20,000 plant species altogether, whereas the nation of Colombia, covering about one-eighth the area, contains a quarter again as many plant species. Great Britain holds 1,443 native plant species, whereas Costa Rica, with a fifth the area, contains five to six times more species.

Aside from thousands of plant species, a "typical" four-square-mile section of tropical forest may have 125 mammal species, 400 bird species, 100 reptile species, 60 amphibian species, 150 butterfly species, and — as a very rough estimate — perhaps 42,000 insect species. Southeast Asia altogether supports 656 mammal species, whereas all of Europe, with four times the area, sustains only a fifth that number. Southeast Asia is home for a third of the world's amphibian species. Peninsular Malaysia has twice as many bird species as are found in all of Europe west of the Soviet Union. The La Amistad National Park in Costa Rica supports, in 840 square miles of forest, more bird species than exist in the full 7.7 million square miles of North Amer-

*Forty percent of five to ten million species may be a conservative estimate. Some experts believe there may be as many as thirty million species on our planet, with 90 percent of them living in the tropical forests.

ican continent. Amazonia alone holds a fifth of the world's bird species. Less than half a square mile of forest in Liberia, West Africa, holds as many butterfly species as does the entire United States. The Amazon River contains 2,000 known (and possibly another 1,000 unknown) fish species, eight times the number swimming in the Mississippi River, ten times the number swimming in all the rivers of Europe. Slightly more than ten square feet of leaf litter on the floor of a tropical forest was discovered to be home for fifty different ant species. A single tree sloth from Panama was found to carry in its fur three species of algae, and three beetle, three moth, and six mite species. From big to little, gigantic to microscopic, living matter swarms multitudinously in a tropical forest, displaying an astonishing diversity found nowhere else on earth.

Dispersion. The coin has two sides. The great, green feast of the tropics nourishes such a remarkable diversity of type, of species, in part because the tropics provide such a vast pantry of continuous nourishment and in part because the individuals of each type are remarkably dispersed.

If we look at temperate-zone forests, we see vast stretches of sameness. Northern America's forests, for instance, feature much the same tree species across the entire continent. By contrast, the Amazonian forest can be mapped into roughly eight distinct forests or *phytogeographic zones,* large patches of distinctive floral and faunal assembly. Throughout the world's tropical forests, in fact, plant and animal species are distributed in patches of concentration. On the broadest scale: even though the lowland rain forests of Zaire, Brazil, and Malaysia are structurally identical, they have almost no plant or animal species in common, and not one common tree species. Nearly half the birds of Papua New Guinea are found only there, and half the mammals of the Philippines exist only there. Nearly a fourth the 1,480 bird species in Indonesia exist nowhere else. Two thirds of the 12,000 plant species existing in Madagascar are entirely local. More than 800 flowering plants in the tropical forests of Sri Lanka are found only there.

So part of dispersed diversity is the reality of local concentrations. Looking more closely, however, we find the species dispersion in tropical forests to be even more profound. For even in a small area, tropical forest species are often thinly represented. If you compare a typical 2.5-acre area of Amazonia containing perhaps 100 different tree species with a same-sized area half a mile distant, fully half the tree species may be different. Corcovado Park in Costa Rica has eight species of *Heliconius* butterfly, but seldom more than one adult of each species per acre. Sixty acres of forest in peninsular Malaysia were found to contain 381 tree species, but 157 of the species were represented by only a single tree.

Mutualism. I first began to appreciate the concept of tropical forest mutualism as Rita Mesquita told me about her ongoing study of *Clusia grandiflora* in the Amazon. The fruit of that tree is an important food source for the bird species she is studying, and the bird is an important seed disperser for the plant. In other words, they are mutually dependent. Actually, several animals, including monkeys, bats, rats, and ants, may depend upon *Clusia grandiflora.* And because it seems to fruit when most other fruit-bearing plants do not, *Clusia grandiflora* may be the keystone in an arch of mutualism. Because it provides fruit when fruit is scarce, its disappearance from the forest could be accompanied by the collapse of a whole arch of fruit-eating species.

Biological diversity and dispersion arise when great and regular nourishment, sun and rain, interact with the powerful force of time — a great and crowded and *extended* evolutionary expression. But extraordinary time, the tens of millions of years of evolutionary continuity that molded the world's tropical forests, has added yet another dimension to that complexity. Species climbing on species are also species depending on species. And many, many tropical forest plants and animals have coevolved, that is, have adapted to each other's crowded presence — both threat and sustenance — so that the matrix of tropical forest life becomes a linked matrix, consisting of immensely complex chains of specialized mutual dependence. Examples of mutualism are many, but let's briefly consider one modality: plant reproduction, the province of seeds and pollen.

Most plants spread their genetic messages through seeds and pollen, and it is important that these messages be cast some distance from the parent, so that offspring do not directly compete with parents for the same resources. Winds cut through temperate forests because they are comparatively open, and many a temperate forest plant takes advantage of this fact by allowing the wind to bear away its seeds and pollen. But tropical forests, insulated by an enveloping, continuous canopy, are virtually windless. A seed falling in a tropical forest will drop almost straight down. So very few tropical plants send their genetic messages on the wind. Rather, they depend upon active, living messengers: birds, insects, mammals such as primates and bats, and (especially in seasonally flooded parts of Amazonia) fish. No one does anything for free, of course, and all those seed and pollen messengers require a certain payment for their services. The payment for flower pollinators often is the food of nectar. The payment seed dispersers require often is edible fruit, and thus tropical forests abundantly produce large and tasty fruits.

These payments, fruit and nectar in exchange for seed dispersal and pollination, are part of a barter economy that has gone on continuously for tens of millions of years. And in very many instances the barter has become quite specialized. More than 900 fig species in the tropics depend upon wasps for pollination. In turn, the wasps' larvae eat figs. But — and here is the critical nub of mutualism — each individual fig species provides for and is pollinated by its own specialized wasp species. The Brazil nut tree depends upon a single bee species to pollinate it and a single, squirrel-sized rodent, the agouti, to crack through its hard seed covering — the nut shell — to allow the seed to germinate. The crucial bee species, in turn, depends upon material from particular orchid species to stimulate its mating activities. The Southeast Asian durian tree produces a very large fruit that is considered superbly tasty by humans, who have eaten it since prehistoric times. According to the aesthetic proclivities of human noses, the durian fruit also happens to stink — but that same powerful smell serves as a floating trail of attraction for the tree's *single* specialized pollinator, the bat species *Eonycteris spelaea*.

Fragility. Both of the field assistants in our camp in the Amazon had brought along their own boombox tape players, and at night they would click in tapes of Brazilian sambas and North American rock music. That is to say, sometimes the forest was noisy and sometimes we were.

I generally think North American popular music is OK, although not as good as it was when I was younger. Once the Brazilians played Stevie Wonder, whom I consider excellent. But I began to like the sambas and dislike the rock music, even Stevie Wonder. The sambas, with their lazy, drifting rhythms, their mild syncopations and soft repetitions, seemed in phase with the intricate and regular sounds of the forest. The rock music, with its driving beat and rebellious ruckus, seemed aggressively foreign, a barbaric yawp from the north. It didn't belong there in the forest, and it sounded more and more like stupid banging, a tormented hammering from inside the dungeons of adolescence. I began, then, to believe that the Amazon, for all its immense size and ceaseless noise, possessed its own subtle fragility, in sound and, by analogy, in form. If the forest is an extravagent engine of growth, powered by immense solar energy, it is yet finely, subtly meshed and ultimately fragile.

Tropical forests have emerged in conditions of great environmental stability and have in fact increased that stability by creating their own insulating roof and envelope, the canopy. However, a forest evolutionarily attuned to a 75-million-year summer is deeply vulnerable to a single quick season of disruption. Temperate forests simply close up shop during the winter. Leafy trees drop their leaves and sleep. Animals migrate or hibernate or adapt in other ways. By contrast, tropical forest animals rarely migrate. Food stands before them the year round. And tropical forest plants are not adapted to seasonality. They fruit and flower throughout the year. Temperate forest seeds may lie dormant during freezing winters and hot summers for up to ten years before germinating. But tropical forest seeds can survive only three or four weeks of dormancy before either germinating or dissolving. More generally, tropical forest seeds can be highly sensitive to even minor oscillations in environmental conditions. For instance, seeds from the hundreds of species of Southeast Asian *Dipterocarpus* trees can germinate only between 73 and 80 degrees Fahrenheit or so. When a logging operation removes even a few canopy trees in a Southeast Asian forest, allowing the sun to penetrate, the forest floor temperature may rise beyond the usual zone of fluctuation, and the dipterocarps will cease to reproduce.

Diversity in itself suggests resilience, but the dispersed diversity of tropical forests actually contributes to their fragility. A localized dispersion of species means that species are easily extinguished by local disasters. The 8,000 plant species found only in Madagascar can easily be extinguished by the rapacious deforestation now occurring on that island. The more than 800 flowering plants found only in Sri Lanka are highly threatened by the speedy destruction of that island's 700 square miles of remaining forest. Likewise, a thin dispersion of species, rareness, means that even minor perturbations of a tropical forest can extinguish species. If a tree or other plant, a mammal or bird, a butterfly or other insect is represented by only a few individuals in an area, then seemingly trivial alterations of that area, eliminating merely a few members of the species, can result in a breakdown of the entire kind.

A most fundamentally fragile quality of tropical forests is the quality of mutualism. Since tropical species depend upon each other in a highly specialized manner, the tropical forest actually expresses itself as a complex and living chain or rope, with links that can be broken, strands that can come unraveled, in a pattern of unimaginable intricacy. Destroy a bat and you destroy a fruit tree. Eliminate a wasp and you eliminate a fig tree. In turn, the fig tree may have provided crucial sustenance for some other flying messenger, which may have been a crucial pollinator for another tree. Clear a flood-plain forest in Amazonia, and you may be clearing dozens of species of fruit-eating fish.

Cut down a particular flowering tree in Central America, and you may be cutting down several species of hummingbird that depend on that tree during a few weeks each year when no other tree flowers. Such a labyrinthine linkage of life forms makes it impossible to modify simply a tropical forest ecosystem. Extinctions bloom into extinctions in unforeseeable directions and dimensions.

Beneath such foci of fragility lies yet another: poor soils. Soil consists of air, water, decomposed organic matter, and decomposed rock. The quality and proportion of these four elements determine the quality of a soil, but most often the predominant element is rock. Unfortunately, most soils in the wet tropics are based on very old geologic formations. Amazonia, for instance, sits atop two massive plates of rock, the Brazilian and Guyanan shields, that are billions of years old. Over that vast time, surface minerals important to vegetative growth have been leached out by rainfall, so the soils are unusually infertile. It is true that the Andean mountain range, marking the western boundary of Amazonia, is comparatively young, formed only a couple dozen million years ago, and still nutrient-rich. Through erosion, the Andes have cast rich sediments into a foothill strip of western Amazonia. And the rivers originating in the Andes, including the Amazon itself, have distributed sheets of Andean sediment on several wide floodplains across the basin. It is also true that elsewhere in the tropics, river sedimentation and recent geologic events such as volcanic eruptions have strewn and spewn strips and pockets of unusually fertile soil.

Still, nearly two thirds of soils in the wet tropics are of similar types: red- or yellow-tinged, acidic, infertile soils technically known as *oxisols* and *ultisols,* which we can simply call tropical red soils. Tropical red soils usually have good physical characteristics. They are deep and well-drained, for example. But chemically they are impoverished, even hostile: highly acidic, with high aluminum toxicity, and crucially deficient in phosphorus, potassium, calcium, magnesium, sulphur, zinc, and other minerals. Temperate-zone farming may yet be possible on several of these soils, but it would be efficient only with sophisticated planning (with specific crops and particular rotational schemes) and the heavy addition of fertilizers. Moreover, the fertile and well-drained soils that cover 15 percent of the wet tropics altogether are, in many instances, already cleared and under cultivation, and they occur primarily in wet tropical Asia.

The great paradox, then, is that the world's most productive ecosystems are growing on some of the world's least productive soils. How do tropical forests sustain their almost unrivaled productivity on such inhospitable soils?

Any forest partakes of an economy of nutrients that are both poured and stored — that is, added to the forest and recycled within it. An important source of added nutrients for forests in the wet tropics is rainwater. In one part of central Amazonia, rainwater adds, per acre per year, almost two pounds of iron, under three pounds of phosphorus, and about nine pounds of nitrogen. And whereas in temperate forests falling rainwater actually draws nutrients away from high leaves and decaying organic matter, in tropical forests growing on poor soils the process is reversed. At the top of the forest immense, multilayered leaf systems act in concert with leafless plants such as lichens and algae to filter and drink the water and to scavenge nutrients.

In forests with rich soils, up to nine tenths of the system's minerals are stored in the soil. But tropical forests on poor soils present a complete reversal of that picture, with the great majority of nutrients stored in the living plants themselves. A test patch of Venezuelan forest, probably typical of poor-soiled tropical forest, was found to contain three fourths of its principal minerals above the ground, in the plants. Adding ground-level organic matter — litter and humus — this forest maintained 92 percent of its principal nutrients in organic form.

A tropical forest economy functions by means of extremely efficient recycling, a very miserly exchange from death to life. Miserly? Nutrients from dying plants are neither lost nor left out of the living economy for long. Instead they are very quickly recovered. Leaves that fall to the ground, for example, decompose and are recycled in about six weeks (compared to the year-long decomposition of leaves in temperate forests). Decomposition is assisted by steady heat and humidity, by fungi and bacteria, and the feeding and fragmenting activities of grasshoppers, katydids, crickets, ants, termites, cockroaches, earthworms, millipedes, and the like (all of which in turn provide food for such predators as spiders, centipedes, beetles, ants, frogs, lizards, and small snakes).

Recycling directly from plant to plant is accomplished most impressively by roots and certain fungi. Whereas in temperate forests, root systems specialize in drawing minerals from the earth and thus tend to propagate downward, tropical forest roots direct their attentions elsewhere. They spread widely, bifurcate finely into broad and shallow, very dense mats that are extremely adept at absorbing minerals near the surface. Some feeder roots even climb up convenient tree trunks to drink dribbling moisture and minerals. And when a leaf or other litter falls on top of the ground-level root mat and begins to decompose, tiny root-tips actually reach out and attach themselves to the detritus and directly draw forth nutrients. In addition, the roots of many tropical trees, especially in poor-soiled forests, are aided by extensive colonies of cooperative fungi. These fungi grow on and around roots (the symbiotic combinations are called *mycorrhizae,* based on the Greek words for *fungus* and *root*) and critically increase the roots' abilities to absorb several minerals. Mycorrhizae may be essential for the growth of several or most tree species in poor soils.

Impressive though it is, however, this efficient economy of nutrition masks the fragility inherent in poor soils. Quite obviously, clearing such a forest clears away most of the stored minerals. Burning such a forest scatters, in the form of ash, a thin, rich layer of some of the plant-stored nutrients, which will be washed away and leached down into subsoils within a season or two. Either burning or opening the canopy of a tropical forest to the sun may destroy the root mat and kill the crucial mycorrhizae, making forest regrowth difficult or impossible.

OUT OF THE FOREST, driving in the jeep back to Manaus, we crossed cattle ranchland once again. This time I asked the driver to stop so I could get out and look around.

The sky was an immense blue dome crossed by wind-driven clouds of great size. Wind blew across the open land, and in the far distance I could discern the green wall of forest we had left. But between me and the forest stretched a quiet, deathly moonscape pasted with sunlight: a ghost forest of old tree stumps, weathered to a bony, dirty white, sometimes fire-blackened, often topped with the mushroom-shaped nodules of termite nests; random scatters of fallen, cracked logs, like piles of huge old bones; and here and there the desolate skeletons of hugely grand trees still standing, yet stripped, dry, charred, with a few high limbs still stretched out or broken and dangling, and the occasional dark bird sitting on the occasional limb like a dirty rag. The ground was baked and reddish, irregularly covered with a waving mat of rough grass.

I was looking at two important versions of the Amazon: in the distant background a forest as it has existed since before the birth of Orion; in the foreground a ranch as it has existed for perhaps ten years. The forest could easily be changed into ranchland, but everything I have read about the subject suggests that the ranchland may never again become tropical forest. Which is better?

From the perspective of a biologist, and from the perspective of most plants and animals a biologist studies, the ranchland was now practically a desert. I didn't even see cattle. The only evidence of life I could distinguish, looking all the way to the forest wall at the horizon, was grass, many large termite nests, and a few birds. From the perspective of a rancher, of course, the land was not at all desert; it was appropriated and cultivated soil, from which might emerge personal wealth and perhaps some measure of fulfillment. At best, Brazilian ranchers imagine a wealthier nation in which the nation's present-day poverty is overcome by a new prosperity for all. Yet the record of most ranching enterprises in Central and South America is not encouraging.

As of 1980 some 950 million beef cattle were dreamily transforming vegetation into meat on this planet, mostly in the United States, the Soviet Union, New Zealand and Australia, Central and South America. That is the supply side. And the supply has recently been dwindling: per person, world production of beef declined from 25.5 pounds in 1976 to about 22.2 pounds by the early 1980s. In contrast, the world demand for beef, at least in the industrialized world, seems ever-increasing.

Demand outstrips supply, resulting in higher prices and expanded markets. The United States, the world's biggest producer of beef, is also the biggest consumer and biggest importer. As recently as 1960 the United States imported almost no beef, but today around one tenth of the beef eaten in the United States is imported — most from New Zealand and Australia, but a significant portion from Central and South America. North Americans prefer beef raised on high-quality grain, which laces fat into flesh. Central and South American cattle feed mostly on rough grasses, which makes for leaner, less desired flesh. Once it reaches North America, therefore, grass-fed Central and South American beef is usually turned into fast food and processed meat: hamburgers, hot dogs, sausages, baby food, pet food, and so on. Southern meat costs about half the price of home-grown beef, so when a North American goes down to the local fast-food outlet and purchases a burger, he or she may be saving a nickel, thanks to what tropical forest expert Norman Myers has dubbed *the hamburger connection*.*

What does the hamburger connection mean to Central and South Americans? Clearly, some people derive a profit from it. But since most land is owned by the already rich and politically powerful, that profit is not well distributed. While Costa Rica, for instance, expanded its beef production fourfold from 1960 to 1980, local consumption

*Since Myers popularized this notion, some North American fast-food outlets have stopped buying imported beef, while others insist they never did. McDonald's 3 billion yearly hamburgers are extracted from only 300,000 home-grown animals.

of beef actually declined by nearly half, to 34.6 pounds per year, which is less than what the average North American domestic cat purrs over.

What does the hamburger connection mean to Central and South American tropical forests? Since tropical forests have commonly been considered wasteland in the American tropics, wealthy ranchers easily acquire large tracts of forest land. The rancher may then hire local labor to cut and burn away the forest and plant rough grasses, or alternatively he may allow landless peasants to clear and farm the land for a season before taking it over. Once the land has been cleared and planted with grass, it will sustain a few beef cattle — for a while. But within five years the soil will already be severely depleted of nutrients. Without added fertilizer, the soils continue to decline rapidly until they are worthless as ranchland, and the rancher must move on to destroy new forest. All told, ranching in the last two decades has destroyed fully one half of the tropical forests in Central America and is now removing perhaps 7,700 square miles of additional forest in Central and South America each year.

Brazil has actively promoted large-scale cattle ranching in its share of the Amazon. In 1973 Brazil was behind only the United States and the Soviet Union as a home for cows and bulls, and Brazilian planners foresaw that beef exports could knock a nice dent in the nation's huge foreign debt. Thus, prospective ranchers in the Amazon were offered large land concessions plus a generous income tax write-off (50 percent) on other projects in Brazil.

Cattle ranching may not be inherently profitable in Amazonia, even though the land is cheap (not counting the usually ignored environmental costs) and can be cleared cheaply. Nonetheless, the tax incentives tempted large numbers of Brazilian and foreign investors to put on hats and spurs and become ranchers. A United States consortium, Brescan-Swift-Armour-King Ranch has taken control of 280 square miles in eastern Amazonia, for example. Other U.S. groups, including Twin Agricultural and Industrial Developers, Anderson Clayton, United Brands, Gulf and Western, and Goodyear, have likewise become Amazonian ranchers. George Markhof of Austria is an Amazonian rancher. From Japan, Mitsui and several other groups hold land. Liquigas of Italy has a 2,000-square-mile ranch. And Volkswagen of West Germany, the car manufacturer, holds 540 square miles of Amazonia in southeastern Pará State, of which half is supposed to be ranchland for 120,000 beef cattle.

Several recently opened Amazonian ranches, however, have already declined significantly in productivity, and some are now abandoned. Of almost 10,000 square miles of ranchland created since 1960 along the Belém-Brasília Highway, one fifth was, by the end of the 1970s, in an "advanced stage of degradation." Altogether, over 30,000 square miles of Amazonian forest were denuded to create 336 ranches between 1966 and 1978, and an uncertain but large additional amount of forest has been cleared for many thousands of other ranches of varying size. Today clearance for cattle ranching may be the biggest single form of forest destruction in Brazilian Amazonia.

While Latin American tropical forests are being cut down to provide a few years' grazing for beef cattle, largely in deference to the tastes and appetites of eaters in the developed nations, the tropical forests left standing could provide a continuous supply of wild animal protein for local consumption. In many tropical nations, people habitually obtain half or more of their animal protein from wild mammals, birds, reptiles, and amphibians. In West Africa many people depend upon the meat of cane rats; although planners seldom consider this fact when evaluating commercial forest exploitation, the habitat of the cane rat thus possesses a clear, if hard to quantify, value. Similarly, the agouti and paca of South America, as well as the bandicoot and deer mouse of tropical Asia, are valuable food resources. Wise conservation of the forests would preserve such important local sources of animal protein.

Latin American ranchers manage to profit from raising cattle at a low level of efficiency largely because it has been easy to acquire new forestland. Ranchers have found it cheaper to clear and burn new sections of forest for new grazing land than to use more efficiently the land they already have. Yet cattle ranching in Latin America could become a less wasteful enterprise. Brazil has lately eliminated the tax benefits once given to wealthy corporate ranchers in Amazonia, thus forcing them to examine more closely the economic realities of what they are doing. Other Latin American governments could legally discourage the quick expansion of ranches into new forestland. The United States could simply stop buying Central American beef, which fosters major ecological disruption *there* but constitutes only 1 percent of the beef market *here*. Or the United States could help persuade Central American ranchers to do what they will eventually be forced to do anyway, that is, diversify and improve their efficiency.

Improving the efficiency of a ranch means getting more meat, milk, and leather per unit of land over time — using the land more intensively, using it intelligently enough that it needn't be abandoned in a few years. Improving efficiency might include acquiring better cattle breeds, controlling pests and diseases more thoroughly, and improving the grasses on which cattle feed. Finally, tropical ranchers might increase their efficiency simply by switching. That is, they could expand their repertoire, and raise some of the meat-, milk-, and leather-producing species native to the tropics: new domesticates.

Just as most of the world's commercial plant foods come from an astoundingly small number of cultivated species, so most of the world's commercial meat comes from just a handful of domesticated mammal and bird species: cattle, goats, sheep, pigs, chickens, turkeys, and ducks. Some of these domesticated species originated in the tropical forests. For example, two of today's most important domestic birds, chickens and turkeys, include in their genealogies tropical forest fowl. Domestic chickens, around five billion egg layers and meat producers today, all trace their family tree back to the red jungle fowl of India and the guinea fowl of West Africa. Domestic turkeys hark back to *Meleagris gallopavo* of South America, which may have been domesticated first by the Aztecs. Domesticated water buffalo in Southeast Asia are descended from the wild *Bubalus arnee*. The Bali cattle of Southeast Asia are descended from wild bantengs. The humped zebu cattle of southern Asia are descended from wild koupreys of the forests bordering Cambodia and Thailand. A few wild koupreys may still remain; if they could be found and caught, they could be cross-bred with their domesticated relatives and provide a critical immunity to the disease of rinderpest.

My point is not that we should be grateful to the tropical forests for providing past domesticates, but rather that we can yet look to the forests as a source of new domesticates. Several tropical forest species might profitably be domesticated or semidomesticated in the future without causing the kinds of ecological havoc associated with beef cattle. Animals that already thrive in the tropics ought to be less susceptible to tropical diseases than ordinary domestic cattle. Wild water buffalo, for example, produce substantial quantities of milk and about the same amount of meat as beef cattle. Because they do well in wet land, they might be excellent domesticates for the floodplains of Amazonia. The capybara, a South American rodent the size of a large dog, is said to be entirely palatable. It efficiently puts meat on its bones by eating coarse weeds and swamp plants, covers itself with an excellent leather, and reproduces six times faster than domestic cattle. Capybaras could be grown domestically on what is otherwise swampy wasteland; actually, they are already ranched in Venezuela. The manatee, a very large aquatic mammal of the New World tropics, likewise feeds on water plants that have no direct value to human nutrition. It might be farmed in tropical lakes and deep swamp areas. Seven aquatic turtle species of Amazonia, which also feed on what humans consider waste vegetation, might provide an annual edible meat supply of five tons per acre of lake, roughly thirty-five times the yield of domestic cattle grazing on the same area of cleared forest land.

SCIENCE FICTION: a nuclear power plant in New Mexico blows up and melts down, broadcasting a cloud of radioactivity that renders half the state uninhabitable within a year. The plant contains an unusually large amount of fuel, and heroic efforts to stop the steady dispersal of radioactivity are quite unsuccessful, so that within two years the entire state has to be evacuated. Well over a million people are moved elsewhere, but radioactivity continues to rise above the plant and be swept into the atmosphere, where it is carried farther and wider each year. Some eminent scientists are publicly stating that radiation from New Mexico now spreads across almost 80,000 square miles of additional United States territory each year, the area of Indiana and Ohio combined. Some of the radioactive elements have relatively short half-lives, and it is expected that people will be able to return to some of the affected areas within a few years or decades. But an area of about 36,000 square miles — the size of Indiana — is being dusted with an extremely slow-decaying radioactivity, so that for all human purposes, some 36,000 square miles of the United States each year is rendered permanently unfit for habitation. A monumental disaster! How many *People* magazine celebrities will lend their figures and physiognomies to the cause? How many tough-talking corporation executives will hold hands with sequined rock stars? How large a portion of the Gross National Debt will be committed to stopping the further spread of radiation?

Science fact: the habitat-nation of most of the world's nonhuman primates — that belt of forest stretching across the tropics of Central and South America, West and Central Africa, and South and Southeast Asia — right now covers perhaps 3.3 million square miles, approximately the land area of the United States. Although the tropical forests express 75 million years of evolutionary stability, nearly half their original area has been eliminated in the last two centuries, mostly since World War II. Today people are altering the tropical forest cover at the rate of roughly 80,000 square miles *each year*, equivalent to more than half the area of New Mexico, or the combined areas of Indiana and Ohio. Some of those forests are only partially degraded and, given the right conditions, could return to something resembling primary forest within a few years, decades, or centuries. But each year about 36,000 square miles of tropical forest is totally eliminated. What was tropical forest is now permanently something other than forest — permanent ranch or farm, permanent grassland or scrub, or eroding wasteland.*

One reason that the forests are being destroyed at this rate is simply because we now possess the technological magic to do it. Bulldozers, chainsaws, and pulpwood chippers appeared in almost the same historical instant as computers and may even more profoundly determine the shape and form of the world to come. Important as the new tools and machines are, though, it might be most enlightening to consider the social and economic forces that drive them. In a geopolitical sense, the tropical forests belong to the developing nations: that's where the forests are. But by and large, both developed and developing nations are today blindly reaching into the tropical forests with bulldozer, chainsaw, chipper, and match, and drawing out, once and for all, a few decades' worth of products to satisfy immediate demands.

The world's demands on the tropical forests are vast, absolutely unprecedented, and accelerating very rapidly. In the developing world, the size and acceleration of those demands is caused by several factors, but most pervasively by one: human population growth. While the industrialized world is not quite exploding in population, it more than compensates for that moderation by the immoderate and expanding appetites of a consumerist lifestyle. Together, the developing and developed nations approach tropical forests in pursuit of four major commodities: fuelwood, commercial wood, farmland, and ranchland. Worldwide, ranchland is the least of the commodities taken. But in Central and South America it is the greatest.

We can take the story of Noah's Ark as the literal telling of a literal truth, or we can see it as a myth or parable. In either case, it is a significant and inspiring story of worldwide ecological disaster in which some humans, by reasoned choice, seized an opportunity to preserve global wildlife for future generations. But the Deluge of our time, threatening a large portion of the five to ten million living plant and animal species, will finally require an Ark that cannot easily be measured in cubits, that must be deeper and more integrated than the desperate cages of captive breeding, that must be larger and more secure than even the biggest and best of national parks. The Ultimate Ark is the earth itself, that cloud-strewn, blue and green sphere turning through the lonely vacuum of space.

*These figures are based upon a thorough study sponsored by the United States' National Academy of Sciences, carried out by Norman Myers, and reported in his *Conversion of Tropical Forests* (1980) and *The Primary Source* (1984). Some other reports on deforestation have been more conservative, but nearly all agree on the general scale of loss.

The common and less significant changes on our Ultimate Ark occur right before our eyes — daily politics and local economic fluctuations, weekly dramas and disasters in the cities and countrysides, yearly developments — but the most significant and enduring changes are sometimes too large and expand over too great a time to be seen directly: profound and sometimes irreversible changes within the full mantle of life.

We have a limited perspective, and the deepest function of education, science, and the accumulated wisdom of culture is to expand that perspective. Yet most people and nearly all governments act as if the loss of wild species is a secondary problem, an occasion for occasional sobriety, perhaps, but not alarm. Few people are aware that soon after this century's end we may have destroyed, in large part through short-sighted and crude attempts at exploitation of the tropical forests, up to one million wild species. Even fewer seem particularly alarmed by today's global spasm of human-caused extinctions, except where it affects the larger and more dramatic species. Yet many of those in the best position to know the facts and their implications, that is, botanists and zoologists, are very alarmed. As Professor Edward O. Wilson of Harvard University expresses the matter: "What event likely to occur in the 1980s will our descendants most regret, even those living a thousand years from now? My opinion is not conventional, although I wish it were. The worst thing that can happen — *will* happen — is not energy depletion, economic collapse, limited nuclear war, or conquest by a totalitarian government. As terrible as those catastrophes would be for us, they can be repaired within a few generations. The one process ongoing in the 1980s that will take millions of years to correct is the loss of genetic and species diversity by the destruction of natural habitats. This is the folly our descendants are least likely to forgive us."

Much as we now peer back at fourteenth-century Europe and see superimposed over a multitude of smaller events the deep scar of the bubonic plague and its implications for the political, economic, social, and psychological tenor of the period, so future historians may describe our small cusp of time at the end of a millennium as the Era of Extinctions: up to one million species erased by human design. Long after we have dropped our momentary preoccupations as expressed in the Kingdom of the Best-Seller, long after we have forgotten our deeper concerns as measured by thousands of miles of newsprint, we may be forced to remember what we hardly knew: the violent dismemberment of the world's tropical forests and the quiet eclipse of a million species.

Three

PIECES

BACK IN MANAUS, I took a room in the cheapest hotel I could stand and spent a few days looking around.

One day I entered the most prosperous tourist shop in town and observed an excellent summary of every tourist item I had seen anywhere else in town, although the prices were doubled and tripled: stacks of Indian baskets, piles of necklaces, genuine Amazonian arrows and blowpipes, straw human effigies with hyperbolic penises, rows of small latex animal figurines, rough scales from the pirarucu fish, rough and bony tongues from the pirarucu, as well as postcards and T-shirts with Amazonian themes. On the wall was a World Wildlife Fund sticker with a picture of the muriqui and the slogan, "Save Brazil's Endangered Primates."

I noticed that the most expensive necklaces, displayed inside glass cases, were made of beads and small teeth — flat, rectangular incisors about the right size for the tooth fairy. Monkeys' teeth. And then, in one corner of the shop I saw something I had read about: monkey-tail dusters, a pair of them. Monkeys are killed, their tails cut off. The vertebrae are drawn out, and wooden sticks are thrust in to make the tails rigid, resulting in wooden-handled household dusters, functional and attractive. One reason this shop was so expensive was that the salespeople spoke English, more or less, and so I caught the attention of one and asked what kind of monkey the two tails were from. He said, "Tail of a monkey." I said, "I know, but what kind? What type?" He turned to the muriqui sticker on the wall, pointed to it, and said, "Thees one." "But that's a muriqui," I said. "It's very rare, and not even an Amazonian species." He insisted it was a muriqui tail.

I asked to talk to the shop owner, and eventually was greeted by an American. "I wanted to ask you about these monkey tails. One of your workers told me they were from muriquis." "I think so, yes." "It can't be — the muriqui isn't even an Amazonian species." "Maybe not, then. That's what the person who sold it to me said."

As it turned out, the shop owner and I had a mutual acquaintance, a person I'll call Gordon Zeudel, an anthropologist who had been in Manaus several times during the past few years. "Zoodie! God! How is the old boy? Married? I'm so embarrassed! I knew he got married, but I never sent him a wedding present. Tell you what: I'll fix up a little something for you to take back to him, a wedding present. All right? Oh, Zoodie is great. He's really nuts about animals, especially monkeys. I love animals too — that's one thing Zoodie and I have in common. Want to see my pet margay?"

I followed him upstairs to his barren, bachelor apartment above the shop. The walls were completely bare, the floors bare wood, and the living room had no furniture except an old wooden table, two wood chairs, and a ratty couch. The whole place stank of cat piss, which was from the margay, which was sitting on top of a refrigerator in the kitchen. The margay was impressive. He looked rather like a domestic cat, but three times bigger, spotted, very lean and powerful-looking, with a thin taut face and huge eyes. He allowed me to stroke his head and paws.

Before I left the shop, the owner handed me a small, wrapped package. "This is a little wedding present for Zoodie. Monkey-tooth necklaces. Very fine work. Tell Zoodie, I know he loves monkeys, but tell him he shouldn't be upset: Indians eat the hell out of monkeys." And, since I had asked so many questions about the two monkey-tail dusters, he gave them to me. "These are a gift for you." I took them and left the shop.

The necklaces and dusters preoccupied me a good deal during the next several days, and I went through a complicated internal discussion about what I should do with those pieces of dead monkey, until I finally resolved it in the following way. The monkey-tooth necklaces I mailed to Zeudel. They were a wedding gift from friend to friend, and I had agreed to deliver them. The monkey-tail dusters I took, one night, down to the edge of the Amazon River, tossed them in, and watched them float down and then disappear beneath the black, sliding surface.

The tails I threw into the Amazon were probably taken from woolly monkeys, inhabitants of a large piece of rain forest around Manaus and to the west. One reason I was distressed to find the dusters, though, was that the critically endangered Amazonian monkey I had come to find, the southern bearded saki, happens to be blessed, or cursed, with an extraordinarily full and bushy tail, which is occasionally also transformed into a household duster.

The southern bearded saki inhabits a piece of rain forest about 750 miles downriver from Manaus, south of the eastern end of the river. So one afternoon, I bought a hammock and attached its ends to two hooks fixed beneath the roof of a wooden boat, next to the hanging hammocks of three dozen Brazilians, local people mostly, who were bound for small towns and villages downriver. A horn blasted, an old diesel engine was flung into action, generating motion and electric light, and we disconnected from the wharf at Manaus just as a yellow trace of final sunlight had begun to erase itself from the black sky. We rocked and plowed out into the middle of the river and turned downstream, moving through the black water with a soft tearing sound, our good downriver speed quickly raising a pleasant wind that cooled faces and bodies and rippled the tassels of three dozen and one cotton hammocks. We were following now and would soon overtake, so I imagined, those two monkey tails I had days before tossed into the water.

At first the sky was entirely black and overcast, while in the distance fragments of lightning quivered and sputtered. Later the sky opened up to reveal, dimly, half a universe of stars. People stayed up late and chattered earnestly. Drinkers laughed, domino players slapped dominoes down on a table beneath generated light, two lovers stood hip to hip in the dark stern of the boat. But gradually people climbed into their hammocks, turned in the edges and mummified themselves inside, and went to sleep.

At dawn a breakfast of coffee, eggs, and rolls was served on a wooden table in the middle of the boat, underneath hammocks. And after breakfast I climbed onto the roof, which was railed and decked, and observed a sky of great possibilities: clouds filmy and puttering, smoky and solid, flat crushed pillows and billowing mashed potatoes, white and gray and blue, purple and pewter and, at great distances, a hazy seam between water and sky. In the distance also, before us, behind us, I could see cloud islands with rain underneath, dark hats dropping gray veils — and on occasion during that day and the river days that followed, we would run into one of those hats and veils and be struck by a rush of wind and pouring water. Blue tarps along the sides of the boat would be unleashed and dropped; we would pass through the storm in blue light, with rivulets of water running across the deck beneath our hammocks.

The river itself cast a milky brown surface that tore open before our prow into sheets of white lace.

The river was a wide brown highway that first morning, and the forest on either side remained remote. Pieces of the forest floated along with us down the river, of course: logs, branches, whole trees, and many floating islands of vegetation. One of these islands transported a half dozen complacent white herons; another carried, as I discovered when we approached it, an obscure symphony of frogs. But mostly the forest passed us as a distant and silent green ribbon. Later long islands appeared, the river turned and narrowed, we approached the forest more closely and could see its complex variations on green: upward fulminations of giant trees, lower penetrations of thick vegetation, palm fronds hanging like split paper, featherform palms with dense bursts of green, tree trunks thickened and covered with a waterfall of climbing vines, a snare of plants reaching out at the water's edge with leathery leaves. Sometimes we came close enough to see white or red or yellow flowers within the green; the curved elegance of a heron, standing bright and white within green grass; the sudden burst of a flock of small birds; the slow undulations of large birds pumping through the air; the wheel and glide of lone, long-winged predatory birds and vultures. On occasion we moved right into the shadow and reflection of shore, close enough to hear the forest: a fluting and crying and whistling of birds, a sizzling of insects.

We rolled down a river of epic dimensions — eleven times bigger than the Mississippi, crossing almost an entire continent, draining with the assistance of 10,000 tributaries one fifth of all the fresh water in the world. The forest along the shore vibrated deeply; the river below us rolled deeply to the sea. And we rolled with it, talked, drank foamy beer, dreamed fragile dreams, dropped pieces of civilization off the stern: fruit fragments and bone bits, bottles and cans, plastic cups, cigarette wrappers, papers, excrement. We rolled past people on the shore — Amazonians dark and lean, many of Indian ancestry, who farm the shore and fish the river, who stood before or within their thatched and stilted wooden cottages. Smoke might be rising from the back of a cottage, laundry flagged out in front, and typically a small wooden boat was tied up to a post or pier at the water's edge. Sometimes, too, you would see signs of placid domesticity: red flowers in pots before an open door, a tame green parrot perched on a porch railing, a white cat half asleep at a window ledge and half lifting its head, five children sunning themselves at the river's edge. We passed gardens and small plantations, often pigs and dogs and chickens, sometimes goats and cows and horses. We passed several small villages, also: a cluster of houses and sometimes a white-washed, one-room mission chapel with a wooden cross on top.

We stopped at the village of Santa Cruz, situated on a small grassy elevation, which included a white stucco chapel, a school entitled Escola Rural Santa Cruz, and a blue wood pavilion, palm-thatched, open on all four sides with a herringbone wooden railing: the Centrosocial Santa Cruz. Behind the Centrosocial wheeled excited white birds, and what I took at first to be scattered pieces of a dead tree in the grass, debarked and white, turned out to be lying-down cattle. A mahogany horse stood near them, eating grass. Under the gaze of a dozen schoolchildren and fewer adults, we tossed a plank down to the wooden dock there, discharged two passengers, and took aboard three big fish the size and shape of pillows, lead-colored with a smear of red. We stopped at actual towns, too, unloaded passengers and cargo — steel pipes, boxes of Tiger Head flashlights from China, bags of dead chickens, small perforated cartons of live baby chicks — and then continued on downriver, departing with a blast of the horn and a roar of the engine.

Meanwhile I was experiencing a breakthrough with Portuguese. Portuguese is not so very different from Spanish after all, I discovered, and I happily watched my functional IQ rise from 50 to 80. I became facile enough to share a couple of beers and half an afternoon with one young man, a twitcher and leg shaker with curly hair over a brief forehead. He wore flip-flops and a nylon basketball suit, and told me he was returning from Manaus to his home in Santarém, downriver. "Manaus stinks," he said, "but Santarém is beautiful. You can know Manaus in three days, but Santarém will take six months to know." He glanced to the right, to the left, crackled his plastic beer cup, shook his legs, and continued. "Tourists come to Brazil, stay a few days, go. The samba, samba, samba. They say, 'Brazil is beautiful, Brazil is good.' They don't see the poverty, how difficult it is to live. Brazil and Ethiopia: the same. Many hungry people in Brazil."

The second evening, I sat on the roof of the boat and shared a couple of beers with a young woman from São Paulo. She had short black hair, dark eyes, black eyebrows, a broad regular face — she was, to summarize, quite good-looking — and was stylishly dressed entirely in red, including red canvas shoes and a red scarf. A hole opened in the overcast sky, and stars appeared. "The stars!" she said. "Very beautiful. You understand?" "Yes: stars. Beautiful. I understand," I said. She waved toward the water and a forested shore somewhere out in the blackness. "Nature: the purity, the simplicity of nature. I love nature. Nature is very beautiful. You understand?" I understood. She was drawing out her vowels now, emoting a poetry in her mind. Some time and more conversation passed before she leaned back, touched my arm, and said, "You're very beautiful. Your hair, your eyes, your beard: very beautiful. You understand?" It was a lie, of course, but I was surprised enough to pretend I didn't understand, and so she repeated herself. Soon after, she said something I actually didn't understand, so she wrote down on a piece of paper the sentence I comprehended as "My name is Jzabel." Jezebel! By the time I recognized my mistake — an initial *I,* not *J* — it was too late: my mind had created and was shamelessly reveling in a Jezebel, dressed in scarlet and behaving seductively.

Soon after, Jezebel's younger sister and brother appeared from down below. Jezebel told me she was twenty-two, her sister fourteen, and her brother eight years old. She began to explain why they had been in Manaus and why they were heading toward the family home in a small village downriver, but the explanation became very complicated. She sent her sister away, and soon her sister returned with a newspaper, which was unfolded in front of me under a deck light. The newspaper told of a government official who had been murdered a few days ago by an unknown assailant with a pistol. There were photographs, one a close portrait of the dead official lying in his casket, another of the mourning family around the casket. She pointed to the dead man: "My father." She pointed to a woman of the mourning family: "Me."

In the night the boat swerved into swamp and headed for a certain grounding on shore, then swerved back out again at the last minute. Later we turned into a channel and pulled up at a dock and a small town, lighted by a few white light bulbs. Some people stood at the dock beneath swooping brown bats, while we unloaded bags of rice, boxes, a refrigerator, as well as Jezebel and her sister and brother, who waved good-bye.

So I rolled along downriver, and so pieces of the river and the life on and along it scrolled before me. Halfway down I rested for a few days at the town of Santarém before boarding another boat.

This second boat almost lost a grandmother one night. We came to a town: a few white lights in the black, a few flat edges of buildings illuminated by the lights. We plowed toward a large pier with a rounded roof, under which men stood, insubstantial in a pool of bluish light. I stood next to the boat's wheelroom, and the captain stood beshadowed inside. I heard the captain ring twice on a doorbell, signaling the engineman in the lower stern to cut the engine. Nothing happened. The captain rang again, ring, ring, but still nothing happened — was the engineman asleep? — and we continued plowing through the water toward the pier. The captain rang a third time, ring, ring, then a fourth. Finally, the engine was disengaged, but we were still moving fast, gliding now into shallow water. Three rings: the engine rumbled into reverse, slowing our forward motion. The captain twirled his wheel; the boat slowly maneuvered sideways, sideways toward the dock on high pilings. Two rings, the engine neutralized again, the captain still twirling his wheel, and the prow of the boat settled heavily into a piling, cruuuuuuunnnch! Then it slowly bounced back. With the engine softly switching forward and reverse, without tying up at the pier, the captain shouted orders for his crew to begin discharging passengers. Seven passengers and their baggage. But the pier was significantly higher than the prow of our boat, so the passengers had to climb onto the prow, toss their luggage up onto the pier, and then hoist themselves up. The last passenger to leave was a frail old woman in a cotton dress, who, crouched and teetering, carried a carpetbag up to the prow of the boat. Someone tossed her carpetbag onto the pier, and she prepared to follow it. Just then, however, the boat began rolling back from the pier, and she leaned off-balance over the black water and crumpled into a fall — this was fast becoming a nightmare — until ten hands from above, from below, caught her and hauled her up like a fish onto the pier. Then, without a second's hesitation, the captain rang three hard rings, the boat churned into reverse, and we moved back into the rolling river.

The next morning we entered a thick white cloud filled with rain, and eventually out of the cloud a big city appeared: boats, ships, wharfs, old buildings, antique churches, and everywhere the rectangular spines of concrete high-rises. It was Belém, near the mouth of the Amazon, where I left the river and ran through sheets of rain to find a taxi.

In Belém I telephoned a friend of a friend of a friend, an American ethnobotanist who works at the Emilio Goeldi Museum, the most important tropical botany, zoology, and anthropology research institute in Amazonia. His name was Anthony Anderson, and I hoped he would help me figure out how to find the southern bearded saki. Anderson's office, at the Goeldi research campus, was on the edge of town, but the museum proper included a piece of rain forest in microcosm right in the middle of Belém, so we agreed to meet one noon at the museum entrance. I stood at the entrance to the Museo Goeldi, plastered by an equatorial heat and sun, and waited, waited — no Anderson. After nearly an hour, though, the American comedian Steve Martin appeared on the sidewalk, walking my way. Steve approached and introduced himself as Anthony Anderson. "Has anyone ever said you look like Steve Martin?" I asked. He smiled. "Yes."

We had lunch, and I unfolded my maps and explained that I had come to find the most endangered monkey of the Amazon. The full species of bearded saki will be found both north and south of the Amazon River. The northern subspecies inhabits a large piece of Amazonia, north into Venezuela and the Guyanas, and it is not endangered. But the southern bearded saki inhabits a small and rapidly disappearing piece of Amazonia south of the river, delimited to the east by palm swamps toward the Atlantic Ocean, to the west by a major tributary of the Amazon: the Xingu River. A short article plus some correspondence I had in my pack suggested that southern bearded sakis would be found in the Gurupí Forest Reserve, about 150 miles south of Belém. Although he had never seen the monkey himself, Anderson had several good ideas about how I might get to Gurupí and find it, but his best idea, as it turned out, was to introduce me to another American at Goeldi, an ornithologist named David Oren.

Oren's office was filled with books, baskets, and birds. He led me into the museum's specimen vault, where we found a few drawers full of southern bearded saki pieces, primarily skins and skulls, smelling of mothballs. The skins seemed about the size of a domestic cat skin, with a bushy tail at one end and a bushy beard at the other. The back fur varied in color from black to chestnut to tan, and Oren informed me that this monkey varies in color across its range: from full black at its eastern extremity near the Atlantic Ocean to grayish tan at its western extremity (defined by the Xingu River). In the middle of its range, along the Tocantins River, the southern bearded saki is mainly chestnut-backed. He suggested that I would be most certain to find members of the subspecies — these would be chestnut-backed — up the Tocantins River on the edge of a large reservoir recently created by the hydroelectric dam at Tucuruí.

A week later I stood on top of the fourth largest hydroelectric dam in the world: Tucuruí Dam. I was a guest of a subcontractor for the dam's owner and builder, Electronorte, and was accompanied by a Brazilian biologist employed by the subcontractor. The biologist, Marion Meyer, spoke English.

One side of the dam was noisy, as the Tocantins River poured out of a rectangular aperture at its base, rose in a trembling white arc of water and fell into the lower river with a continuous roar, raising a white plume of mist that turned to gray, then became invisible higher up, and finally turned into a gray cloud bottom. The other side of the dam was quiet, a long lake stretching to the horizon, blue and polished under a mostly blue sky, ruffled thinly in places: lacquer accidentally scratched by sandpaper. On the lake a speedboat was waiting. Marion and I sat in the boat and were driven south for a bumpy hour.

The lake was a reflected blue in the distance, milky brown as you looked down into it. Land on both sides of the lake was mostly covered by forest, although we passed several spots where new farms and ranches announced themselves with clear-cut swaths of stubble, grass, and red earth, often reaching right down to the water. Many of the trees under the reservoir had been killed with poison and defoliated with Agent Orange before the dam was completed, but most of them had not been removed; thus, except across the deepest parts of the reservoir, the speedboat navigated through a maze of dead treetops: gnarled and twisted, bare of bark, colored silver. At last we tied up at the dock of Base 4, an environmental monitoring station maintained by Marion's employer.

Marion and I walked up a dirt trail to the camp: several wooden buildings on a high, sandy clearing overlooking the lake. A regular breeze picked up cool air from the surface of the lake and ran it up to the camp.

Workers at the camp agreed that, yes, we would be able to find southern bearded sakis in the forest there — they called this monkey *cuxiu*. So, after a brief lunch, five of us walked along a dirt road and then turned onto a narrow path into the forest.

The forest arched as high as a cathedral, and shards of light pierced and plummeted through green into green. Both high and low, palm trees spread huge, featherform fronds some twenty or twenty-five feet long. A breeze frequently moved from across the reservoir into this forest, stirred high leaves and vibrated the fronds. From my time in rain forest north of Manaus, I had prematurely concluded that the Amazon was a very noisy place. This piece of Amazonia, however, turned out to be surprisingly quiet. In places, it is true, we heard the agonized groans and roars of howler monkeys. But mostly we listened to a few insects shaking maracas and a few birds. In the distance some birds made modest piping noises: toucans I was told. Nearer, some obscure bird flung a most regular call: a bright, piercing whistle, wheep wheeyo. Other birds fluted tyoo, tyoo. On occasion woodpeckers knocked a random rhythm, and one time two pairs of parrots burst out of the trees and pawwked.

Butterflies of many colors and patterns appeared silently, sometimes singly and large, pumping through the air with a loopy motion, sometimes small and in dueling pairs, flickering and rolling.

I heard a hiss, a crash, and saw a flash of tan. "Did you see?" asked Marion. "Do you know this word for Bambi?" she asked. "Deer?" I said.

We came upon a big tortoise on the trail, with a brown, tessellated shell; its head was hidden behind two knees; its legs were folded in and covered with a pebble-surface skin, orange-yellow pebbles in black. It was the size of a large mixing bowl. During that first afternoon we discovered half a dozen others, that big and smaller, one as small as a teacup. "The people around here," Marion said, "they like to eat tortoise."

We walked away from the trail, stumbled over crazy piped networks of roots, ducked under twisting vines and lianas, avoided thorned trees, eschewed tarantula holes that pierced the earth everywhere. We observed several small signs and samples of life: ants carrying leaf flags, for example, and little frogs with bright orange backs, as shiny as glazed ceramic. We stooped over the remains of a sloth — tufts of fur, claws, a skull, some bones. We discovered a tree containing four black howler monkeys, silent and solemn, which barely stirred when they saw us. We came upon a giant toad, olive colored, with a triangular head and black rivet eyes, his skin gleaming with the dull sheen of rubber. It was huge, five or ten times as big as any toad I'd ever seen before, as big as a Saint Bernard puppy, and it looked fat, fearless, and fixated: Henry VIII waiting for a big burp.

My companions said we would soon find the bearded sakis. Two of them left our group and disappeared in opposite directions, but they returned an hour later without results. Then the sky darkened, thunder rolled over, and a wind began pushing and pulling the tops of trees. The wind intensified, pulling leaves away from trees, flinging them down. The leaves fell fast at first and then settled into a lower cushion of air and started slipping, flipping, and settling down to the forest floor. Rain appeared soon thereafter, poured into the forest, and encouraged us to call it a day and return to the camp.

Next day, however, we found them in the forest soon after dawn. First I heard a high-pitched whistle, like a squeak, then another, and another. Next I heard splashes in the leaves above, and I looked up to see a dark, thick-tailed monkey the size of a domestic cat, about twenty feet above me, leaping from one thick cluster of leaves to another, landing with a soft splash: pssssssssh. "Cuxiu!" Marion said to me, reminding me of the Brazilian common name for this monkey — but I already knew what we saw: not merely from noting that dark, bushy tail, but also from observing a full beard hanging like a Brillo pad at the chin and two high knobs of fur at the forehead.

I heard more squeaks and slowly began to distinguish the full group, as it hid and moved in the trees high above us. One particular tree seemed full of them, almost like an apartment building, and I looked at perturbations in green there, then noted dark shapes inside — a tail here, half a body there — moving slowly. I watched them walk across branches, svelte and sinuous, saw them jump and jump and jump, and listened to the splash, splash, splash, soft and surfy in the leaves. Then they moved out of that tree into another in follow-the-leader fashion, with a run-run-run-run-leap and a run-run-run-run-leap. Their leaps were stunning fifteen-foot excursions that usually ended lower than they had begun; the monkeys flew and dropped with front hands extended, bushy tails drawn out behind and switching like rudders. I watched eight of them, one after the other, as they walked along the same branch corridor, sinuously, like cats, and then one after the other entered a leaf cluster and drop-leapt out of it, into the air, falling and then breaking the fall with a quick grab on an outcurled vine before splashing into a lower leaf cluster. Each leapt from precisely the same spot, each grabbed the same vine, each dropped to the same lower cluster before scrambling and hopping upward along the same leafy pathway into the new tree. There were thirty-three animals in this group, my companions told me, including about eight fully adult males and one infant still riding piggyback on its mother.

I concentrated on their tails, which reminded me of waving banners. Sometimes the bearded sakis ran with their tails extended behind, curled slightly up or slightly down; sometimes they scurried across branches and through leaves with their tails flipped up and curled back over their heads, as if to serve as umbrellas. They sat with their tails hanging down, curled like commas, and sometimes swung them back and forth like scythes. And later on I noticed that they would sometimes sit on a branch and lean forward, perhaps to grab some leaves to eat, and steady themselves by draping those full tails across a branch behind: almost a grasp.

Their heads and faces were astonishingly extended with extra hair. Both males and females had full Smith Brothers beards drooping from the chin, just about doubling the length of the face. Their faces were extended upward, too, by two bouffant elevations of fur at the forehead, bulbs sometimes tall enough to look like extra uphanging beards. You could tell the fully adult males by noting pink sacks with big marbles inside, which hung loosely beneath their tails; but pretty soon I was able to tell the adult males by looking front on as well, where I saw a tomcat's bulk and swagger in the body and especially long beards and dense forehead knobs around the face. Sometimes the knobs reminded me of an insect's bulbous eyes; sometimes I thought of a bearded man with two bowls placed upside down on his head.

As we wandered beneath them all that day and the next, I became very familiar with their usual call, a high-pitched whistle-squeak ending with an abrupt chuck: squeeeeeeechuck! They made other noises too, including squabbling sounds, but when they disappeared, we could usually locate them again by listening for that call; I believe the call serves them similarly, enabling individuals to locate the traveling group.

Our group down below consisted of five: me, Marion, and three workers from the camp, Baiano, Expedito, and Pedro. As we passed that day and all the next in each other's steady company, sweating in the heat, wandering everywhere, anywhere, stumbling across the forest floor with no other purpose than the free and delightful one of chasing monkeys, I came to like them all. Particularly Pedro — Pedro Pimentel was his full name — who, until the dam was built, lived in a small village near the river and fed his family by hunting. He was short, with square shoulders, a short neck, black hair, a reddened leathery face, high cheekbones. He dressed simply in a cotton shirt, old Levi's, old leather boots, with a machete strapped to his belt. A hunter, he approached the forest with a poet's intuitive gentleness and spoke with a scholar's quiet sureness. Glasses on his face added to the scholarly effect, but the hunter's eyes behind them seemed to miss nothing: birds I never saw; ground animals I hadn't imagined were there; leaves that were really moths; hummingbird nests impossibly hidden behind palm fronds; a fragile bird's nest containing two jellybean-sized eggs, white with maroon splotches like bloodstains, cupped preciously in the nest like jewels on a pillow. He drew forth forest pharmaceuticals, too, and once with his machete whacked open a length of liana that poured out good drinking water. Pedro used to hunt everything, including bearded sakis, yet he spoke of them now with respect: "When you try to catch him, hunt him, and he falls down, he will attack. He has courage!"

We followed the monkeys as they went from tree to tree and food to food. Sometimes they sat — squatted, rather, knees up and holding on with their toes — in new-leafed trees and reached out, grabbed handfuls of tender new leaves, and gobbled at them. Much of the time, though, they seemed to prefer nuts, often thick-shelled ones, which they ate in a swarm, raining detritus into our faces and hair, dropping leftover pieces of their food, split husks and shells, to the ground where we stood. Once we stood beneath them as they broke open green and heavy liana fruit, and the leftover husks plummeted down as sudden and heavy as grenades, arriving in the lower foliage with a startling splinter-splash.

So they traveled through the trees and ate as they traveled for most of the morning. Around noon, when the sun above the trees became most intense and heat settled heavily into the forest, they settled in too, stopped moving, stopped calling, and became difficult to see and hear. It was siesta time in the trees, and we on the ground took our own siesta. We lay down on the forest floor and looked up into a swaying green jigsaw puzzle, figures embedded in a matrix, swaying shadow and light, green into green, opaque into luminous, a high and hypnagogic world of hidden monkeys and exposed leaves, out of which a single leaf would regularly be released and fall, slow and fluttery, like a sleepy moth. Pedro said that Indians lived in this forest before the dam came. When they didn't want you to go into an area, they'd tie two adjacent small palms together, making a looped knot. When you could go in again, they'd untie them. "When you meet an Indian on the trail," he said, "and you carry a rifle, you have to break it open and put it down. Otherwise he thinks you will shoot him."

That night a full moon appeared near the horizon, as big and soft and orange as a cheese pizza, and the next day when we followed the sakis they appeared lethargic, less active and more inclined to disappear and sit quietly in treetops. Pedro thought they had been eating during the night, when the moon was out.

"TELL ZOODIE, I know he loves monkeys, but tell him he shouldn't be upset: Indians eat the hell out of monkeys."

In Brazilian Amazonia, Indians have for centuries hunted wild animals for meat. Indians still hunt in Amazonia and for the most part still rely on the traditional tools: bow and poison-tipped arrow, blowpipe and poison-tipped dart.

The traditional blowpipe is fashioned from two pieces of wood, each nine to ten feet in length, each scooped out down the middle to form a cross-sectioned half. The two halves are placed together, wrapped spirally with strips of bark, and sealed with beeswax, forming a hollow tube that tapers toward the far end and is finished at the near end with a cup-shaped mouthpiece. The dart is long and needle-shaped, sharpened by knife or tooth from the leaf stalk of certain palms, given air resistance in the pipe with a small ball of "fluff" taken from seed vessels of the silk cotton tree. When a target is sighted, the hunter places the cotton ball and a dart, poisoned with curare, in the mouthpiece, aims, and puffs a sudden breath, silently propelling the tiny missile with impressive accuracy for up to 160 yards. The dart sticks, the poison penetrates, and the surprised target drops.

According to the account of Henry Walter Bates, a nineteenth-century British naturalist and collector, a single group of around 200 Tucano Indians in the upper Amazon consumed the flesh of 1,200 woolly monkeys yearly. Today in parts of Amazonia, particularly where Indians are settled near mining camps or religious missions, traditional hunting tools have been replaced by shotguns, increasing the impact of hunting in those areas. Nevertheless, the fewer than 200,000 indigenous people living in the forest today probably place only marginal pressures on wildlife. Hunters among the 7.5 million neo-Brazilians in Amazonia, generally mobile and carrying modern weapons, represent much more of a threat to the wildlife.

The present effects of hunting on primates vary widely from one area of Amazonia to another. In some regions fish, turtles, and larger mammals other than primates are the preferred sources of protein. Wherever primates are hunted, though, the larger types — spider monkeys, woolly monkeys, howler monkeys, capuchins, sakis, and bearded sakis — are nearly always the objects of pursuit. The endangered white ouakari is also occasionally hunted, although many people living within its range express disgust at the idea of eating a monkey that looks, in its own peculiar and bald-headed way, rather human.

To put meat in the pot is probably the major motivation for hunting in Amazonia, but primates are also killed for other reasons. Perhaps most significant is the killing of larger primates for bait to catch fish, turtles, and spotted cats. Occasionally necklaces fashioned from monkey skulls, bones, or teeth are marketed in tourist stopovers, including, as I discovered, Manaus. Occasionally too, as I've mentioned, the tails of bearded sakis and other monkeys are made into dusters. Skins of howler, woolly, and spider monkeys are cut up and fashioned into ornamental browbands for horse bridles in Colombia. The large and interesting hyoid apparatus (throat bone and tissue) of the howler monkey sometimes serves as a drinking vessel in Colombia — part of a folk treatment for goiter. In some parts of Brazilian Amazonia, drinking from a hyoid cup is considered a good way to ease the pains of childbirth, while in Surinam it is a cure for stuttering. In Peruvian Amazonia drinking the ground and boiled hair of a howler monkey is one treatment for a cough.

Finally, though, eating monkeys in Amazonia, dismembering them, using their pieces, is comprehensible. We dismember what we eat, we eat what we need. But now, in our own fragment of time, we have begun dismembering the Amazonian Basin itself.

The drainage basin of the Amazon and its tributaries covers roughly 2.3 million square miles of land and includes the largest single piece of tropical forest in the world. Amazonia extends into Bolivia, Colombia, Ecuador, French Guiana, Guyana, Peru, Surinam, and Venezuela, but about two thirds of it lies within the borders of Brazil.

At first glance Amazonia may appear to be a homogeneous environment. The land is comparatively flat or undulating; from Peru to the Atlantic, the Amazon River descends a total of only 230 feet. The region's warmest and coolest seasons differ on the average by only about 3 or 4 degrees Fahrenheit, and average annual temperatures hover between 75 and 80 degrees Fahrenheit. The vegetation of Amazonia can be neatly divided into three important types: *terra firme,* the vast majority, poor-soiled, nonflooded forest; *igapó,* permanently flooded swamp forest; and *várzea,* rich-soiled, seasonally flooded forest. Yet in spite of its general climatic consistency and stability, Amazonia in actuality presents a great diversity of living systems.

Amazonia is commonly thought of as a super-wilderness, a vast jungle, untamed, hostile, impenetrable. But in fact the region has for centuries been inhabited by large numbers of Indians subsisting on hunting and slash-and-burn agriculture. European colonizers long ago penetrated Amazonia's wrinkled maze of waterways by riverboat, harvesting such forest products as rubber, Brazil nuts, fruits, fibers, and skins, and farming along the great bands of rich várzea floodplain. For very practical reasons — transportation and soils — the development of Amazonia has until recently been concentrated along the rivers, leaving generally intact the vast regions of interfluvial terra firme forest. The exploitation of Amazonia was largely defined by river barriers, ancient as the forest itself.

In the past two or three decades, however, new pieces have been cut out of the Amazon. A new pattern of exploitation has appeared, mostly based upon a single technological marvel: the bulldozer. Post-Bulldozer Amazonia began in the late 1960s and early 1970s, as plans for an 11,000-mile system of highways were etched into reality. Since then a continental mesh has been cast over Amazonia, broadly sectioning the terra firme forests, opening them for the first time in history to easy penetration and development.

Although satellite imagery in the mid-1970s showed the developing highway system as a mere hairline fracture, Brazil had already begun to develop the highway peripheries by means of an ambitious colonization scheme. In the words of former President Medici, Amazonia was "a land without men for men without land." To rectify that situation the National Institute for Colonization and Agrarian Reform began giving away 250-acre lots of land in a twelve-mile-wide band along the highway network to settlers from the impoverished northeast. Original plans for massive colonization, though, were soon scaled down to a supposedly manageable settlement rate of 100,000 northeasterners a year (a tenth of that region's annual population growth). Yet only 50,000 colonists arrived altogether, to face weeds, pests, diseases, poor soils, harsh climatic conditions, and threats from hostile Indians. Many of the pioneers quickly abandoned their allocated plots and took to slash-and-burn cultivation. Fires set by the colonists burned over unexpectedly large areas, government support dwindled, and some of the construction of highways meant to support colonization was stopped.

Cattle ranching has also taken big pieces of Amazonia, as I mentioned in the last chapter, and logging cuts out more. If Brazil were to cut down all of the trees in its portion of Amazonia, it could sell the hardwood for $1 trillion, more or less. So far Brazilian Amazonia has provided very little timber for the world market, but according to a late 1970s projection by the Brazilian Forestry Development Institute (IBDF), the industrial wood taken out of Amazonia should increase ninefold by the year 2000, to 35 million cubic yards — the result of moderate harvesting from 75,000 square miles of forest. Thus the IBDF asked that pieces of forest totaling about 150,000 square miles be set aside for future logging; another Brazilian agency suggested that twice that amount be earmarked for logging. Meanwhile multinational timber corporations — including National Bulk Carriers, Georgia Pacific, and Atlantic Veneer of the United States; Eldal of Japan; and Bruynzeel of the Netherlands — are already at work in Amazonia.

Parts of Amazonia, such as the igapó forests along the Negro River, remain uninhabited and will be difficult to develop. Yet with the rapid movement of people into Amazonia and with development — highways, colonization, ranching, logging, mining, railroads, and hydroelectric construction — large pieces of forest have already disappeared. How much? It is estimated that in the past two decades between 45,000 and 100,000 square miles of Brazilian Amazonia have been cleared.

Bearded sakis live almost entirely in the upper and middle levels of the forest canopy, and they prefer untouched primary rain forest. Their specialized food preference — seeds — and other factors probably limit their natural density in any area, and the southern bearded saki is more limited by the size of its original habitat piece than most other Amazonian primates.

Recent field studies have shown bearded sakis to be somewhat more adaptable to habitat disturbance than previously believed. Heavy logging, in which at least half the trees are taken out, almost entirely eliminates the species. Yet light selective logging, in which no more than 15 percent of trees are destroyed, seems to result in only minor reductions in resident populations, even when some of their favorite food trees, such as the massaranduba, are removed. Even light logging, however, opens up previously inaccessible pieces of forest to this primate's most powerful predator, *Homo sapiens,* so that over the long term its habitat must be protected both from heavy or repeated logging and from the incursions of hunters into lightly logged regions.

Unfortunately for the southern bearded saki, its habitat, small to begin with, coincides with that piece of Amazonia where the Deluge now rushes in most suddenly and powerfully. The Belém to Brasília Highway, completed in 1960, completely slices its range north to south; the TransAmazon highway slices east to west; and two other highways, from Belém to São Luís and to Rio de Janeiro, also cut through this region. Farming, ranching, logging, and general colonization had by the late 1970s eliminated about a third of the forests in its eastern range. Today about half of those forests are gone. The remaining half now provides 95 percent of the timber used for construction and furniture in southern Brazil. Except for one protected piece, the Gurupí Forest Reserve (which has been set aside for controlled, supposedly sustained logging), the eastern forests are now fragmented and discontinuous. The human population in that region steadily increases; subsistence hunting is common; forest clearance continues in a manner largely uncontrolled, unplanned, and illegal.

The rain forests in the central and western parts of the southern bearded saki's habitat — around the Tocantins River and west to the Xingu — have until recently remained intact: immensely rich and ancient reservoirs of biological diversity.

In 1967, however, geologists discovered that the Carajás Mountains, between the Tocantins and Xingu rivers, contain immense mineral deposits. Those mountains, in fact, include the richest piece of iron ore on earth — an estimated 18 billion tons of 66 percent pure ore — as well as a billion tons of copper, 60 million tons of manganese, 124 million tons of nickel, 40 million tons of bauxite, 35,000 tons of tin, and considerable gold. Thus in 1978 a project was begun to develop Carajás: to extract 35 million tons of minerals per year from open pit mines; to move those minerals by railroad from the mountains 550 miles through dense rain forest east to the Atlantic port of São Luís; to acquire electric power for that railroad and related development by building a major dam on the Tocantins at Tucuruí.

The mines at Carajás are now in operation, and a permanent community of 11,000 inhabitants has sprung up around them. The electric railroad now runs between Carajás and São Luís, carrying minerals and, twice a week, passengers too.* The Tucuruí hydroelectric dam is finished also, and now sends power along a corridor of high-voltage transmission lines south to the Carajás railroad.

The Tucuruí Dam on the Tocantins is just one of several large and small hydroelectric projects in Amazonia, all very recent developments. The first Amazonian dam, Brokopondo, was completed in 1964 in Surinam, north of Brazil; the second, Curua Una Dam, opened for business in 1977 in Brazil.

On the surface, at least, Amazonian hydroelectricity seems like a good idea. After all, a fifth of the world's running fresh water flows freshly through Amazonia, and if all that water — counting only the tributaries, not the Amazon River itself — was turning turbines, an electric force equivalent to 5 million barrels of oil a day would be created. Hydroelectricity has its limitations, however. It is not, as some people imagine, one of those wonderfully endless and renewable sources of energy. Dams last only as long as their reservoirs, and reservoirs fill up with silt within several decades or less. Amazonian dams may not even help reduce Brazil's dependence on foreign oil. Industrialized, non-Amazonian Brazil, to the southeast, is even now consuming less electricity than it can generate, and the Amazonian dams are meant to power industrialization and other development in Amazonia itself, not elsewhere. Hydroelectricity, moreover, may be comparatively inexpensive only if one ignores the environmental costs.

But the environmental costs can be considerable. Large dams alter an environment in two important ways. First of all, they flood land. The Brokopondo Dam in Surinam, for example, flooded 570 square miles of rain forest. Flooded vegetation rots, and the decomposition can produce a peatbog effect, with such byproducts as methane, hydrogen, hydrogen sulfide, and carbon monoxide. (The rotten-eggs

*The railroad actually may have created greater environmental disturbance than the mines and miners. The building of the railroad affected twenty-one Indian reserves; 12,000 Indians were compensated and in some cases relocated. Within a few months after the railroad was completed, Brazilians from elsewhere in Amazonia had settled along the line and now occupy the entire right-of-way strip between Carajás and São Luís. Elsewhere in Brazil the conglomerate responsible for the mining and railroad construction has, in the past, dealt commendably with environmental matters. And the conglomerate promised to spend $10 million over five years to mitigate the project's direct environmental impact. Strangely, however — weirdly, even — it spent $6 million of that money to build a zoo at Carajás. The zoo supposedly protects native fauna, but many of the species represented in the zoo are not local.

stench of hydrogen sulfide from the Brokopondo Lake could be smelled forty miles away; workers at the dam had to wear gas masks during the two years of construction.) The increased acidity and lowered oxygen level of such an environment yields dead fish. In many areas of Amazonia, furthermore, dam reservoirs have displaced Indian tribes and other settlers.

Dams alter the environment also by changing running water to still water, which provides excellent breeding sites for malaria-carrying mosquitoes and schistosomiasis-transmitting aquatic snails.* Still water also provides superb opportunities for certain water weeds, including the beautiful but notoriously fast-growing water hyacinth. A year after Brokopondo Dam was finished, for instance, fifty square miles of reservoir were covered by water hyacinth; a year later, the hyacinth carpet had grown fivefold. Water hyacinth blocks sunlight from the water, thus starving other vegetation and fish, and so the Surinam government was moved to spend $2.5 million spraying Lake Brokopondo with Agent Orange, which may have produced its own environmental consequences.

When Electronorte linked together the last pieces of Tucuruí Dam in 1984, a lake began to grow. It grew and flooded more than 900 square miles of virgin rain forest, destroying many unique local plant and animal species. It flooded a fortune in marketable timber, about 36 million cubic yards' worth. As the water rose, it flooded many barrels of Agent Orange, brought in for a defoliation operation and accidentally left behind. It flooded six river towns, and so 6,000 families, people who had for generations depended upon fishing, had to be relocated away from water. It flooded one Indian reserve entirely and two others partly, requiring the relocation of the Indians onto new reserves.

The lake flooded a large piece of faunal habitat, and since many animals became trapped on temporary islands in the rising water, Electronorte supposedly spent $30 million rescuing them, including 450 southern bearded sakis. Ultimately Electronorte released the animals into five pieces of forest along the lake's final edge, five large areas that were designated as "protected" and were, in fact, protected by environmental monitoring stations. The five stations were thoroughly manned, and as the lake flooded, a flood of international journalists was transported to them by helicopter and a fleet of speedboats to witness this remarkable humanitarian gesture: the animal

*Schistosomiasis is a hard-to-treat, potentially fatal infection that proceeds from the larvae of flatworms living on the snails to the human bloodstream, liver, kidneys, and spinal cord.

rescue operation. Once the dam was completed and the journalists were gone, however, Electronorte lost interest. By the time I came to Tucuruí, the helicopter was gone; the fleet of speedboats had declined to a single boat; the five monitoring bases were essentially abandoned except for one, Base 4. And much of the land within those large "protected" areas where the rescued animals were released had been given, sold, and traded away. One large piece of the original protected area at Base 4, for example, is now held by an Electronorte subcontractor, BelAuto, which provides cars and vans for Electronorte employees at the dam. BelAuto is now a rancher at Tucuruí. It might be said, of course, that Electronorte is in the dam and power business, not the conservation business; but in fact, protecting the dam's watershed would have been financially prudent. Now that farmers and ranchers are clear-cutting forest right down to the edge of the lake, Electronorte itself will pay for the shortened lifespan of Tucuruí Lake.

The mines, the railroad, the dam, plus their larger spheres of disruption, represent pieces of a Deluge suddenly appearing in the middle of the southern bearded saki's forest. But the fabulous mineral wealth at Carajás is now exciting a much larger plan for southeastern Amazonia, the Grande Carajás Program, which places 310,000 square miles of Amazonian rain forest — including this primate's entire range — under a $60 billion map of integrated exploitation. Grande Carajás will create a whole series of hydroelectric dams on the Tocantins and Xingu rivers, as well as ore-processing factories, agricultural and colonization projects, and forestry exploitation schemes.

Already the government has approved construction of seven pig iron factories, two metal alloy factories, and two cement factories — with little apparent consideration of even simple and practical environmental issues. The pig iron and metal alloy factories, for example, will need considerable water to operate, but five of the nine planned factories are to be built in areas with minimal water supplies. All eleven of the factories now planned will also require charcoal to operate, which will increase local charcoal demand elevenfold. Coupled with the requirements of nine more factories being planned for the Grande Carajás Program, regional needs for charcoal could surpass two million tons per year, a demand that, according to one authority, "spells doom for the remaining native forests."

While the Amazonian Basin is being dismembered into a thousand pieces for a hundred purposes — the Deluge — a few important pieces are being set aside as Arks. Amazonian nations other than

Brazil have defined several pieces of forest as their own network of Arks, a total of over 27,000 square miles — the size of Ireland. The individual areas are, however, widely dispersed, and only three of them are very large: Venezuela's Canaima National Park and Peru's Pacaya-Samira National Reserve and Manu National Park. All the rest are much smaller. Moreover, these non-Brazilian conservation areas do not begin to represent the full diversity of Amazonia.

The bulk of Amazonia is in Brazil, and the bulk of the responsibility for conservation rests with that nation. Brazil has possessed — still possesses — a unique opportunity for conservation. Brazilian Amazonia remains the largest single block of tropical forest anywhere in the world. Compared to the major forest formations in Central America and Southeast Asia, Amazonia is still relatively intact. In other words, opportunities for systematically planned conservation still exist. Brazil also holds the key to world primate conservation, since it has more primate species than any other nation: forty-one species from all sixteen New World genera; within Brazilian Amazonia, thirty-two species from fourteen genera.

It might be said that organized conservation in Brazilian Amazonia began in 1965, when Brazil's Forestry Code called for the establishment of rain forest parks and reserves and prohibited the clearing of certain crucial forest areas along rivers and lakes, on the tops of hills and mountains, and on steep slopes. It also stipulated that any private individual or organization purchasing forest land could clear only 50 percent of it. But forestry practices in remote Amazonia are almost impossible to monitor, and such legal restrictions proved ineffectual. Pathetically, as of 1975 all of Brazilian Amazonia included only two significant conservation areas: the Araguaia National Park in the State of Goiás and the Amazônia National Park in Pará State. In 1975, however, Brazil's Second National Development Plan called for a large network of conservation areas in Amazonia, and the Brazilian Forestry Development Institute set out to create a conservation map of the region, to define those areas of greatest ecological importance.

One interesting way to draw a conservation map for the present would be to dig up a conservation map from the past: to define as precisely as possible Amazonia's Pleistocene Refuges. The theory of Pleistocene Refuges was first promulgated in 1969 by Jürgen Haffer, a German ornithologist, who had noticed that about a dozen regions of Amazonia were especially rich in endemic bird species. During the Pleistocene epoch (1.8 million to 10,000 years ago), the polar ice caps expanded and contracted four or five times, the ocean levels rose and fell, and the topography of temperate regions altered in various ways, as did global climatic conditions. The earth's tropical forest belts were subjected to great periods of dryness, during which, according to Haffer's scenario, the forests retreated to vast refuges, great islands of wet forest surrounded by dry savanna. The periodic forest contractions isolated and divided previously widespread species, over time allowing for genetic divergence, while the periodic expansions reconnected isolated forest pieces and produced zones of hybridization among previously divergent species. By closely and cleverly examining present-day patterns, considering both areas of concentration and zones of hydridization, one can hypothesize the locations of the Pleistocene Refuges. And since the areas of these past refuges, created by natural events, are still remarkable for their high levels of species concentration, they remain obvious candidates for the artificial creation of refuges in the present.

By the end of the 1970s the Brazilian Forestry Development Institute had identified eighty-three pieces (including twenty-three possible Pleistocene Refuges) as especially worthy of preservation in Amazonia. Based on a close consideration of the eighty-three recommended areas, Brazil in 1979 began defining a huge network of conservation regions throughout Amazonia. Altogether the official conservation pieces make a very large Ark, which at the moment consists of seven national parks, five biological reserves, and twelve ecological stations. The total area of almost 45,000 square miles, bigger than the island of Cuba, is about 2.3 percent of that portion of the Brazilian Amazon legally defined as Legal Amazonia (a smaller, more precisely circumscribed area than ecological Amazonia).*

Is Brazil's Amazonian Ark properly caulked and pitched, mortised and tenoned, ready for the Deluge that approaches? José Lutzenberger, an outspoken Brazilian conservationist, thinks not. As he recently told an American journalist: "All our parks and reserves are in a shambles; there is not a single one that is protected."

Generally, Brazil's conservation plan is designed to preserve rep-

*Brazil defines three major sorts of conservation areas for Amazonia: national parks, national biological reserves, and ecological stations. The national parks are established theoretically for educational, recreational, and scientific purposes — that is, they promote at least temporary and limited human use. The biological reserves are designed more specifically to protect particular ecosystems or certain plants and animals — in other words, they discourage human intrusion. The ecological stations were created to preserve specific pieces of representative ecosystems and also for ecological studies. Beyond those three major types of conservation areas, the Brazilian Indian Foundation administers Indian reserves in Amazonia, which are supposed to protect indigenous peoples and their traditional ways.

resentative ecosystem pieces, largely defined by their vegetation, so many animal species or subspecies are still not included in protected areas. That's one problem. Most critically, however, both the Brazilian Forestry Development Institute, which runs the national parks and national biological reserves, and the Special Environmental Agency, which oversees the ecological stations, are remarkably underfinanced. As a result, only two areas — one national park and one biological reserve — actually exist in reality, with officially marked boundaries and actively employed staff members. The rest are yet scratches on someone's map: paper parks.

As for the southern bearded saki, described by the best experts as "undoubtedly the most endangered primate in Amazonia," it remains unprotected throughout its range east of the Tocantins River; west of the Tocantins, it receives complete protection only within three areas recently established on mining concession property in the Carajás Mountains.

I LEFT BASE 4, left the reservoir and the dam on the Tocantins, went down to the old village of Tucuruí just below the dam, and caught a local boat, the *Gerland de Jesus,* which rolled slowly downriver for a day and a half, with delays and stops at a dozen hamlets and ramshackle villages that half leaned on stilts over the river with wooden stalls and shops and small lunch counters and dreary bars, down the Tocantins to the Amazon and then into Belém.

The time at Base 4 was the center of my journey into the Amazon, not only because it was my actual destination — where the bearded sakis were — but because I felt most healthy, happy, whole, in my power there. A single ill effect of that time in the forest could be seen on my legs, which were bitten in a hundred different places by an insect Marion described as the *carrapato do fogo,* or mite of fire. The hundred bites were actual holes, fiery red and intolerably itchy. Back in Belém, I walked into the first pharmacy I saw, pulled up a pant leg, and pointed to a calf full of red and blossoming holes. The pharmacist divined my meaning and sold me a tube of ointment.

I went to my room at the Central Hotel, took off my pants, sat down, and thoroughly rubbed that ointment up and down my legs, covering all one hundred holes. While I was sitting there with my pants off, I happened to glance at the locked wooden door connecting my room to an adjacent room. It had holes too, five little peepholes, obviously drilled by a nefarious voyeur. Luckily, I had bandages in my pack, so I took out five of them and bandaged the five holes in the door.

A day later, holes appeared in my stomach. That half-cooked hamburger at the Cockroach Diner, I thought. I was overwhelmed by stomach cramps, diarrhea, weakness, a mild fever, all of which readily osmosed from soma to psyche and drained my intent, will, direction, spirit. I became a solitary zombie in Belém and spent a week wandering, going to movies, and lying on my bed, where I was comforted with the text of and irritated by the critical comments on *Jane Eyre*, in the Norton Critical Edition.

During those days I saw pieces of Belém from a vagrant's perspective, arrived at a nodding acquaintance with prostitutes, beggars, cripples: the flirtatious teenage sisters who stayed with a young American in my hotel and breathily asked me to light their cigarettes; the haggard young woman and her four children, who all lived on a sheet of cardboard on the sidewalk's edge up the block from my hotel and looked for coins to drop in their bowl; the handsome young man with a general's profile who, lacking legs and hips, having stumps for arms, spent ten-hour days and seven-day weeks situated like a statue in the middle of the sidewalk, looking for paper money to be placed between his arm stumps. I also saw pieces of Belém from a tourist's perspective and found it a tolerable and sometimes delightful city full of architectural surprises, expansive parks and ornamental pools. An unexpected turn in the street tossed you into a new pattern — a new square, a statue, dark trees leaning over, men playing cards on an overturned box as dusk gathered, a cool wind suggesting something brewing in the sky. Antique buildings were scrolled, domed, arched, tarnished with age and moisture, garnished with grass, vines, ferns, or even small trees rising from a gutter, a niche, a nook at roof's edge. Belém is nearly on the equator, and the daily heat could beat on body and brain and rob you. But when the heat became most criminal, the sky would lower way down, white and woolly, and rain would pour in, a cleansing rain washing the rags and litter, the scattered garbage and stink in the gutters, the dust and grit of a day's city grind, the walks, the streets, washing the heat as well, and the hot faces of pedestrians and merchants and schoolchildren, of shoppers and strollers and Michael Jackson look-a-likes, washing too like the tears of Christ the tired faces of beggars and cripples and whores, washing loose pieces of all, everything, down the sloping streets, down to the power and fullness of the river, pouring into the river, where they joined all the pieces of Amazonian civilization already floating down, the fruit fragments and the bone bits, the bottles and cans, the plastic cups, cigarette wrappers, excrement, and two monkey-tail dusters. The pieces floated, then sank and drifted underwater as alien artifacts into a dense world of cries and scratches, pulsing gills and stroking fins, of silent benthic undulations: a murky nether world where Grendel wakes as giant catfish and 650-volt electric eels and twenty species of bright-toothed piranha, where ragged-scaled crocodiles sink to ragged beds of leaves and pale river dolphins rise to mirrored beds of air, where palisaded bacu pedra and armored pirarucu sleep in slow and steady suspension: a woven wet world, whole and eternal, rolling, forever rolling to the sea.

After a week, my stomach reconstituted itself, and I was ready to move on.

Four

HUNGER

I HAD EATEN only a single small croissant that morning in Senegal, and by midafternoon, as I stood on the ferry, crossing the bay from Lungi airport to Freetown, the capital city of Sierra Leone, West Africa, I felt hungry. But I also felt energetic and optimistic, and I found myself surrounded by bright images of plenty. Out to sea, I saw big patches of water boil with a violent meeting of birds and fish. I saw a long wooden fishing canoe carrying nine men and a hill of fish gleaming silver in the sun. Near me on the ferry a woman wrapped in bright robes squatted next to a whole basketful of fish. Those fish had turned into brass, and she was selling them a few at a time. Behind me the owner of a shiny green Mercedes had cracked open the trunk of his car to give the poultry inside some fresh air, and a rooster in the trunk was broadcasting his presence.

During that morning's trip from Senegal to Sierra Leone, I had met a young Frenchman, an employee of the French embassy in Freetown, who was carrying a diplomatic pouch from Dakar. He kindly offered me a ride from the airport (and later provided me with a place to stay while I was in Freetown). Thus on the ferry I was also standing not far from a Peugeot with diplomatic license plates, against which leaned a young Frenchman wearing a red sportshirt, tan slacks, and well-shined penny loafers, holding a canvas sack with a red wax seal.

Two young girls approached me and my new friend. They smiled coquettishly and begged playfully. "Give us money." My friend smiled and replied in a teasing, playful tone, "What makes you think we have money to give you?" One of the girls responded: "You have car. I think you have money. You gentry." My French friend: "That's not for you. Why should I give you any? No. I have nothing." He shrugged dramatically.

Later an old woman with a basket of green bananas on her head approached us. "Ya, good bananas." She wanted twenty leones, about a dollar, for a bunch. My friend pulled two bananas out of the basket and asked, "How much for these?" "Ten leones," she said. "Twenty leones for a bunch and ten for two?" he said. And as he playfully bargained her down in price, I looked down to see her swollen, calloused feet and up to see her tired eyes and face. We finally paid I don't remember how much for the two bananas, and I ate one and felt less hungry.

The ferry shuddered into dock and dropped its front gate. The woman with the basket of fish rolled a piece of cloth into a circle on her head, placed the basket on top, and began walking. The man with the green Mercedes closed his trunk and climbed into the driver's seat. My French friend and I climbed into his Peugeot, I in back, he in front next to the embassy driver. And after hordes of pedestrians had poured off the ferry, the car began slowly moving.

"No petrol in town," my friend said. "That's why everybody's walking." Indeed, I saw only two or three other cars penetrating the crowds and crowds of walking people.

Freetown seemed, at first, a comfortably small town. The buildings lining the streets were ancient and weathered two-story clapboard houses with falling-down gingerbread trim, and shanties patched from scraps of corrugated tin. Crowds of people milled outside the houses and shanties, and inside (as I saw through wide-open windows) more crowds of people sat, cooked, played dominoes. As we proceeded, Freetown began to seem bigger than I had first thought, sprawling, with long winding streets, dirt sidewalks, ravinelike gutters separating street from sidewalk, and interminable rows of shanties. We passed into the center of town — at first, two or three blocks of buildings from Chicago of the 1930s, then another couple of blocks with several modern-style, multistory hotels and commercial buildings — and then we turned back onto a winding street lined with shanties and dilapidated clapboard houses.

During the ride through Freetown, my friend described his economic theory of Sierra Leone. "Sierra Leone should be the richest country in West Africa. It has diamonds, second biggest deposit in Africa, behind South Africa only, but 80 percent of the diamonds mined leave as . . . contraband. Is that the word? It has gold. It has trees. It has fish. In fact, people from Senegal go here to fish. It has water, which could provide enough hydroelectricity to power the country, but no one has built the dam. The government is 100 percent corrupt. Everyone is corrupt, from the highest minister to the lowest civil servant. There are many millionaires in Sierre Leone. You see them in their Mercedes." He imitated someone behind a steering wheel. "Mercedes on a bad dirt road. But they are Lebanese and don't put their money back into the economy. It goes out to banks in Switzerland."

On the far edge of town we turned off the road, frightening a couple of goats, and sped past a gate and guardhouse into a walled compound, where my friend's California-style ranchhouse was situated in the middle of a green carpet of lawn and shaded at the front door by a large tree.

We said good-bye to the driver, who left the car there and walked away. My friend introduced me to his house guard, who wore an inverted cloth bowl, a Muslim cap, on his head. He had been sitting on a small rug in the carport and praying, I think. My friend then introduced me to his cook, whose name was Abdul. Abdul was a small, older man, barefoot, wearing ragged clothes, with a bald spot in the middle of his head, and a soft and pleasant but slightly obsequious manner.

Abdul had already prepared a superb dinner, enough for two, and so my friend turned on the house fans, and we sat down to eat. My friend had not been in Sierra Leone long, but he had already developed theories about how to treat the servants. "You've got to be strict with them," he explained as we ate. "Otherwise you find, a month later, your radio missing." After a second helping of dinner and then dessert, my hunger disappeared.

I like to think that the African monkey I had come to see, the western black-and-white colobus, need never go hungry so long as there are trees around. For, like the other members of the *Colobus* genus in Africa, the western black-and-white is a leaf eater. Although all of the Old World leaf-eating monkeys will eat fruits, flowers, and seeds when those foods are available, they have evolved to survive on leaves. They prefer young leaves, which are easier to digest, more nourishing, but harder to find than mature leaves. However, several species can do quite well on a diet of mature leaves.

The main advantage of leaf eating should be obvious, since leaves are probably the most abundant food in a forest. Leaves also provide enough water that the leaf eaters don't need additional water sources. But leaf eating has three disadvantages: leaves are not very nutritious; they consist primarily of cellulose and are therefore hard to digest; and many trees have evolved chemical and other defenses to protect their leaves from being eaten. Not to be outdone, however, the leaf-eating monkeys have evolved particular physiological features to overcome these disadvantages. Leaf-eating monkeys eat large quantities of leaves, up to a quarter of their body weight a day, to obtain sufficient nutrition. And they digest this bulk very slowly through a labyrinthine-chambered alimentary system, which harbors plentiful bacteria to break down the cellulose. Finally, leaf-eating Old World monkeys have evolved resistances to many of the chemical defenses of trees, and they are smart enough to avoid the most poisonous leaves.

The name *Colobus* derives from a Greek word meaning *mutilated* — and all these African leaf eaters either lack thumbs or possess only small protuberances where their thumbs would be. Missing a thumb is probably an advantage when these monkeys occasionally travel by swinging from one branch to the next, and they can still grasp objects by pressing them between fingers and palm.

Africa contains several *Colobus* species, including four closely related black-and-white varieties. These four species discontinuously span the entire midcontinent, from Senegal in the far west to western Kenya in the east. They are spectacularly beautiful animals, with bare, black faces, long tails sometimes terminating with a dramatic, shaggy tuft, and a full, billowy mantle of very long fur across the back. The four species are most accurately separated according to anatomical features, including skull shape, the overhang of the nose, and the protrusion of the jaw. But they are most quickly distinguished by their patterns of black-and-white coloration. One species, the severely threatened black colobus, is in fact entirely black. The species distributed farthest east, the eastern black-and-white colobus (or guereza), is most notable for an extraordinary, long and pure white fringe around its cape of black fur. The southerly Angolan black-and-white colobus displays long, flowing white epaulets on its shoulders. And the western black-and-white colobus, which lives within a narrow band of coastal forest between Senegal and western Nigeria, possesses an entirely white tail, terminating in a bushy tuft of white.

To get to western black-and-white colobus territory, I jumped into the back of a pickup truck in Freetown one morning. Along with me the truck carried two native Sierra Leoneans, a half dozen American Peace Corps volunteers, our personal bags and baggage, plus a fifty-gallon barrel and several large jugs of extra gasoline.

Not long after the truck left the city limits of Freetown, a century or two disappeared, and we entered a peasant society powered by human muscle, with almost no electricity and entirely no telephones. Houses were built of mud brick or mud plastered over sticks and roofed with corrugated tin or thatch. It was hot, and both men and women in the villages along the road tended to deal with the heat by going shirtless, although some of the more fashionable women wore lacy bras and held up open umbrellas for shade. The sky carried Georgia O'Keeffe clouds in the early morning, but they soon became dumplings in white gravy and then dissolved into a cotton ceiling of haze.

After nearly two days, late in the second evening, the truck arrived with two people — me and a Peace Corps volunteer named Anne Todd — at the small village of Kambama, on the edge of the Moa River in eastern Sierra Leone. Anne parked the truck in a mud-wall, thatch-roof garage in the village, and we were warmly greeted by several villagers.

Lacking electricity, the village was dark and wonderfully quiet, glazed by a mild moonlight, spotted by the glow of one of two dying fires and the flicker of candles here and there inside mud houses. Anne took me to meet the village chief, Chief Duwai, a lanky old man with a chin stubble of white whiskers, who was sitting on his front porch playing checkers by candlelight. After exchanging introductions, greetings, and gifts with him, Anne and I carried our supplies down to the river and placed them in a motorized rubber raft. Anne started the motor and told me to shine my flashlight onto the river so that we could watch out for rocks and the resident crocodile. I swept the river before us, back and forth, with the flashlight beam; it penetrated a thin film of mist and flashed off the surface as a second beam, which terminated in a broad circle of light moving back and forth in the trees on the far side. We turned into the current, slid through a rock-lined chute of water, never saw the crocodile, and half an hour later beached the raft at our destination, Tiwai Island.

Next day I went into the island's forest looking for western black-and-white colobus monkeys. The forest of Tiwai Island is mature secondary forest, about sixty years old, and even though many of the trees are now huge — broad and tall — still the forest exhibits large areas of dense undergrowth characteristic of secondary forest, and the higher forest is often obscured by lower vegetation. A grid of trails has been cut into part of the forest, yet in places the lower vegetation becomes so thick that dappled trails turn into dark tunnels.

When I first entered the forest, I was struck by its comparative quiet — none of the racket of the Amazon near Manaus, I thought. Cicadas rattled castanets; some other creatures steadily produced soft percussive bell tones, lightly tongued flutes and piccolos. A soft melancholic whistle rose and fell in tone; a soft hooting sounded like someone blowing across a bottle top at a distance. Once in a while I was mildly startled by a break in the pattern, a chuffing sound, loud at first then disappearing, that reminded me of the chugging of a steam engine; eventually I learned to identify that sound as the wing beating of a departing hornbill.

Creatures other than humans use the trails of Tiwai, and the first time I entered the forest, I was entranced by the quick flicker of dragonflies and the slow rowing of a white, flowerlike butterfly, opening and closing its wings, slowly bobbing in the air along the trail in front of me, flying no faster than I walked. Spiders with flat round bodies like crab shells maintained webs across the trails to catch flying insects. And at night, large fruit bats careened along the trails — as I discovered to my great discombobulation one night, walking alone through the forest, when a dark object like a big dishrag whacked through the air right next to my ear.

Still, the forest is comparatively quiet, the noises mostly musical and steady. And your first observation of monkeys at Tiwai is likely to begin, as mine did, with an unexpected, heart-stopping shift in sound, the sudden crash of a large body jumping through or into leaves, a rustle-burst, followed by quiet and a little rainstorm of dislodged leaves falling, and then a warning cry in the trees, kree, kree, kree, followed by the thump, thump, thump, thump of feet galloping across branches, and more rustle-bursts and more falling leaves. You may look up to see through a screen of leaves the silhouette of a rather humanlike body with a tail, casting off in a long, diving leap.

But eventually at Tiwai you'll find a spot where the lower vegetation opens up into a vista of high trees, and you'll see and then watch a whole huge family of monkeys. Tiwai Island is stiff with monkeys, holding one of the densest concentrations of monkeys in that part of West Africa, with nearly a dozen species, including the western black-and-white colobus. At night it's possible to find a nocturnal primate, the galago, by chasing its sporadic call, which sounds like an old sewing machine, and then shining a flashlight into a thicket until two small eyes flash back. Chimpanzees also live on Tiwai, and in the early morning their characteristic hooting is heard in the distance. They're bold, but smart enough to keep to themselves, and they seldom show themselves to people on the island.

When at last I was able to get a good look at the black-and-white colobus monkeys, I was impressed by their tails, which hang like thick white ropes with thicker white tassels at the ends. I found it easiest to recognize the members of this species by those distinctively white hanging tails, and I found it convenient to count them by counting those highly visible tails.

These monkeys sometimes sit side by side, three, four, even five abreast on a thick branch, thirty to seventy feet above the ground, knees up, hands on knees or around ankles, tails dangling. Other times they rest even more closely clustered, turned to one another, one or two lying stretched out across a thick limb and undergoing serious grooming by their companions. They may sit straddling a large tree limb like riders astride horses; they may lie flat and sprawled out on a limb, arms and legs hanging down, tail draped, looking like tuckered-out dogs.

In motion, they sometimes swing themselves beneath branches, but usually they walk flat-footed and quadrupedally across branches, sometimes slow in a follow-the-leader style, sometimes scattering in fast gallops. They cross gaps between trees with great diving leaps, stabilizing their flight with that long, tufted tail, landing with both hands and feet extended. On occasion they miss their branch of choice, usually then catching themselves in any handy cluster of foliage and quickly dashing into more supportive territory. In motion, their dramatic coloration is highly visible and may serve to flag other members of the group. But when these monkeys are still, that coloration serves as camouflage, breaking up the visual pattern into some semblance of sun-dappled shadow.

The sight I remember most vividly is this: one late afternoon I looked up to see a single black-and-white colobus sitting at the top of a monumental tree. As I peered up through my binoculars, I saw that he was plucking off and opening up giant seed pods that looked like huge brown peapods. After eating the seeds, he would discard the empty pod halves, which would spin to the ground. It was unpleasantly hot where I stood, but I soon recognized that where this monkey sat — one knee up, forearm casually resting on that knee, looking for all the world like a person sitting in the upper tier at a baseball game — the wind was blowing wonderfully cool. And his white tail fluttered in the wind like a banner. I wanted to be up where he was, and then I thought that one seldom-considered advantage to real estate in the high trees is the great view and pleasant breeze. He soon saw me, however, and began leaping down the tree. When he moved, suddenly, other monkeys started moving too, and I ultimately counted five black-and-white colobuses in that tree. They were leaping and diving great distances, and they soon disappeared into other parts of the tree and other trees.

Black-and-white colobus monkeys live in coherent social groups of from three to more than a dozen individuals, which inhabit and defend forest territory from other groups of their own species.

The ratio of adult males to adult females in the social group averages about one to three. Many groups include only one fully adult male, with perhaps one or two adolescent males. Other groups may include two or more fully adult males. But the adult males don't seem to like each other very much; they coexist in a group with some degree of tension. One, usually the biggest and oldest, is dominant: subordinate males will move away as he approaches, look away when he stares, or occasionally greet him with a "chewing-and-lip-smacking" gesture of appeasement. The females coexist in comparative harmony, and if there is a dominance hierarchy among them, it is relaxed, perhaps based upon some habitual deference to older females.

Generally, the full social group coheres through shared experience and mood, sleeping, moving, foraging, resting, sunbathing, and grooming together. Mothers and infants share the most powerful bonds, but all members of the group express some degree of mutual attachment, especially during grooming.

Grooming may take place at any time during resting periods. But in the morning, after the group arises, there occurs an extended grooming ceremony. Grooming, of course, has a hygienic function: with both hands, an individual presses down and parts the fur of another, and with tongue and lips cleans and removes loose particles of skin and foreign matter. Yet perhaps more important, grooming reaffirms emotional and social ties. During the grooming ceremony individual animals may approach each other, seemingly at random, and mutually groom. Or one individual will deliberately seek out another. An individual who wishes to be groomed will approach another, lie down, and wait. If the other resists or refuses — by ignoring, looking away, or even slapping with the hand — the approaching animal may still lie there and persist, eventually persuading the other with several brief slaps to the cheek. An individual who wishes to groom will either approach another from the back and begin grooming, or, approaching from the front, grab the other's fur at the forehead, pull down the head, and begin working from that position. Adult females do more than their share of grooming, while infants, juveniles, and adult males get more than they give. Still, all or almost all members of the group will groom and be groomed during the morning session.

All primates can be said to have a home range, an area where they live and forage. When a home range or part of one is actively defined and defended from other members of the same species, however, it is called a *territory*. Black-and-white colobus territories are roughly thirty-five acres in size, and good forested habitat for this primate may include a whole series of adjacent territories, a regular suburbia in the forest, each territory inhabited and defended by a single social group. The territory should provide sufficient food for the group more or less permanently, as well as good favorite sleeping trees.

The dominant male is always the definer, or signaler, of territorial possession. In the early morning, before the rest of the group has left the sleeping tree, or late in the evening when the rest of the group has already turned in, or even during the hot part of the day when the others sit or lie languidly in a shaded portion of the forest canopy, he is busy. He climbs to the top of a high tree and watches for neighboring groups. At dusk, during the night, and in the early morning, this dominant male announces his group's occupation of territory with a loud "roaring" that carries more than a mile through the forest, at the same time jumping violently up and down on a branch or from one branch to another in his tree. Neighboring groups usually cannot see him jumping — so roaring is the important territorial signal. Still, the leader's violent leaps lend an additional pulsing quality to the roar and may impress members of his own group. Once begun, the roaring becomes contagious, answered by the roars of dominant males from nearby territories, and soon a chorus develops.

In spite of the roars, often one group will try to invade the territory of another. Members of the trespassing group may gather first at the border, perhaps in a tall lookout tree, and pause to consider the situation and gather their courage by engaging in a quick and intense bout of mutual grooming. Then subordinate males slowly cross the border, followed cautiously by other members of the group. The home group of monkeys may also have gathered near the border, psychologically preparing for battle with their own brief grooming session. When the trespassers appear, they are met by adult male and female members of the home group, who crouch in branches, glare, open their mouths wide, and click their tongues. The males may further threaten the intruders by sitting, legs extended, and displaying erect penises toward the objects of their displeasure.

Should the trespassers persist, the dominant male of the home group may at last approach with violent, noisy leaps, slapping on branches, breaking off twigs and dead branches. He may then proceed with a leaping display, jumping twenty to thirty feet down into lower vegetation, landing with a dramatic crash, then ascending and jumping again. Meanwhile the subordinate males may chase, wrestle with, or attack the intruders. When finally the trespassers are driven out — the usual outcome — the home group reassembles and the dominant male emits a triumphant territorial roar.

SCIENTISTS HAVE STUDIED the eastern black-and-white colobus more thoroughly than the western species I pursued at Tiwai Island. Later in my African trip, I was able to observe a group of eastern black-and-white colobus at the edge of Lake Naivasha in Kenya, and I quickly hypothesized why the eastern species is so much better known. Simply, East Africa is savanna, not tropical forest. The trees are lower and farther apart, so the monkeys there are much easier to find and watch.

Moreover, the nine individuals I observed in East Africa were entirely inured to seeing humans with binoculars sticking out of their faces, so I was able to sit in an aluminum lawn chair and watch them for hours at a time. Yes, I saw that these eastern black-and-whites had long black tails with white tufts at the end and flowing capes of long black fur with beautiful white fringes. When they leapt, the white fringe lifted up, swirling, and floated down as they landed. Their faces, solemn and expressionless, seemed to me gray rather than black. The white fur around their faces looked much like a man's hair and full beard, except that it flowed up into a straight and narrow band across the forehead. And the black fur on top of their heads turned to rise thickly up, resembling a thick flat-top haircut with a low part in the middle.

For me, the most interesting member of this eastern family of nine was the baby. Black-and-white colobus monkeys start life with a full white coat of fur, which they begin to lose after the first couple of months. This baby, a female just starting to shed her natal coat, was still remarkably awkward. While most of the family relaxed in the lower part of the tree, scratching and grooming, the baby kept busy higher up in a maze of small twigs and branches. Unsteady and unsure of herself, she kept rushing about, making small leaps, losing her foothold, hanging desperately by one hand or two, trying frantically to locate a foothold for her two kicking feet. Sometimes she would land on top of a thin branch, grasping with all fours, and then suddenly flip upside down and find herself hanging by one hand. She slipped and tripped and stumbled and tumbled up there, yet kept moving and climbing constantly, forty lethal feet above the ground. It was unnerving to watch, and several times I thought she was about to fall all the way to the ground and die. But she never did, and I noticed that at least one and often two or three older members of the family group were positioned right beneath the baby, ready to catch her. I came to think of the tree as a large house and the baby's twiggy portion as the nursery.

Once, as I watched those monkeys in Kenya, my fantasy engine accidentally slipped into high gear, and I began seeing them as a tribe of glum clergymen dressed in clerical black and white. Then I perceived that even their dramatic coloring merged with the natural colors around them: black and white, branch and trunk, earth and cloud. And I looked at my own clothes — red shirt, blue pants, yellow pack — and observed how my own careless dress had interrupted the flowing assemblage of earth colors around me.

An animal's fur reflects, as camouflage, the exterior world of nature. Human clothing, often a forced compilation of artificial colors in angular patterns, refracts from the interior, expressing minds that in coming to cerebral consciousness, to ego and time, have foregone or forgotten the peace of a meditative attention in nature. (I'm thinking of contemporary temperate-zone styles. People living traditional lives in the tropics often cover themselves with cloth patterned like leaves, ripples, scales, feathers, in colors soft or bright, yet never brighter than a bird's.) It is thus an interesting if perhaps atavistic twist of habit when humans drape themselves with animal skins. And it became a twist in my thinking to recall, as I saw the tribe of glum clergymen in clerical habit, that black-and-white colobus monkeys are the only primates in the world that have been intensively hunted to provide fancy garb for people.

Skins from the eastern black-and-white colobus have been used in East Africa for centuries to create warriors' caps and capes, ceremonial headdresses and costumes, anklets and shield coverings. Their tails were occasionally converted into fly whisks. The export trade in black-and-white colobus skins is also centuries old. Skins of the eastern species were exported into Central Asia at least by the time of Marco Polo (late 1200s); and as late as the 1960s, thousands of pelts were exported each year from East Africa (through Somalia) into the Middle East.

For more than a century, moreover, a substantial trade to the markets of Europe and North America has waxed and waned, involving pelts from all four black-and-white colobus species, but primarily from the western black-and-white. The pelts were usually graded according to glossiness, and the western species happens to have the most glossy fur. The eastern species' pelt, with its long white fringe, has sometimes been valued for the dramatic mixture of colors, but often furriers dyed the white portions black to simulate the more valuable western pelt.

The first "black monkey" pelts of the recent European trade appeared in London during the Great Exhibition of 1851. A furrier at the exhibition fashioned the black pelts into muffs and short capes.

Later furriers cut the skins into strips to line the edges of coats and dresses or used the whole pelt to create hats, capes, and other fancy apparel. By the time the fashion declined near the beginning of this century, well over two million black-and-white colobus monkeys had been slaughtered to provide skins for the wholesale fur markets of London, eventually to be retailed in Germany, Italy, the United States, Canada, and elsewhere. The market expanded again after World War I: 30,000 to 40,000 skins were traded in 1923, with sales continuing for perhaps a decade. And for about eighteen months in 1972 and 1973, it once again became fashionable to drape the skins of dead black-and-white monkeys across one's body. Furriers in London were selling up to 10,000 pelts monthly, mostly to Italian, French, and German dealers. As in the past, western black-and-whites provided the bulk of the skins, although many from the eastern species were also sold.

By the early 1970s a significant tourist trade in rugs and wall hangings fashioned from black-and-white colobus skins had developed as well. In East Africa the market in such skins to tourists was enormous. A 1972 survey of tourist shops in Nairobi and Mombasa, Kenya, turned up rugs sewn from the triangular back skin pieces from five to forty-nine animals. Altogether the Kenya shops were displaying skins from over 5,000 monkeys; adding an estimate of undisplayed stocks, the shops may have accounted for about 27,500 dead monkeys at that time. One dealer in Mombasa claimed, furthermore, to be exporting rugs containing about 5,000 skins annually to American dealers, mostly in Baltimore, Maryland, and to German dealers. A second survey in Nairobi in 1974 counted far fewer rugs for sale, about half of which were displayed in a single shop, "Jewels and Antiques" on Kimathi Street. It has been roughly estimated that Nairobi dealers were selling up to 20,000 skins yearly at that time, while Mombasan dealers may have been selling 700 monthly during the tourist season. Kenya finally banned the sale of black-and-white colobus rugs to tourists in 1978 (but gave one to Pope John Paul II during his 1985 tour of Africa).

In West Africa the extent of the recent tourist trade in western black-and-white colobus skins is poorly documented. We know that colobus skins have been sold to tourists in Sierra Leone, Ghana, and Nigeria, but probably not in great numbers.

SEVEN MEN from local villages worked at Tiwai Island, clearing trails, repairing buildings, and acting as field assistants for the occasional visiting scientist. I didn't need a field assistant, but one morning one of the staff, a young man named Minah, offered to lead me to several of the different monkey species. Like most of the African men I met in the countryside, Minah was lean and perfectly muscled. His handsome face had three bulges on it, two for his cheekbones, one around his mouth; elsewhere the skin stretched tight. He was healthy, energetic, spirited — lean but not emaciated. He seemed terribly shy and seldom spoke.

Minah found monkeys by ear. He would stop, listen intently until he heard the calls of the species we sought, then follow the sound. We once stopped and listened to nothing I could figure out. "You want to see Diana?" he said. I indicated yes. He said quietly, "That is the voice of Diana." And off we went, soon to discover four or five high trees packed with more than forty Diana monkeys. The Dianas, it seemed to me, squeaked, squealed, and emitted rasping groans.

We found groups from several different species in that fashion, but I was growing hungry. So, as we walked through the forest, my stomach emitted a rasping groan. Immediately Minah stopped, cocked his head, and waited with great attentiveness for the sound to repeat itself. It didn't, and we began walking again. But once again my stomach raspingly groaned, and so again Minah stopped and listened, intent and wordless. Finally I said, "Let's go have some lunch," and we returned to camp and ate lunch.

Lunch was a big dishpan full of white rice covered with a stew of, if I remember accurately, boiled greens, tomatoes, peppers, and peanuts, thickly mixed with a foul-tasting palm oil. Nine of us — the seven staff members, Anne Todd, and I — sat down, and we ate communally, reaching with our hands into the large pan and scooping out big gobs of the mixture, which we pushed into our mouths. The lunchtime conversation was quick and animated, although I understood almost nothing of it. (The Africans spoke among themselves in a local language, Mende, and talked to Anne in the national dialect, Krio.) But whenever I wanted to join in the conversation, Anne served as translator. I took the opportunity to ask my lunch companions whether they ate monkey meat. Muslim tradition forbids eating monkey meat, they told me, so few of the older people in the villages around there eat it. But younger people don't care much about such traditions, and they'll eat monkey meat whenever they can get it. The preferred species are the spot-nosed monkey, Campbell's monkey, and the mangabey, but other types are also consumed.

In West Africa game meat has traditionally been a major source of animal protein, and even today around three quarters of the people in this part of Africa include wild meat in their diet. The Senegalese ate nearly 380,000 tons of meat from wild mammals and birds yearly in the early 1970s. In the Northern Province of Ivory Coast, it was estimated in the 1970s that each person ate about an ounce of wild meat a day, which accounted for most of the meat consumed. In some rural regions of Ghana, perhaps three quarters of all meat eaten is from wild game. And during the mid-1960s in southern Nigeria, domestic meat accounted for only 21 percent of the total animal protein consumed. Yet rural consumption is only part of the picture: in many areas of West Africa, professional hunters make a living selling game to intermediaries, who transport and resell it in urban centers. In Accra, the capital of Ghana, one market alone sold at the very minimum 155 tons of wild animal meat over a period of seventeen months between 1968 and 1970.

West Africa as a whole now doubles in population every twenty-five years, and domestic supplies of meat have failed to keep pace with the exploding demand. Although people in many areas once excluded various kinds of meat from their diet, as wildlife has declined, the list of acceptable foods has expanded. Throughout the region all of the wild ungulates (hoofed mammals) are eaten, as are wild cats, and in some areas domestic cats. In Western Ghana, we are told, people "are starved for meat, and they will eat almost anything." The chief game and wildlife officer of Ghana, E. O. A. Asibey, has written that Ghanians — and they can be considered representative of the larger region — consume hares, giant rats, cane rats, porcupines, small rodents including common house rats (which supposedly have an additional medicinal value for children suffering from whooping cough), all types of squirrel, most types of fruit bats, anteaters, most birds, all tortoises and turtles, monitor lizards, the African python, the Gaboon viper, the puff adder, the agama lizard, ants, maggots, giant snails (which in some areas are a major protein source), as well as bush babies, chimpanzees, and all available species of monkey.

This expanding appetite and genuine need for animal protein has resulted in overhunting in many parts of West Africa and the depletion of most West African animal species. In Ivory Coast, for instance, monkeys have been pretty much eliminated from all areas except officially protected sanctuaries. In Ghana the pygmy hippopotamus and the manatee are gone, while the olive colobus monkey is going. Western black-and-white colobus monkeys had already been extinguished in many parts of Ghana by the mid-1950s; although officially protected by the Wildlife Preservation Act of 1961, a decade later they were still being killed at the rate of about one per day in the forest reserves of western Ghana. They were reported to be "heavily poached" outside of protected areas in 1978.

In Sierra Leone flocks of wildfowl and game birds, as well as predatory birds, noticeably diminished in a single decade, the 1960s. Overhunting of certain predatory animals in Sierra Leone, including leopards, civet cats, and pythons, brought about an explosion in the number of cane rats, which began feeding heavily in rice fields and cassava plantations. And except in devoutly Muslim parts of the country, where religious tradition still somewhat protects them, all primate species — including the western black-and-white colobus — were "relentlessly hunted," according to a 1970 report. Hunting in Sierra Leone's neighbor Liberia has already reduced the numbers of primates enough that teams of Liberians, armed with semiautomatic weapons, now buy their way across the border and hunt monkeys in eastern Sierra Leonean forests. The meat is preserved by smoking, smuggled back into Liberia, and sold at urban markets.

The problem will not easily be solved: West Africans are hungry, and West African economies are not even beginning to keep pace with the unprecedented growth in human population. The wages of ordinary workers in Sierra Leone are minuscule by Americans' — make that anyone's — standards. Abdul, the cook at my friend's house in Freetown, was paid about eight dollars a month, a little more than usual for a domestic cook, but less than the cost of the rice he would eat during that time. The workers at Tiwai Island get coffee in the morning and a full meal at noon, but Anne Todd told me that when she first came to Tiwai, the staff weren't offered food or drink at camp, yet were putting in a full day's work. "That was during the hungry season, too, which means that they were probably only eating one day out of every two at home," she said. "No wonder these guys look lean," I said.

Before the domestication of animals, beginning around nine millennia ago in the Middle East, any person who ate meat ate the flesh of a wild animal. Today most people in the industrialized nations eat the meat of domesticated animals, neatly packaged, its bloody origins remote and forgotten. Yet many people living in the tropics continue to rely on and often prefer the meat of wild game as an important source of animal protein. For these people, primates are simply one type of wild game. No one would argue that hungry people should not hunt. Nonetheless, hunting in many parts of the tropics has recently become overhunting. The game is declining, so hunting contributes not only to the occasional extermination of species, but more steadily to the diminishment of an important food resource.

I was in Sierra Leone during the month of March, which is the end of the dry season. One day while at Tiwai Island, I emerged from a swim in the river and looked up to see on the other side a massive plume of dark smoke, expanding upward in a great explosive billowing of white and gray, sickly blue and black. The fire continued all day and into the night, filling the air with a dark haze and the smell of burning, and continued even into the next day. Another day, when I walked several miles west from Tiwai to the Gola Forest at the western edge of Sierra Leone, I saw more fires, as well as leveled secondary forests and farm bush ready to burn, and already burned high forests, transformed into eerily quiet graveyards of tree skeletons and wilted weeds. Later yet, when I left Tiwai and traveled across the country back to Freetown, I saw fires and dark plumes of smoke regularly, nearly everywhere it seemed. It was slash-and-burn time in Sierra Leone.

Tropical forest farming is a very different enterprise from temperate-zone farming. Most tropical forests grow on ancient soils from which the nutrients have long ago been leached. The forests have responded to this fact, over tens of millions of years, by developing distinctive nutrient cycles. Since the bulk of crucial nutrients energizing a tropical forest resides not in the soils but within the living body of the forest itself, when a tree or a leaf dies, separates from the living whole, and falls to the ground, both climate and ecosystem conspire quickly to decay the dead vegetation and recycle its nutrients back into the living forest. Impoverished soils mean that temperate-zone agriculture, stationary farming, doesn't work. What does work is slash-and-burn.

For centuries, traditional forest dwellers around the world have practiced in moderation a style of slash-and-burn farming we might describe as *rotational.* A patch of forest is cut down and entirely burned over. The burning transforms the forest into ash spread across the ground, which provides a burst of nutrients substantial enough to sustain farming for a few seasons. After a short period of farming, the forest patch is abandoned and left fallow for two, three, or more decades, and then cleared, burned, and farmed once again. In other words, rotational slash-and-burn agriculture is not based upon the progressive invasion of forests but rather — with small numbers of farmers per area and long enough fallow times — upon sustained use.

But in recent decades traditional forest farmers, themselves expanding in numbers, have been joined by newcomers to the forest: landless colonists, culturally unprepared for sustained rotational agriculture, who approach the forest as an enemy. If the traditional farmers still practice a sustainable, rotational slash-and-burn agriculture, the newcomers practice what we might call *progressive* slash-and-burn. They arrive in large numbers, hack away at a forest's edges, farm cleared areas until they are entirely depleted, then progress more deeply into the forest. Instead of a tolerable matrix of fallow patches that will return to secondary forest and ultimately to primary forest, they leave behind a swath of scrub and degraded cropland that in many cases will never see forest again.

Unchecked population growth combines with a direct dependence on farming (almost 70 percent of the third world lives in rural areas) to turn a certain sector of people in tropical nations toward the great forests, where they see progressive slash-and-burn farming as their one option for staying alive. Thus well over 200 million forest farmers — nearly one out of every twenty people on this planet — are now clearing 65,000 square miles of primary tropical forest per year, converting perhaps half that amount into permanent scrub or grassland, and disrupting the other half severely enough that, while some trees may eventually grow back, the full forest will never return to its original state of complex integrity and exuberant diversity.

The western black-and-white colobus habitat is a long, narrow belt of tropical forest reaching into eleven relatively small nations on the southern coast of West Africa, from Senegal in the far west through Gambia, Guinea-Bissau, Sierra Leone, Liberia, Ivory Coast, Ghana, and Togo, to western Nigeria in the east. Throughout this region forests are being destroyed by slash-and-burn farming — as well as by cash-crop farming, logging, and other developments. Sierra Leone, which may have been three-quarters forested in recent times, now includes primary forest on only about 3 percent of its full area.

IT WOULD BE HARD to overestimate, I believe, both the current extent of ecological devastation in Sierra Leone and its suddenness. A generation or two ago the country was a land of great forests and considerable wildlife. Now it is not, and the next decade has been described as "the last chance." Many people I talked to are pessimistic about the future of conservation in Sierra Leone. One person said: "They're just not ready for conservation here. When they think of a plant or animal, it's only in terms of what it can be used for. Any money coming in will never go to conservation. And since the purpose of the government is to rape the country, you're not going to get much through official channels."

However, John Waugh, a Peace Corps volunteer who is also one of the few non-Sierra Leonean members of the Conservation Society of Sierra Leone, is a good deal more positive: "When I first came to this country, I was told, 'We're glad you're here: No Sierra Leoneans care about conservation.' But they do. There are people here who care a great deal. The government is very interested in conservation — it's just that economic constraints are so great. The government would like to develop an extended system of parks and reserves, but they just don't have the money to do it." Waugh believes that conservation in West Africa must be integrated with development, and he argues for conservation according to a concept of *sustained use* — the idea that wilderness and wildlife are important resources that can be preserved *and* exploited, if the exploitation is moderate enough to sustain, not deplete, the original resource base.

Sierra Leone right now maintains two important Arks for West African wildlife: Tiwai Island to the west and Outamba-Kilimi National Park to the north.

Sixty years ago Tiwai Island was a slave island, a large plantation farmed by slaves who had been acquired in wars between the local Mende people and other ethnic groups of the larger region. The British, who managed Sierra Leone as a protectorate, outlawed internal slavery in 1928; the island's slave plantations were abandoned, and the land began to revert to secondary forest. Small parts of Tiwai have been farmed since then, but its natural isolation limited the extent of farming and provided a natural sanctuary for about 130 tree species, 140 bird species, and 10 primate species. Currently Tiwai Island is defined as a wildlife sanctuary and permanently protected by the Barri and Koya chiefdoms of western Sierra Leone. It is the only locally (instead of nationally) established sanctuary in all of West Africa. Farming is still allowed on parts of Tiwai, and local people have complete access to the island, but hunting is forbidden. Tiwai could become a tourist attraction, but right now it serves as a research site for scientists.

Within Sierra Leone's northeastern sector lies a region of open savanna and deciduous forest fragments that happens to be rich in wildlife, particularly two contiguous areas known as Outamba (285 square miles) and Kilimi (90 square miles). Lions are gone from Outamba and Kilimi, and elephants nearly so, victims of European sport hunters carrying modern weapons. But twelve primate species, including such endangered types as chimpanzees and western black-and-white colobus monkeys, remain. The two areas, part of the larger Tambakha Chiefdom, are inhabited by the Susu people, who practice slash-and-burn farming.

In 1974 Outamba and Kilimi were officially designated as game reserves, the first in the nation. Meanwhile, during the 1960s and 1970s, Sierra Leone had become a major exporter of live chimpanzees for the European and American research industry. After American pharmaceutical companies applied for permission to purchase 275 chimpanzees from Sierra Leone, though, three Western conservation organizations sponsored a survey to find out just how many of those intelligent apes actually were left in that country. The survey, soon expanded to include other mammals, confirmed both the decline of chimpanzees and the great value of Outamba and Kilimi — and ultimately encouraged the central government to convert the two reserves into a national park.

In considering Outamba-Kilimi as Sierra Leone's first national park, perhaps the most difficult problem was how to approach the Susu people occupying the land, who have lived in the region for around two centuries. (Luckily, Outamba and Kilimi are among the most lightly settled areas in the country.) The Susu, as converts to Islam, supposedly do not eat the meat of primates. Nonetheless, anthropological studies have shown that they feel a profound alienation from animals and nature and will readily kill primates and most other animals as pests or potential pests. They disdain herding and keep no animals as pets, while the few domestic animals they do keep — chickens, goats, and sheep — are left to roam and feed themselves as they can. Traditionally, the Susu consider agriculture to be divinely ordained, a sacred activity, while the bush and forest are regions of the profane, inhabited only by wild animals and malevolent spirits. The fragments of forest that are the most important ecological refuge for the primates and other large mammals of Outamba and Kilimi are viewed by the Susu as the best areas for farming: the trees are an indicator of moisture in the soil, and they stand on land not already depleted of nutrients by the annual grass fires that sweep across open savanna. In short, the human and wildlife inhabitants of Outamba and Kilimi compete for the same few pieces of forested land. The first plans for a national park called for resettling resident Susu farmers into other parts of the Tambakha Chiefdom; but that idea proved controversial, so the Susu farmers remain within the area.

In any case, Sierra Leone's president approved plans for Outamba-Kilimi Nationa Park in 1981, and the following year the World Wildlife Fund–U.S. began financing construction of park facilities and training local people to staff the park. Finally, in July of 1984, the World Wildlife Fund turned the park over to the Sierra Leone government.

One hopes that the few western black-and-white colobus monkeys inhabiting this new and very important West African Ark are now protected. But the future of Outamba-Kilimi is unclear, for it must somehow, and someday soon, I believe, prove itself economically.

Sierra Leone, blessed with some of the best beaches in the world, populated by an unusually warm and friendly, handsome people, could become a tourist Mecca, but it is not now. Moreover, the road between Freetown and Outamba-Kilimi does not cater to the convenience of ordinary tourists, and building a good road would probably cost much more than the park could ever return in tourist revenues.

ONE MORNING at Tiwai Island I opened the food box — big cockroaches inside slowly backed away from the surprise of light — and took out a loaf of bread: my breakfast. The African staff had already arrived, and were drinking their morning coffee. But as I started chewing on the loaf, I suddenly felt I was being inconsiderate, and so broke the bread into several pieces and passed them around.

Not only at Tiwai, not only in Sierra Leone, not only in Africa, but everywhere in the tropics, I found myself often humbled by the sense of how easy it was for me to open the food box, when that box was closed for so many people around me. I was Marie Antoinette, touring the provinces in a glass coach.

For no good reason that I can remember, I once fasted for twelve days. During my travels into primate worlds, circumstances often led me to go hungry for half a day or even a full day. I once climbed an entire mountain, up and down, powered only by two Coca-Colas and a piece of toast. For middle-class Americans like me, in other words, hunger is a hobby. And as I traveled through the many hungry spots of the tropics, I always trailed behind me the invisible umbilical cord to a wealthy temperate-zone economy. I carried near my solar plexus a small sack containing all the necessary jujus — American cash, credit, passport — to palliate the rasping groans of my own hunger.

I possess, in short, no personal expertise on hunger, but I have lately acquired the following belief. We are witnesses to a war, global in extent, motivated by human hunger. Individual soldiers of hunger enter the tropical forests and pluck out the moving fauna. Much more devastatingly, whole battalions of hunger move right into the forests permanently and level them entirely, top to bottom. But the great tragedy and paradox is that the tropical forests, left standing, could serve as a permanent larder that would indefinitely provide sustenance, whereas the forests cut down will be empty, a lost larder. When the animals are gone and the soils depleted past renewal, the war's ill-considered strategy will have been fulfilled, yet a vast army of warriors will be even more hungry.

I left Tiwai Island in a wobbly dugout canoe with a fast leak.

Five

FOOD

WEST AFRICANS are not sentimental about wildlife or forests. The lingua franca of Sierra Leone, Krio (derived from a mostly English-based Creole spoken by free former slaves from the Americas who came to Africa and established a settlement at Freetown in 1787), expresses a frankly utilitarian attitude toward nature: Krio words for tree and animal are pronounced "stick" and "beef." Sierra Leoneans, moreover, have several good reasons not to be sentimental about nature. Nature there can manifest itself as malaria, sleeping sickness, river blindness, and a dozen other terrible diseases; as biting insects and enervating heat; as cobra, mamba, viper, and half a dozen other supremely venomous snakes. Tiwai Island is a wildlife sanctuary, where the killing of any wildlife is forbidden. Thus the research station has two latrines — the first was abandoned because a viper had taken up residence there. And sometimes when I visited the second, I was spontaneously inspired to think about vipers sliding in dark places and to consider well a utilitarian view of nature: a vision of forest as food.

People now commonly eat and enjoy around two dozen fruits — including avocado, banana, breadfruit, guava, jackfruit, lemon, lime, mango, orange, papaya, and pineapple — that originated from tropical forest species. Collectively, we now eat 40 million tons of bananas each year, 13 million tons of mangoes, 1.5 million tons of papayas, and 1.5 million tons of avocados.

One might argue that since these wonderful tropical fruits are already thoroughly cultivated, we face no particular loss, other than a sentimental one, in converting their ancestral trees into timber or fuelwood, or ash to fertilize the ground for a season or two. However, more than a dozen important commercial fruits still grow wild in tropical forests, as well as 250 commonly eaten but not commercially exploited species. Ultimately, the world's tropical forests may contain 2,500 different fruit species that could contribute to the world supply of food, compared to the approximately twenty fruits that grow naturally in the temperate zone. Among the underexploited tropical fruits, the mangosteen of Malaysia is considered unusually delicious. The pomelo of lowland Southeast Asian forests is one of the very few citrus fruits that will grow in the wet tropics; it is also the largest of the citrus fruits, bigger than a grapefruit, and it will grow even in saline marsh areas.

Other than fruits, tropical forests have already given the world such important foods as cane sugar, millet, okra, peanuts, rice, sweet potatoes, taro, and yams. These, and in fact all of today's widely cultivated food crops, are descended from a handful of choices made a few millennia ago, when Neolithic peoples first improved on nature by farming. That handful of major Neolithic crops continues to define today's world food resources largely through the persistence of memory, habit, and taste, while the tropical forests and other genetic reservoirs offer an immense cornucopia of possible new foods; at least 1,650 plants have already been identified as potential vegetable sources.*

Perhaps most important, the tropical forest genetic reservoirs may contain new, cultivable sources of green protein. World agriculture is mostly a carbohydrate industry. While adult men need about 14 percent protein in their food, and while children and pregnant or nursing mothers need 16 to 20 percent, most of the world's extensively cultivated crops, such as cereals, potatoes, and yams, yield from 1 to 12 percent protein, dry weight. Thus people who can afford it supplement their vegetables with such high-protein foods as dairy products and meat, which are comparatively expensive and represent an inefficient use of increasingly scarce resources. A beef steer, for example, probably consumes over seven times more protein from vegetation than it delivers as meat.

Soybeans are an exception in agriculture, providing high protein

*N. W. Pirie comments on the persistence of taste in Western cultures: "Even in Europe and the U.S. a flavor and appearance unacceptable in an egg is acceptable in cheese, a smell unacceptable in chicken is acceptable in pheasant or partridge, and a flavor unacceptable in wine is acceptable in grapefruit juice. These things are a matter of habit, and habits, although they will not change in a day or even a month, can readily be changed by suitable example and persuasion."

yields — but soybeans are a temperate-zone crop. Might we find a tropical soybean? Growing wild in the tropical forests of New Guinea and Southeast Asia, and already providing food for forest tribes in New Guinea, the winged bean plant contains far more protein than any of the more commonly cultivated food crops of the tropics. In fact, the bean itself is nutritionally equivalent to soybean, containing 40 percent protein, as well as a high content of edible oil and several other valuable nutrients. Additionally, the leaves, shoots, flowers, and pods of the winged bean plant can be eaten as greens, and its underground tubers taste like early-season potatoes but contain roughly 20 percent protein, several times the protein content of potatoes. Presuming that the wild winged bean can be cultivated widely and economically, this plant — until recently obscure and thought worthless — should prove a most valuable commodity for people in the tropics.

Tropical forests have already provided us with such nonessential but highly valued beverages and flavorings as coffee, tea, cocoa for chocolate, and vanilla. What they seem about to offer us includes some important alternatives to sugar and the carcinogenic synthetic sweeteners. Unlike the standard natural sweeteners in use today, which are all based on naturally growing sugars, these new natural sweeteners are both protein-based and unusually powerful, and should not result in the sorts of health problems caused by excessive and habitual sugar consumption. The miracle fruit, actually a berry, of West African forests is a powerful protein-based sweetener, as is the serendipity berry of West Africa, discovered only in 1965 by an American scientist. The serendipity berry is said to be 3,000 times as sweet as ordinary sucrose, while sweet proteins derived from a third recently recognized product of West African forests, the katemfe fruit, are supposedly 1,600 times more potent than sucrose. Sweeteners based on katemfe proteins are already marketed in Great Britain and Japan.

Aside from having provided the genetic material for several of today's important cultivated plants and domesticated animals, as well as offering the potential of several new major crops and domesticated animals, the tropical forest genetic reservoirs also provide continuing genetic support for the stability and health of today's farming and husbandry.

The world food supply is based primarily upon a small number of cultivated plant and domesticated animal species. Yet several of these species are, in turn, derived from a very narrow genetic base, a very small number of wild founders. For instance, the entire North American soybean crop originated from only six plants taken from one area of Asia. Almost three quarters of the United States potato crop derives from only four wild varieties. Furthermore, because cultivated crops and domestic animals have long been bred to express a few particular qualities, many of them have come to possess even a smaller spectrum of genetically transmitted qualities than their original wild founders. Their genetic narrowness is a double-edged sword: while cultivated crops and domestic animals may be unusually productive, they are also unusually vulnerable to disease. By contrast, a wild species will typically proliferate in many subtypes, some of which may not be vulnerable to particular diseases. In short, the world's cultivated and domesticated food resources deeply depend upon the continuing presence of their wild relatives, many of which are still living in the tropical forests and other declining genetic reservoirs. Let me give some examples.

Rice. Many of today's high-yielding domestic rice crops were recently threatened by a disease called grassy stunt virus, until they were cross-bred with a wild rice strain from Africa that provided critical resistance.

Sugarcane. During the 1920s aphids carrying the sugarcane mosaic virus devastated Louisiana's crop and threatened the entire North American sugarcane industry. By 1926 the virus had reduced the annual crop in Louisiana from around 180,000 metric tons to 43,000 tons. Fortunately, a sugarcane growing wild in Java was found that possessed resistance to the mosaic virus. Breeding in that wild variety probably saved the United States sugarcane crop. Other wild sugarcanes have from time to time provided the domestic crop with critical resistance to red rot, gummosis, and other serious diseases.

Coconuts. By 1978 a lethal yellowing blight in Caribbean coconut trees had destroyed 375,000 of the trees in Florida and was killing 1,000 per day in Jamaica. By 1982 the disease had spread into the Yucatán. Fortunately, Malaysian coconut trees were found that possessed resistance to the blight; they may replace the American trees.

Peanuts and tomatoes. Resistance to leafspot, provided by a few wild species of peanut found in Amazonia, has recently increased the productivity of cultivated peanuts by an amount valued at $500 million per year. Two species of wild green tomatoes found in Peru during the early 1960s have improved the coloring and soluble-solids content of commercial tomatoes enough to be considered worth $5 million yearly.

Coffee. Latin American coffee almost entirely derives from a species of African origin, the *Coffea arabica.* Just as South American rubber trees are commercially farmed in Southeast Asia to avoid certain coevolved native parasites, so this African coffee species was grown in Latin America partly to separate it from its native parasite, coffee-leaf rust disease. By 1970, however, thirty different strains of the coffee-leaf rust disease had appeared in the coffee plantations of Brazil, threatening to ruin that important domestic crop. Within six years the disease had spread far north into the coffee plantations of Central America, which were at the time producing $3 billion worth of coffee beans a year. Happily, that particular crisis had been anticipated. A team of scientists had journeyed to the highland forests of Ethiopia in 1964, where, although only one eighth of the original forests remained, they found enough wild *Coffea arabica* trees to take out a few specimens. Preserved in a plant bank in Costa Rica, they finally provided resistance to the leaf rust.

Actually, all cultivated crops are vulnerable to diseases that either already threaten or can be expected to. Yet in the tropical forests and almost everywhere else, the supply of wild cultivars is declining. In North Africa wild species of coffee, okra, and pea are declining. In Asia wild cucumber, eggplant, radish, squash, black-eyed pea, and yam are diminishing. In the Mediterranean wild asparagus, cabbage, beet, lettuce, wheat, and turnip are disappearing. In Latin America wild bean, lima bean, pepper, corn, potato, sweet potato, pumpkin, and tomato are being lost. Corn, potatoes, rice, and wheat — these four cultivated crops feed more people than the next two dozen most important crops combined. A single major disease sweeping through the fields of any of those four genetically limited species could cause a disaster that would make the Irish potato famine — in which a single plant disease resulted in the starvation of two million people in two years — seem trivial by comparison.

Wild-growing plants also offer the potential to improve qualities other than disease resistance. For example, an extremely fine-tasting type of wild cocoa has been discovered within the less than one square mile of forest that is the Rio Palenque Biological Reserve of Ecuador. A wild variety of coffee naturally free of caffeine has been found in some forest fragments of the Comores Islands off eastern Africa. A wild variety of teosinte, the original source of domestic corn, has recently been found in a small mountainous forest of south-central Mexico. This wild corn ancestor possesses the same chromosomal structure as domestic corn and thus might be cross-bred with domestic corn, but it is also a perennial plant that thrives in a wet and chilly climate at high elevations. Crossing wild teosinte with the domestic corn crop, therefore, might establish a good domestic species yielding a perennial harvest (without annual plowing and planting) in "corn orchards," and might also allow the domestic crop to be grown in higher, wetter, and cooler regions. The wild teosinte, furthermore, is resistant to several diseases that regularly threaten domestic corn.

The tropical forests' vast genetic reservoirs also offer new opportunities for pest control. If the tropical forests can be described as a 75-million-year feast, they can also be described as a 75-million-year battleground, where millions of plant and animal species, concentrated in fecund profusion, have competed voraciously. The result of such crowded and intense competition is an extreme concentration of predators and parasites — pests — and defenses against them. Hundreds of thousands of plant species, for instance, have coevolved with millions of insect pests, and in that extended coevolutionary struggle, tropical plants have developed an extraordinarily broad armory of pest defenses that one day may prove useful to human agriculture.

Everyone hates mosquitoes, but people who are not farmers tend to underestimate the problem of insect pests. In fact, insects are said to reduce the world food supply by two fifths. In the United States one fifth of the food crop is regularly lost to insect damage, and $3 billion worth of poisons is used each year to save an estimated $12 billion worth of crops from the ravages of tiny mandibles.

Spreading synthetic poisons over food crops is a comparatively new idea. In the United States the practice became widespread during the 1940s with the use of several poisons, including the very effective but persistent DDT. When Rachel Carson warned of the serious environmental consequences of this practice in her book *Silent Spring* in the early 1950s, American farmers were casting 200,000 pounds of poison over their crops each year. Today American farmers pour on more than 5,000 times that amount, or 1.1 billion pounds per year. It is true that the DDT-style poisons once favored by American farmers have been replaced by several less persistent poisons and that the traces of DDT in our environment are gradually diminishing. It is true that some scientists believe the concentrations of pesticides now found in our ground water are too attenuated to worry about. Yet it is also true that many of today's agricultural poisons are more mobile than the 1950s poisons, more soluble, and more toxic. It is also true that most fungicides now used commercially are known to cause cancer and birth defects in laboratory animals. Even though the Environmental Protection Agency (EPA) was required by the U.S. federal government in 1972 to review the possible health and environmental effects of 600 active ingredients in 50,000 already approved agricultural poisons, it is also true that the EPA had not by 1986 analyzed even one of those ingredients.

Some people will argue that the dangers of synthetic pesticides have been exaggerated and that increased agricultural productivity in a hungry world is worth the price. Yet at least in the United States, we have no certain evidence that massive poisoning has really improved farm productivity. The same proportion of crops, one fifth, is lost to insect damage now as before the widespread use of poisons. Why? First, the great majority of pesticides randomly eliminates both target pests and those beneficial insects that either compete with the pests or prey on them. Second, insects are notoriously fast breeders and can quickly evolve resistances to particular poisons. Thus, whereas before World War II only seven species of insects and mites threatening American agriculture were considered to be resistant to insecticides, today around 450 species possess resistance to at least one insecticide, and some 20 species are now resistant to *every* known insecticide.

Synthetic insecticides are clearly not the full answer. Many plants in the tropical forest genetic reservoir produce *organic* toxins that kill or debilitate insects, that repel them, that inhibit growth, or that obviate reproduction — yet are biodegradable and often not toxic to larger mammals. In fact, insecticides have been made from naturally occurring tropical forest plants for centuries. Central American Indians, for example, have long used powder from the seed of the sabadilla lily to discourage lice. Amazonian and Malaysian natives have traditionally extracted the very powerful rotenoid compounds from certain legume species to poison fish and insects. One of these rotenoids requires only a trace concentration (one part per 300,000 parts water) to kill fish. Taking as their cue the traditional uses of rotenoids — which we now know can be extracted from at least sixty-seven plants in Amazonia and Southeast Asia — Western scientists have already successfully produced an insect poison apparently harmless enough to larger mammals that it can be dusted over livestock. Other insecticides already taken from the tropical forest genetic reservoir include the pyrethrins, extracted from several plants related to the chrysanthemum. Scientists have identified an Amazonian tree species that manages to repel the monstrously voracious leafcutter ants; its natural chemistry may yield the secret of a new, powerful, and target-specific insecticide. Many other natural insecticides based on the weaponry of tropical forest plants remain to be found and used.

Ultimately, modern agriculture may benefit most from an approach known as Integrated Pest Management. Integrated Pest Management would introduce natural insect predators in combination with emergency — as opposed to routine — spraying of pesticides. This practice, it is said, could reduce the amount of pesticides used on modern crops by three quarters without significantly affecting crop production.

Natural insect predators? The tropical forest genetic reservoir contains an army of them: pests' pests, insects that will displace, or that specialize in preying on, the eggs, larvae, pupae, or adults of particular other insects. Unlike the shotgun approach of most insecticides, which eliminate both beneficial and harmful insects at random, insect predators can work like a rifle, sniping away at only the target species. As with natural insecticides, the use of natural insect predators is actually an old art. Ancient Chinese placed nests of a predatory ant, *Oecophylla smaragdina,* into citrus groves to discourage vandalizing caterpillars and boring beetles. In recent times the U.S. Department of Agriculture has identified 1,000 species of insects, mites, and pathogens that could economically improve American agriculture; and we now have around 250 sample cases of successful pest control through the introduction of pests' pests. In Florida an introduction of parasitic insects in 1973, at a one-time cost of $35,000, has been saving the citrus industry an estimated $40 million yearly. In Hawaii a tachinid fly from the sago palm swamps of the Molucca Islands has helped inhibit the sugarcane beetle borer, which was wreaking $750,000 to $1 million worth of damage each year. In California, several pests' pests introduced over a fifty-year period increased crop yield and reduced pesticide use for an estimated net gain of $1 billion. Since the tropical forest genetic reservoirs contain millions of different insect species, one imagines that they contain many, many insects and other small predators that can benefit modern agriculture by controlling insect pests.

To quote from an otherwise thoughtful statement to a symposium for botanists a decade ago: "Only a mentally unbalanced person would say that plants and animals are more important, more useful, and in more need of care and love than the millions of people throughout the world — particularly those in the so-called 'third world' — who are dying before their time because of poverty, disease and hunger." That line of argument is hard to resist, since no one will admit to being "mentally unbalanced," since we all place the highest value on human welfare, and since we are all moved by the image of people "who are dying before their time because of poverty, disease and hunger."

We may all agree that people are more important than plants and animals. The speaker subtly implies, however, that the welfare of plants and animals is somehow in direct opposition to the welfare of people. Yet people are not usually dying because plants and animals are living. As the wildlife in West Africa declines, are West Africans progressively less hungry? Now that Nigerians are forced to import wood for construction, are they better off? Obviously not. On the contrary, life in the tropical forests sustains human life; and people everywhere, in the tropics and temperate zones, are being impoverished as we impoverish the tropical forests. When the forests are gone, or so altered as to be minimally useful as genetic reservoirs, ecological buffers, and sources of wood, fuel, and food, people everywhere will be poorer and that much closer to death.

In the long run, no one will benefit from the destruction of an ecosystem containing two fifths of the world's species. And not even in the short run do the usual sorts of tropical forest exploitation benefit most impoverished people of the third world. The massive deforestation of Costa Rica over the last two or three decades may have allowed for tremendous increases in national beef production, but the average Costa Rican now eats less beef. The deforestation of Nicaragua may have made the Somoza family rich, but it made the great majority of Nicaraguans and their descendants poorer.

Once we have disabused ourselves of the notion that today's style of tropical forest destruction pays the greatest possible dividends to hungry people, we are ready to consider alternatives: overall, a new pattern of exploitation that *combines* conservation with development in an intelligent fashion, ideally profiting the people of tropical nations while sustaining their forests perpetually.

As a centerpiece for the new pattern of forest exploitation, let's begin with the following single principle: all of the major forms of assault on or exploitation of tropical forests should be shifted to secondary — already degraded — forests. The remaining virgin forests, still rich in species diversity, should be left intact as critical parts of the Ultimate Ark, a permanent heritage for humankind, serving in perpetuity as genetic reservoirs and ecological buffers. (That's the central principle, although one might modify it in some instances. For example, traditional forest-dwelling peoples living in primary forests would continue to practice their minor and essentially nondestructive forms of exploitation as they have for centuries. Non-forest-dwelling people who live by gathering minor forest products such as rattan from otherwise untouched primary forests would also continue that style of exploitation, although some core areas of virgin forest might be declared off-limits for even such comparatively trivial exploitation.)

In order for this plan and principle to succeed, however, subsistence farming in secondary forests must become more efficient. I cannot pretend that such efficiency will come simply or easily. Nevertheless, improving the efficiency of tropical forest farming probably requires, more than anything else, that farmers begin to rely more on mixed cropping. Mixed crops provide better resistance to erosion (since some crops remain when others are harvested), make less extreme nutrient demands (since different crops require different nutrients), and reduce the losses to pests and diseases (since crops have different vulnerabilities).

The many types of legumes, with their capacity for drawing nitrogen from the air and storing it, should be included in any multispecies subsistence system, since the legumes can help fertilize not only themselves but plants around them. It has been demonstrated that interplanting the leguminous giant ipilipil tree with a wheat crop adds nitrogen, phosphorus, and potash to the surrounding soil and thereby significantly improves the production of wheat. Interplanting ipilipil trees with dryland rice in Indonesia, combined with the use of other natural fertilizers, has more than doubled the rice harvest. And rubber tree seedlings in Malaysia reach latex-producing maturity two years sooner when they are interplanted with low-growing legumes.

Experimentation will no doubt uncover many excellent systems for interplanting at the subsistence level, but we should also draw on the extensive accumulated wisdom of settled forest societies. Some forest-dwelling people have long supplemented their wild foods with multispecies dooryard gardens that imitate the larger ecology of the surrounding forest. The gardens contain crops emerging at several horizontal levels, planted very densely in a manner that maximizes natural plant synergies.

In Central America a dooryard garden may include at the ground level such root crops as malanga, yams, and sweet potatoes. Mixtures of corn, beans, squash, melons, and pineapple come next, planted in complementary combinations and exposed to the sun. Elsewhere in the garden, low woody plants such as coffee, chili, pepper, and pigeon pea will grow in the shade of higher trees such as bananas, cacao, and citrus, which are themselves interplanted with higher mango, avocado, and breadfruit trees. Chayote vines may climb on the citrus trees, while above everything else tower such emergents as coconut and pejibaye palm. Necessary fertilizers may be acquired from mulched crop leftovers, garbage, manure, or more exotic sources such as bird guano or organic refuse stolen from ant heaps. Although it is not done in Central America, such a dooryard garden could be supplemented with a fish and snail pond. The garden could provide fodder for some domesticated animals. The integration of a number of kinds of plants, plus the intermixing of annuals and perennials, buffers against extreme nutrient losses. The very variety of plants in such a garden discourages attacks by individual pest species and encourages the continuing residence of pests' pests. As well, plants known to inhibit certain pests can be planted among the plants that are vulnerable to those pests.

Is such a garden really practical? Members of the Lua tribe in northern Thailand grow up to 75 different food crops, along with about four dozen other plants for medicines and other purposes. The Tsembagas of New Guinea grow as many as 50 different crops at the same time. The Hanunoo of the Philippines cultivate 430 different crops and plant as many as 40 different plant species in a single two- to three-acre garden. The Lacandon Indians of southern Mexico grow up to 80 different crops in a plot about the same size. However, forest gardening at that level of sophistication is based upon culture, learning developed and held and passed on by stable societies. Newcomers to the forests, first-generation farmers living not in stable communities but in cultural diaspora, will not easily acquire the knowledge and patience necessary for such complex gardening. Also, the forest societies that possess as a cultural heritage the capacity for multispecies gardening are today being displaced and obliterated by deeds and leases, bulldozers and chainsaws. Nonetheless, the principle is simple enough, and new farmers could succeed in practice even if they do not attain the superb complexity of the best traditional gardens.

As the models of ecologically sound forest farming in this century fade, we can still examine models from the past. It has been suggested that traditional slash-and-burn farming can sustain around 65 people per square mile, but Mayan societies in the Yucatán practiced a system of forest farming that fed 300 to 400 people per square mile at the height of that civilization, in 800 A.D. According to recent archaeological discoveries, the Mayans may have gardened their crops in small, raised beds of fertile earth, called *pet kotoobs,* that drained swampland and, during dry seasons, allowed moisture to percolate up. Experiments in swampy lowlands of Mexico based upon the ancient raised-bed model have succeeded in demonstrating the high productivity of this method and have additionally shown that canals between the raised beds can host turtles and fish.

Six

ISLANDS

I'M SITTING AT THE RIM of Mt. Visoke, one of the eight Virunga Volcanoes in Central Africa, looking down into this extinct volcano's crater, hearing distantly the soft lapping of water in the crater lake, seeing a green slope — grass, ferns, moss, clusters of swordleaf, and the phallic extrusions of giant lobelia — dropping precipitously into white mist. The mist sweeps thickly before the wind in ghostly waves. It's raining softly, and I listen to the distant, then close, tympani of thunder.

The thunder recedes. The mist draws open like a curtain to reveal the opposite rim of the crater. I see the lake itself, first as dim patches of water, rippling white the color of mist. But the mist further attenuates until the sky itself is exposed as blue, and the ripples on the lake are becalmed into moving sparkles. The air breathes sweetly, rushes across alpine meadows behind and to either side of me, and I listen now to the susurration of wind across grass and the faint lapping of water. I hear wings whipping air: two white-necked ravens land next to me, then leave.

I'm wet, cold, tired, short of breath from the altitude, and huddled beneath a poncho — dipping my head inside to write these words, poking my head outside to make these observations. My African guide lies next to me on the volcano rim, entirely wrapped in his poncho, very still, possibly asleep and dreaming.

The mist thickens again. The thunder returns, rattling right above us now, accompanied by the quick fire of lightning and by tempestuous winds, and then rain hits in a downpour. It's time to go, so we rouse ourselves and begin our descent down the muddy trail.

The mist and rain become a soft wall, and I watch the figure in front of me walk into it and disappear, while all around me a pageant of insubstantial shapes, swords and giant phalluses, emerges from and melts into white. Everything is white or whitish except for some small yellow wildflowers on the ground, whose color appears to be intensified by the mist, so bright now they seem almost sources of light.

The trail has turned treacherously slippery, and we slide and stumble out of the subalpine zone into a forest of *Hypericum revolutum* and *Hagenia abyssinica* trees. The latter: twisted, hulking, ancient giants. Drooping massive trunks and mossy limbs, dripping lichens and ferns and orchids, splashing pale waterfalls of glacial green moss — they are appropriate apparitions in this dim dream world. The rain diminishes, and we pause many times, coming down the slope, to consider the twisted and ponderous presence of *Hagenia*.

Later the mist thins enough that we can look over into a grassy meadow to the west, where two huge buffalo are grazing. They graze like cows, then lie down in the grass like cows. My guide shouts at them, but they're too far away and don't hear. He violently bangs a stick against a tree and shouts more and more loudly until at last they notice him. Mildly alarmed, they stand up and trot away. He laughs uproariously and points at the drama he has just created, two fleeing buffalo.

That meadow, incidentally, slopes into a saddle of forest and meadow between Mt. Visoke and Mt. Karisimbi, both within Rwanda's Parc National des Volcans, and since we are now at an altitude of around 10,000 feet, my map suggests that the meadow must be close to Dian Fossey's old research station in the Rwandan Virungas, called Karisoke, which has been closed to the public since she was murdered. I look up to see the volcano rising dim, dark, and silent above us out of the mist, like an island out of a white sea.

Near the end of our descent, the trail curls across a stream and into a grassy meadow. I look down just in time to avoid stepping on two giant earthworms. They look exactly like the nightcrawlers fishermen use to fool fish, except that each is more than a foot long, as thick as clothesline, perfectly clean, and bright pink. When they move, slowly, their pink skin stretches into bands of opalescence.

The mist is gone now, and a little farther down the side of the volcano I look into sunlight under an azure vault, look across the broad, gently undulating slopes of sediment spread richly around the base of these volcanoes like so many dropped skirts — everywhere cultivated in a patchwork pattern. And I recognize that Mt. Visoke and her seven sister volcanoes huddle together as an actual island: an island of forested wilderness within a patchwork sea of cultivated fields.

APES EMERGED from the primate stock about 25 to 30 million years ago, appearing within the great belt of tropical forest that then stretched virtually unbroken from Africa to Asia and into the Southeast Asian islands. Today's apes are represented by three major groups: the lesser apes (several species of gibbons); the great apes (four species: the orangutan, the chimpanzee, the pygmy chimpanzee, and the gorilla); and a third group comprising the single species *Homo sapiens*.

We can think of the apes as a line of primates that for various reasons became large. Largeness, of course, confers certain advantages. A large animal is more likely to eat than be eaten. Also, larger animals tend to live longer. Their general rate of metabolism is slower, and thus the organs associated with metabolism, including the heart, are less stressed. Their large size, however, meant that the apes could not so easily walk or scurry on all fours across branches as most other primates do. And so the apes became hangers and swingers. Instead of depending for support solely on a single branch underfoot, they came to depend on the support of one or two branches from above. (Today's adult gorillas spend most of their time on the ground, but they, like all the apes, possess an anatomy suited to hanging and swinging, acquired during their evolutionary past.)

Hanging and swinging required some major anatomical adaptations. First, long and powerful arms attached to high shoulder blades, with an unusual flexibility at the shoulders and elbows, providing a superb overhead reach. Second, a broader, shorter torso with a less flexible spine. Third, a pelvis attached to the spine and legs in such a fashion that the legs hang downward, somewhat parallel to the spine, rather than at right angles to the spine. Fourth, a wide pelvis and associated musculature for supporting the internal organs. (For horizontal-walking monkeys, the rib cage serves this purpose nicely.)

But perhaps the simplest way to distinguish an ape from the other primates is to glance at his bottom: no tail. Almost all tree-dwelling monkeys have full tails, which they commonly use for balancing; some of the New World monkeys use the tail as an extra limb. For leaping monkeys, a tail can also serve as a crude rudder to influence the trajectory of a leap. But for a primate that hangs or swings and frequently moves upright or partially upright, a tail — hanging parallel to the spine — would mostly be a hindrance. Anyway, apes are tailless.

Apes represent the extreme of some tendencies in primate evolution. They are extremely oriented to visual rather than olfactory events, for instance. Indeed, apes have very flat faces, with small noses rather than rounded snouts or muzzles. And they are the most intelligent and behaviorally flexible of the primates. They possess the largest brains, ranging in size from about 100 cubic centimeters for gibbons to an average of 490 cubic centimeters for gorillas to an average of 1,390 cubic centimeters for humans. Of all the primates, apes undergo the longest periods of gestation and immaturity, which means that the offspring endure or enjoy the longest periods of dependency and learning. Gestation ranges from 210 days for gibbons to 265 days for gorillas and humans. The offspring, helpless at birth, take from seven to thirteen years to reach sexual maturity.

Humans are not directly descended from any of the modern ape species. Current knowledge, based on the unearthed evidence of several dozen skulls and hundreds of bones, suggests that the last common ancestor of apes and humans may have appeared as long ago as 10 million years. If we consider the measures of chromosomal and biochemical similarity, humans are most closely related to the African apes, the chimpanzees and the gorilla, whereas the Asian apes, the orangutan and the gibbons, are more closely related to each other. Current taxonomic conventions, however, divide the apes differently: orangutans are thought of as great apes and placed in a family with the African apes, while the several gibbon species are given another family, and the single human species has its own. However we categorize the apes, they are, with the exception of *Homo sapiens,* among the most threatened of all the major primate groups: the International Union for the Conservation of Nature identifies four of nine gibbon species and *all* of the great apes as threatened.

The gorilla is the greatest of the great apes and the biggest and strongest of all the primates. Perhaps because it fears no significant predator other than humans, it is also among the most placid. Males grow to be about twice as large as females; at full adulthood males may weigh 350 pounds or more and stand well over five and a half feet tall. Gorilla skin and fur is black or dark brown; the animal's bare face may appear a deep, opaque, shiny black: jet black. Gorillas' eyes are deeply set and widely separated. Their ears are small and flattened against the head. Their noses, consisting mainly of two large, flared nostrils orbited by prominent folds of skin, have been compared to squashed tomatoes. Yet those tomatoes are varied enough that field researchers often use nose prints — line drawings of the nostrils and the folds of skin around them — to identify individuals. Their arms are long and extraordinarily powerful, while their legs are comparatively short and rotated outward at the hip. Their great, broad hands look similar to human hands except for the thumbs, which are shorter and smaller.

The maturing male develops a high, conical crown of bone and muscle atop his head, the *sagittal crest.* Because the fully adult male acquires a broad saddle of white or silver fur across his back, mature males are sometimes called *silverbacks.*

Female gorillas in the wild reach sexual maturity and start breeding at around ten years, then bear a single offspring every three to five years. The newborn infant clings to its mother's chest and abdomen for about the first four months of life, then learns to ride piggyback. For at least the first six months, the infant is nourished primarily by its mother's milk, but gradually it learns to select and eat appropriate foods: leaves, shoots, and other vegetation. By the age of three years, the young gorilla is usually weaned and traveling on its own, though it will remain partially dependent upon its mother until full maturity. Juvenile gorillas are curious, playful, and acrobatic, and they often climb and clamber in trees. The immense bulk of full adults, though, makes such activity a less good idea, and so adult gorillas climb cautiously and generally prefer the terrestrial life.

Gorillas are social animals, spending most or all of their lives in the close company of a family group that can vary in size from two to almost forty individuals, with a median size of five to nine. Beginning its day about an hour after sunrise, the family group moves and feeds leisurely on several different types of vegetation until midmorning, when it stops for a siesta that extends into midafternoon. During siesta time several members of the group may gather loosely around the dominant male, the silverback. If the sun is out, individuals may warm themselves in the sun, lolling luxuriously like sunbathers at a beach, arms and legs casually extended. After siesta the group resumes feeding and continues on its leisurely course until dusk, at which time each individual constructs a rough sleeping nest, either in a tree or, more commonly, on the ground. All told, the gorilla family group wanders a distance of only about five to ten football fields during an average day; over a full year it may cover only several square miles of forest. For the most part, the group stays within a limited and familiar terrain. Some primate species hold strict territorial domains, which they will defend aggressively against incursion from other groups of the same species. But gorilla families often share their ranges with one or more other groups. Groups even occasionally rest or bed down for the night next to each other.

The obvious leader of the gorilla group is the silverback, who frequently leads the group to food, determines the direction and time of travel, and the time and place of resting and nesting. The silverback's mountainous mass may help simplify the complexities of leadership and communication. When he determines it is time to rest, he stops and rests. The others do the same. When he determines it is time to move on, he rises and begins walking purposefully in a particular direction. The others follow. During foraging, individuals will spread out and frequently wander out of sight of each other. But when the group begins to move, the infants run to their mothers and climb on their backs, while the mothers and subadult males arrange themselves behind the silverback to follow him single file through the forest. The silverback maintains mating priority with the adult females of the group and is typically the father of most or all of the young group members. Subadult males sometimes serve as sentries or as a rear guard for the moving group, but the silverback is usually the most active protector of the family. Should a disturbance or threat appear, he will either lead the others away or momentarily defend them with a stereotypical bluff attack.

The bluff attack probably looks like anything but a bluff to the object of the attack. The silverback may precede his charge with such displays as hooting, chest-slapping with cupped hands, and the peculiarly delicate gesture of placing a leaf between his lips. He will then stand to his full height, bark and scream-roar, before crashing toward the threat on all fours, slapping and tossing vegetation aside, charging with surprising speed and force, only to stop within a few yards or feet or inches of the threat — if that person or animal stands firm. Should the object of the charge turn and attempt to flee, however, the attack becomes no longer a bluff. The silverback will pursue, finally to deliver a devastating bite or blow.

Two mid-nineteenth-century missionaries in West Africa, Wilson and Savage, are officially considered the gorilla's European discoverers. Wilson and Savage collected several gorilla skulls and sent them, accompanied by a sensational description of the species, to the European anatomists Jeffries Wyman and Richard Owen. "They are exceedingly ferocious, and always offensive in their habits, never running from man as does the Chimpanzee," wrote Savage. "It is said that when the male is first seen he gives a terrific yell that resounds far and wide through the forest, something like *kh-ah!* prolonged and shrill. . . . The females and young at the first cry quickly disappear; he then approaches the enemy in great fury, pouring out his cries in quick succession. The hunter awaits his approach with gun extended; if his aim is not sure he permits the animal to grasp the barrel, and as he carries it to his mouth he fires; should the gun fail to go off, the barrel is crushed between his teeth, and the encounter soon proves fatal to the hunter." After examining gorilla skulls in the peace of his laboratory, the anatomist Owen felt qualified to add his own information about this animal: "Negroes when stealing through shades of the tropical forest become sometimes aware of the proximity of these frightfully formidable apes by the sudden disappearance of one of their companions, who is hoisted up into the tree, uttering, perhaps, a short choking cry. In a few minutes he falls to the ground a strangled corpse."

Intrigued by such reports, the Philadelphia Academy of Sciences financed Paul Belloni Du Chaillu, a French-born newspaper reporter from New Orleans, to explore Africa and find some gorillas. Du Chaillu spent four years on that continent and returned with dead specimens of gorillas and other mammals, as well as over 2,000 birds. He also wrote a successful book about the trip, *Explorations and Adventures in Equatorial Africa* (1861).

"I protest I felt almost like a murderer," related Du Chaillu in that account, "when I saw the gorillas the first time. As they ran — on their hind legs — they looked fearfully like hairy men; their heads down, their bodies inclined forward, their whole appearance like men running for their lives." But six days later Du Chaillu confronted a large male of the species face to face. "Then the underbrush swayed rapidly just ahead, and presently before us stood an immense male gorilla. He had gone through the jungle on his all-fours; but when he saw our party he erected himself and looked us boldly in the face. He stood about a dozen yards from us, and was a sight I think I shall never forget. Nearly six feet high (he proved four inches shorter), with immense body, huge chest, and great muscular arms, with fiercely-glaring large deep gray eyes, and a hellish expression of face, which seemed to me like some nightmare vision: thus stood before us the king of the African forest." The gorilla clapped his chest, then produced a characteristic roar, which Du Chaillu described quite accurately. "It begins with a sharp *bark,* like an angry dog, then glides into a deep *bass roll,* which literally and closely resembles the roll of distant thunder along the sky, for which I have sometimes been tempted to take it where I did not see the animal. So deep is it that it seems to proceed less from the mouth and throat than from the deep chest and vast paunch." After displaying his large canines, the animal roared once again, providing substantial stimulus for Du Chaillu's imagination: "And now truly he reminded me of nothing but some hellish dream creature — a being of that hideous order, half-man half-beast, which we find pictured by old artists in some representations of infernal regions." The gorilla advanced to within about six yards, and Du Chaillu fired; with a groan the animal fell face down, "shook convulsively for a few minutes, the limbs moved about in a struggling way, and then all was quiet." Although some of Du Chaillu's contemporaries challenged the accuracy of his account, in general those memoirs confirmed all previous notions of the gorilla: that it was a thing most brutish, a human-shaped monster of aggressive ferocity, a mute, untrained, unrestrained Caliban.

For an entire century after Du Chaillu, curious people learned what they could about gorillas by peering into the cages of zoos and the glass cases of museums or by sighting down gun barrels — except for one late-nineteenth-century scientist, who camped inside a steel cage in Gabon and waited for curious apes to emerge from the surrounding forests. Then in 1959 George Schaller and John Emlen, zoologists from the University of Wisconsin, arrived in central equatorial Africa to initiate a long-term field study of the mountain gorilla subspecies. Emlen's work was completed within six months; Schaller remained in Africa for two years with his wife, Kay, and afterward published two superb books about his experiences, *The Mountain Gorilla* (1963) and *The Year of the Gorilla* (1964).

Schaller felt that earlier accounts of the species' ferocity had been exaggerated. He refused to carry a gun, feeling that "firearms have no place in my kind of study. No animal attacks without good cause, except on rare occasions. My inclination is to give the charging animal the benefit of the doubt, hoping that it is merely bluffing." Happily, that intuition proved sensible: during his two years of study, Schaller was not once charged by a gorilla. And his study profoundly altered what we know about this primate. Schaller demonstrated boldly and convincingly that gorillas, if observed with respect and on their own terms, are not to be feared. He discovered an animal that was shy, peaceful, and almost entirely vegetarian, only occasionally feeding on the meat of grubs and snails.

Schaller was deeply impressed by the gorillas' resemblance to humans, not merely in anatomy, but in their gestures, typical body postures, and facial expressions. When they wake up in the morning, gorillas stretch and yawn. They will sit up on the edge of a tree branch and casually dangle their legs. When they lie on their backs, relaxing during midday, they sometimes place their arms under their heads as humans do. And in the expression of feeling, Schaller found them to resemble humans. If irritated or angry, they frown. Uncertainty or an ambiguous situation will cause them to bite their lip. Juveniles throw temper tantrums. Schaller found gorillas to be warm and affectionate with each other, often expressing their closeness in strikingly recognizable gestures of affection. He concluded that gorillas resemble humans in an emotional sense and that he could predict their behavior by considering them as emotional and sentient beings.

In 1967 Dian Fossey went to the Virunga volcanoes to continue the study of mountain gorillas that Schaller had begun. Fossey spent almost two decades with the animals, becoming so accepted by them that juveniles would approach her, touch her, pick up and examine her equipment, and even at times play with her, while their parents looked on with casual tolerance. Fossey's study tremendously expanded our scientific knowledge of mountain gorillas, while publicizing the reality that poaching and habitat destruction could soon extinguish these beautiful giants.

Zoologists today recognize three subspecies of gorilla, distinguishable by a number of traits, including average size, bodily proportions, length of fur, and shape of the head. Western lowland gorillas have broader faces and smaller jaws and teeth, for instance. Eastern lowland gorillas are the largest of the three. Mountain gorillas, perhaps most notable for their thick and glossy fur (an adaptation to the cooler climate of their high-altitude home), are intermediate in size, with silverbacks weighing around 340 pounds and standing about five feet eight inches tall.

The three subspecies are distributed within two large pockets of forest in equatorial Africa, separated by a forested gap of over 600 miles in the Congo Basin. In West Africa, members of western lowland subspecies are virtually gone from the northernmost part of their former range in southern Nigeria. They now inhabit a continually declining range within portions of Cameroon, Central African Republic, Congo, Equatorial Guinea, Gabon, and the Cabinda enclave of Angola. Altogether, roughly 40,000 individuals of this group remain. To the east, across the Congo Basin into the lowland forests of eastern Zaire, a few thousand eastern lowland gorillas inhabit several areas of forest within a diminishing range. Finally, in a tiny island of forest known as the Impenetrable Forest, in Uganda, about 115 mountain gorillas may still survive; and several miles south, in the Virunga Volcanoes region (intersected by Zaire, Rwanda, and Uganda), around 290 mountain gorillas are still alive.

I WAS LED by two Rwandan guides and accompanied by three Americans on a difficult three-hour hike through forest, bamboo, and thick brush at the base of Mt. Visoke, one of the eight Virunga Volcanoes, within Rwanda but not far from the Zaire border. The bamboo itself was forest, sometimes so thick we tunneled through it, but mostly thinner, penetrated with trails and easy to walk through, the ground covered with spear-shaped bamboo leaves and broken bamboo sticks the color of straw and smelling like straw. The thick brush sometimes expressed itself with a sweet smell of mint and sometimes with the surprising sting of nettles: as sharp and painful as a bee sting that persists at its original intensity for hours.

We hiked a muddy trail, gradually climbing the lower slopes of Mt. Visoke, until at last that path disappeared altogether and we began following gorilla spoor: trails of crushed vegetation and fresh nuggets of dung — bluish black, compact, the size of apples with a smell like horse manure, but milder. We pursued these new rough trails into the brush, sometimes hacking our way through difficult spots, or tunneling under on hands and knees, or pushing our way over and stomping through. But mostly we trod a maze, a dozen meandering gorilla trails that diverged and converged through dense brush as irregularly as the rivulets of a rocky river.

One of the guides sounded as if he had a bad cold. He kept clearing his throat in a coughing sort of way. But when the other guide began coughing too, I realized that we were near gorillas and that the coughs were a signal, a way to avoid surprising them. Still I saw and heard nothing until we came to an island of higher vegetation at the crest of a small hill. Here the trees and brush were head height and higher. I heard, then, cracking in the brush — a loud cracking of solid branches. I looked over to one side and saw the top of a small tree trembling, and then I heard a brief and deep rumble-grunt. We continued, and finally I saw through an opening in the underbrush the patch of a large black boulder, furry.

The guides still coughed, and we slowly walked along the trail, which meandered past the opening and the black patch of fur, then doubled back until we saw the entire trail blocked by the black and silvery back of an adult silverback gorilla, sitting turned away from us, eating. He looked as big as a car. He gave us one glance over his shoulder, then ignored us and continued eating. Thick heavy neck, fur a bright black glistening in the sun. He swiped at plants around him and put green wads of leaves into his mouth: chomp, chomp, chomp. We waited. Eventually he raised himself onto all fours and pushed off from the trail, crashing slowly into the brush, creating a new trail as he went.

Cautiously we advanced along our trail until we stood at the top of the silverback's new trail. Fifteen feet away from us, at the end of this new passageway, he was eating, now sitting in profile to us. Huge, with a great, drumlike belly and massive chest and arms, he glanced at us occasionally with only the mildest regard, as if we were an unusually boring television program.

He had a wonderful way of stripping leaves off vines. Snatching a vine with his hand, he would place the near end in his mouth and swipe the whole vine through until only a big wad of leaves was left, which he would then, slowly and casually, proceed to chomp. The wad of leaves at his lips would slowly disappear into his mouth. We watched him quietly for a long time before I noticed that his right hand was entirely missing, severed at the wrist. Lost in a poacher's snare?

We moved on to find other members of the group, who were also wandering in their food, eating everything around them. Sometimes we merely heard them cracking and chomping in the brush. Sometimes we merely saw bushes and trees shaking. Sometimes we saw an arm or a face appear, then disappear. But finally we came within arm's length of an adolescent male who was up to his neck in vegetation, a pit of food, delightedly eating whatever he placed his hands on. His eyes were bright, sunlit to a clear mahogany, and his face seemed to express, not curiosity exactly, but a mild amusement at our presence. Nonetheless, he too essentially ignored us and kept on eating: chomp, chomp, chomp.

What do these gorillas think? I asked myself. What is their mental world like? The answer came back: green. All around them, green. Their food, green. Their thinking green and uncomplicated.

Next day I went out to see another group of mountain gorillas, accompanied this time by five other people, including two Rwandan guides, the assistant director of the park, and a Hollywood producer who had come to scout locations for the film about Dian Fossey's life.

I was prepared for several more hours of walking and crawling through thick brush and stinging nettles, and thus was surprised when we came upon gorillas right near the edge of the park, within a bamboo thicket. Once again we heard them first: deep, rumbling grunts and a cracking of vegetation. Three of us entered a narrow tunnel in the bamboo and crawled single file on hands and knees for several yards until we came to a multichambered cave in the bamboo. Two or three yards away a female gorilla sat facing us, potbellied and dull-eyed, tolerating but apparently not interested in us — our clothes and gadgets, our delicate hands and three little faces. And then from behind her appeared a toddler gorilla with a yellow root in its mouth. The toddler approached us, walking on its four spidery legs, until suddenly, alarmed by its own boldness, it turned back and tumbled into the female's lap, then climbed out again and disappeared into the tunnel behind her. The adult female remained, however, and eventually lay down on her back to relax. Later, with the yellow root still in its mouth, the furry toddler appeared once again, inquisitively stumbling toward us, and then slowly backed away.

We left the bamboo thicket and walked into a clearing, a meadow of grass and thistles and bamboo clumps nearly surrounded by a more continuous bamboo thicket. There, at the base of a small clump of high bamboo, the group's silverback sat facing us, next to an adult female, also facing us. It looked as if we had just interrupted an important tête-à-tête.

We kept our distance as the silverback considered us, looking monstrous, massive, mountainous, his great head elongated by the sagittal crest, that broad miter of fur and bone, his eyes deep. Several times he turned his face away and then back, glaring very directly at us, until at last he stood up on his hind legs, then turned and lowered down on all fours and lumbered heavily away, displaying a broad, bright band of silver fur across his slowly swaying back. The female moved away too, and then I watched the silverback crouch down on his elbows, rump up high, and dig down into the earth with his hands. He pulled out thistle roots, straw-colored and filamental, and ate them.

All his actions so far had been remarkably slow and ponderous, but a few minutes later he was suddenly seized by an impulse: a quick sideways dash, a mighty slam to the earth with his doubled right fist, and a frontal dash toward the female, who in turn dashed away to the edge of the meadow, where the meadow terminated in bamboo.

Not long after, the rest of the family walked out into the meadow single file, on all fours: first a mother with a baby, then two mothers with toddlers, then one lone adult, and one lone adolescent. The baby rode piggyback, clutching the fur at its mother's neck. One of the toddlers also rode piggyback, while the other toddler scampered into the clearing alongside a female I imagined was his mother. Once this second toddler arrived in the open meadow, he stood up on his hind legs, strutted briefly, clapped his chest — a terribly comical miniaturization of grown-up chest-clapping — and then ran over to his peer, who by then had jumped off his mother's back. The two toddlers grabbed each other and fell to the ground wrestling, rolling over and over.

A half dozen people stood still and cautious, watching ten mountain gorillas in a meadow. The gorillas ignored the people, and went about their usual midday business of relaxing, resting, playing, casually eating green vegetation, and once in a while stooping down to dig out thistle roots. The baby and toddlers were beautiful and fascinating, so delicate and finely formed, looking much more human-proportioned than the adults, reminding me so much of children in footed pajamas.

The silverback and his female consort still hung out well away from the rest of the group. Indeed, they were far enough away that I could no longer see them very well and eventually forgot all about them altogether, until I heard several high-pitched cries and saw that they were mating: the female on all fours, the silverback covering her from behind. It was the briefest of passionate spasms, over within several seconds.

The meadow was surrounded by bamboo thicket, except one part that opened directly into farmers' fields. In fact, looking out through that opening, I could see a farmer's house not far away, a round mud hut with thatched roof. While we watched gorillas, I sometimes heard the sound of drums: someone was having a party. We were that close to the park boundary — and I was reminded then of the ancient and continuing warfare between humans and gorillas.

A gorilla carries a lot of flesh, and the western lowland subspecies in particular has long been hunted as game. The Bulu and Mendjim Mey tribes of southern Cameroon, for instance, have hunted western lowland gorillas for food, as have the Bengum, Mahongwe, Bachangui, Sameye, and Eschira people of Gabon; the M'Beti of Gabon and Congo; the Fang of Equatorial Guinea; and some nomadic pygmy tribes.

During the early 1940s the American adventurer Armand Denis stayed with the M'Beti, a people who "from time immemorial" had hunted gorillas as their only source of meat. Members of the tribe spoke of gorillas as a hated enemy tribe against whom they had long ago declared war. Gorillas and gorilla hunting provided not only food but entertainment — a major focus for daily activities among the M'Beti and the principal theme of their communal dancing.

Denis noted that the M'Beti hunted gorillas by creating islands in the forest. A hunt he observed began after individual trackers had located a family of gorillas during the day and noted their sleeping place for the night. Once the animals had constructed their nests and gone to sleep, the trackers sent for reinforcements from neighboring villages. The reinforcements, perhaps twenty or thirty additional men, arrived with spears and long knives, and during the night, as the animals slept soundly, the men stealthily cleared a circle in the underbrush, surrounding the entire family group. By dawn an encircling path had been established, and more reinforcements (eventually around four hundred men and women from several villages) arrived, carrying large, rolled-up rope nets along with weapons. The nets were placed upright and tied every two or three feet to standing trees around the entire clearing, thus forming a five-foot-high fence around the newly created island.

Then workers guarded by spearmen quickly expanded the cleared area inside the nets to a width of about five yards. The M'Beti hunting weapons were primitive: spears and harpoonlike weapons with barbed iron tips fashioned to break off after penetrating an animal's hide — and about twenty muzzle-loading guns, two of them ancient flintlocks provided with hammer-and-cap firing mechanisms, some with split barrels wired together, one constructed of an old iron pipe, all packed with gun powder and loaded with "bullets" — chisels, nails, screws, nuts, scraps of iron, even compressed pieces of tin can. Denis thought he recognized the wadded remains of cans of sardines, Campbell's soup, Heinz spaghetti, and Dominion apricot jam from his own refuse heap. Even with such crude weapons, though, the communal hunt proved terribly effective. The area of the island was diminished by transecting the original circle with new paths through the brush and additional nets, until finally the awakening gorillas were tightly enclosed. The animals, driven by noise to one side of the encircling nets, were dispatched one by one with all handy weapons.

Fred Merfield, a British hunter, guide, and specimen collector, lived with the Mendjim Mey people of Cameroon for five years during the 1930s. Although their women were forbidden to eat gorilla meat, which was supposed to cause infertility, the tribesmen regularly hunted gorillas for food, often kept their skulls and bones as charms, and used gorilla skins as belts. The Mendjim Mey expressed no fear of gorillas, referring to them contemptuously as merely a variety of unarmed bushman. Typically, the Mendjim Mey hunted in groups of two or three men, and they sometimes courageously attacked large male gorillas at close range, using only spears. They preferred to find gorilla families, tree them, and pick them off with spears and wooden crossbows that fired poisoned darts.

Some wild animals can be considered as competitive with humans to one degree or another, but gorillas are particularly so. For one thing, they actually prefer areas of past or present human habitation, where primary forest has been partially degraded, opening up the canopy to encourage the growth of edible ground-level vegetation. For another, gorillas — bold and destructive foragers — have developed a taste for such cultivated crops as peas, maize, carrots, taro, manioc, bananas, papaya, pineapples, sugarcane, peanuts, kola nuts, and palm nuts. Able quickly to ruin large areas of cultivation, gorillas have often been killed as crop raiders in many parts of their range.*

In southern Cameroon, Fred Merfield once witnessed a gorilla roundup by the Yaounde people, carried out in retaliation for crop damage. A gorilla family — not necessarily the group that had destroyed crops — was located one evening in a patch of forest. By dawn large numbers of Yaounde hunters and their women had surrounded the sleeping group and begun clearing a circular band, like the M'Beti, trapping gorillas on an artificial island. Spearmen guarded the work in case the awakening gorillas should attempt to break through, and the women helped keep the animals at bay by marching around the outside of the fenced clearing, singing, pounding on drums and kerosene tins, and shaking bells and rattles. Construction was supervised by the local chief, dressed in an oversized trenchcoat, followed by his personal cook and cupbearer, who supplied the chief with food and drink. At the same time, the local witch doctor solemnly walked around the inside of the circular clearing, blowing on a magical horn and making important gestures with his left hand to keep the gorillas away. When the fence was nearly complete, the witch doctor buried several charms at a spot where the gorillas seemed most likely to attempt to escape. "By now Old Man gorilla was making his presence known by repeated, deep-throated growls," Merfield reported, "and every time he growled he was answered by renewed efforts from the drummers and the shrieking women." But by midmorning the fenced island was complete.

Guarded by men holding spears and a few guns, workers then began clearing away the brush inside the circle. At that point the silverback made several attempts to lead his family group through the fence, but each time he was driven back with spears. Finally, an adult female attempted to break through and was killed with spears. Another female, with an infant clinging to her back, was dispatched with spears, cutlasses, and guns, then mutilated. (Her still living infant and a second captured infant were passed to Merfield's companion, Major P. H. G. Powell-Cotton; both animals survived only a few days of captivity.) Another adult female escaped over the fence with three spears sticking out of her back. A young male was hacked to death with knives, and his arms and legs were cut off. Finally, the silverback was

*Because the vegetation of their forests is low and accessible, and also, perhaps, because local cultivated crops do not tempt them with the most succulent fruits, mountain gorillas are not crop raiders.

found, sitting helplessly in the underbrush, pincushion for about a dozen spears. "Making no attempt to retaliate, he was just sitting there, rocking to and fro as more spears were thrust home at close quarters; his mouth was wide open, crying shame on his tormentors." Merfield attempted to finish the animal off with a bullet, but he was pushed back — the Africans didn't want death so much as revenge. "Three or more spears hit the gorilla and then another man, armed with a heavy club, leaned forward and slowly, methodically, began clubbing him to death."

Because of its enormous size and power, its threatening appearance and boldness, the gorilla has traditionally been regarded by many indigenous Africans not merely as a walking piece of food or a greedy crop thief, but also as an object of fear and hatred. Yet fear and hatred have perhaps quite commonly merged with a powerful fascination for an animal that seems so profoundly humanlike in form and behavior. Thus the M'Beti had declared war on gorillas as an enemy tribe. And thus the Mendjim Mey once claimed (to Merfield) that they hunted gorillas and chimpanzees partly as a good substitute for humans — whom they purportedly ate until the beginning of this century.

The Fang of Equatorial Guinea have professed to abhor chimpanzee meat because the animal is "almost human," and eating it would be a sin. Yet some older members of the tribe told Jorge Sabater Pi — a zoologist from the Barcelona Zoo who during the late 1960s polled local inhabitants of Equatorial Guinea concerning their dietary predilections — that they preferred gorilla meat above all else, particularly for its delicious flavor and fatty texture, also for its natural flavoring of the plant *Aframomum danielli,* whose small red fruits and stalks gorillas commonly consume. Although gorillas were by then scarce in Equatorial Guinea, and legally protected from hunting, at least five were eaten by members of the Fang tribe in the Monte Alen region in 1968. And it has been suggested that the lingering taste for gorilla among older members of the tribe is associated with a past interest in human flesh: the Fang allegedly practiced cannibalism until the beginning of this century, when they were persuaded to stop by missionaries and the Spanish colonial government. The favored parts of the gorilla — the muscles of its face, the palms of its hands, soles of its feet, and the tongue — correspond to what once were the favored portions of human flesh.

The meat and various body parts of this primate have also been valued for their symbolic or magical properties. Du Chaillu reported that his own hunting party, which included several Fang tribesmen, ate most of the gorilla he shot but carefully preserved the brain, to be used later in charms that conferred strength in hunting and success with women.

As for the mountain gorilla of central equatorial Africa, its ears, little fingers, tongue, and testicles have occasionally been used in a traditional potion to endow the imbiber with virility and strength. Nor has the fascination with pieces of big gorilla been limited to native Africans. Briefly during the mid-1970s, a few mountain gorilla hands and heads were sold as souvenir trinkets to Western tourists.

Dian Fossey wrote in detail about an engaging young male mountain gorilla she named Digit, who seemed particularly inquisitive about his human observers and liked to examine their strange possessions — thermoses, cameras, lenses, notebooks, gloves, and so on. He would smell and even lightly touch the clothing and hair of recently arrived humans. But Digit was probably too inquisitive. On December 31, 1977, poachers killed him with spears, then hacked off his head, hands, and feet. Nevertheless, poaching may threaten the mountain gorilla most severely by accident or indirection. Poachers in the Virunga Volcanoes region generally pursue buffalo, elephant, and antelope with spears and arrows, pit traps, and snares. The snares, constructed of hemp or wire nooses, bent bamboo springs, and small stick triggers, are usually intended for antelope or other medium-sized game. When gorillas occasionally spring them, the animals usually break free, but with the wire or hemp noose still binding a wrist or ankle. Fossey observed six instances of young gorillas bound by snares set for antelope. In two cases a silverback bit off the lethal bracelets. But in the remaining four incidents, the victims gradually became weaker and finally disappeared, most probably to die. Fossey believed that a few dozen African poachers were rapidly destroying the wildlife of the Virungas.

During the last century gorillas also provoked the trigger fingers of many North American and European hunters. Du Chaillu's words suggest not merely self-justification but even a measure of self-conferred heroism, as he describes slaying a ferocious "being of that hideous order, half-man half-beast." Du Chaillu donated the proceeds of his pleasure to science. Other hunters likewise found they could enjoy themselves and benefit Western science simultaneously. The British hunter Fred Merfield bagged 115 western lowland gorillas for European museums and exhibits. Merfield also provided occasional ape fetuses for the anatomical studies of Professor Ian Hill of University College and skulls showing tooth decay for Sir Frank Colyer's dentistry studies. And Merfield's companion, Major Powell-Cotton, "world-renowned naturalist and big game hunter," acquired skins and skulls of western lowland gorillas for his own private museum at Birchington, Kent ("the largest collection of big game ever shot by one man").

Specimen collections represent a critical part of zoological knowledge. But in a few instances specimen collecting has pushed rare species and subspecies even closer to oblivion. The mountain gorilla, for example. Not long after the discovery of the subspecies in 1902, the American Museum of Natural History hired Carl Akeley to kill five specimens so that the museum might display some stuffed examples of an already rare great ape. Sweden's Prince Wilhelm managed to kill fourteen mountain gorillas during his hunting expedition in the 1920s, and an American named Burbridge took an additional nine around the same time. Harvard University also sent out a gorilla hunter. Today you and I can stare at the results, a male mountain gorilla standing fiercely poised inside a glass case within the Museum of Comparative Zoology, forever ready to beat his chest with closed fists (taxidermist's error — those fingers should be spread open, palms cupped), forever defending his family against Harold J. Coolidge, Jr., gun-toting nemesis from Harvard.

Altogether, between 1902 and 1946 Americans and Europeans, usually carrying permits issued by the Belgian colonial government, killed at least sixty-one individuals of the nearly extinct mountain gorilla for sport and science, a major fraction of the four hundred members of that subspecies now remaining.

GORILLAS, feeding mainly on ground vegetation, actually prefer partly settled areas of forest, near roads and villages, or places where a village or farm existed in the past — where moderate shifting cultivation has degraded primary forest into a mosaic of primary and secondary patches. Yet gorillas are reluctant to cross open meadows or roads, and they cannot survive extensive forest clearance. Destruction of large forest tracts for lumber, agriculture, and human settlement now most significantly threaten the world's remaining wild gorillas.

Most of the western and central equatorial African nations that still harbor gorillas have created Arks that include habitat for these primates. Yet most of those conservation areas are not adequately protected. The full Ark for gorillas seems comparatively large, but only a third of the protected land actually contains suitable habitat. Altogether, conservation areas in equatorial Africa nominally protect only a few to several thousand gorillas — a large number compared with some other endangered primates, yet perhaps not large enough (particularly given the geographical isolation of the several conservation areas) to maintain adequate genetic representation over extended time.

The mountain gorilla, the rarest but most thoroughly studied of the three subspecies, is also the only one protected within its entire range, which, as I earlier mentioned, consists of two very small pieces of central equatorial Africa: Uganda's Impenetrable Forest and the Virunga Volcanoes region, which is protected by Uganda, Zaire, and Rwanda.

Uganda is essentially outside of the African tropical forest zone. Within recent history, forests covered more than 6 percent of the country, but today they cover only half that, about 2,100 square miles altogether. The 120-square-mile Impenetrable Forest is one of Uganda's final pieces of intact forest, and it happens to contain remnants of a Central African Pleistocene Refuge. Extraordinarily rich in plant and animal species, the Impenetrable Forest is home to ten primate species, including three nocturnal prosimians and two great apes — the chimpanzee and the mountain gorilla. The seven diurnal primate species are thought to include 12,000 individuals in total, of which about 115 are mountain gorillas. So far the Ugandan government has protected the Impenetrable Forest as a forest reserve and game sanctuary. Many Ugandans view the forest with pride, and one day it may become a national park and perhaps a major tourist attraction. Meanwhile, though, the forest is a habitat island, almost entirely surrounded by cleared farmland and a densely settled populace. It has been estimated that 500 to 1,000 people enter the forest each day, legally and illegally, to gather firewood, bamboo, and honey, and to hunt. Poachers have already extinguished the cape buffalo and leopard from this forest, and have reduced elephants to a herd of fewer than thirty animals. Even though poachers seldom pursue the many primates, snares and pitfalls constructed for antelope and pigs are a threat to the ground dwellers.

Just a few miles south of the Impenetrable Forest rise the Virunga Volcanoes. The entire Virunga sanctuary consists of a kidney-shaped piece of high forest and meadow about twenty-five miles long and six to twelve miles wide, undulating across the slopes of an ancient chain of eight volcanoes, two of which are active. This remarkable sanctuary originated as Africa's first national park, the Albert National Park, created in 1925 by the Belgian colonial government specifically to protect the mountain gorilla. Uganda added eighteeen square miles to the protected region in 1930, calling its portion the Kigezi Gorilla Sanctuary. After discarding colonial rule, Zaire and Rwanda renamed their pieces of the park. Zaire's section is now known as the Parc National des Virungas, while the Rwandan section is the Parc National des Volcans.

Gorilla habitat around the volcanoes is roughly 400,000 years old, but within the last forty years it has been under attack from all sides. Soon after June 1960, when Belgium granted independence to the Belgian Congo (now Zaire), a war for control of the newly independent state began. Political and social chaos quickly extended into the Albert National Park. Rangers were threatened by rebel forces, local people around the park moved inside it, fishermen settled along river and lake shores within the park, and Watusi pastoralists from Rwanda migrated with their cattle across the border into the park. More than two dozen park guards were slain during that period, while soldiers and rebel forces burned farms and gardens around the edges of the park, which may have led to heavy poaching of mountain gorillas for food. Many people have attributed the drastic decline of the Virunga mountain gorillas (from 400 to 500 animals in 1960 to between 260 and 290 by 1971–1973) to poaching and shrinking habitat on the Rwandan side of the sanctuary. But a recent analysis of the Virunga gorilla population suggests that most of the decline occurred in Zaire as a result of the civil war.

Uganda still controls its small section of the northeastern Virungas, the Kigezi Gorilla Sanctuary. However, that nation's human population now doubles every 20 years. Early in this century the population of Uganda's Kigezi region had already increased sixfold over fifty years (partly through immigration). During that time forests were cut down, crops planted, and the soil depleted, so that what was originally a vast forested area is now mostly brown and barren grassland. Forest and open land in Uganda became valuable commodities, so, not surprisingly, in 1950 Uganda officially opened about a third of the Kigezi Gorilla Sanctuary to farming, and even more of the area was settled illegally. Today Uganda's sanctuary in the Virungas is so disrupted by squatters and their cattle that gorillas rarely go there.

Rwanda, at the Virungas' southeastern edge, is the most densely populated nation in Africa; in fact, it ranks among the poorest and most densely populated nations on the globe. Rwanda's population now doubles every nineteen years, and by the year 2020, this overburdened piece of land — smaller than the state of Maryland — is projected to hold nearly 22 million people. On the immediate edges of Rwanda's Virunga region live 780 people per square mile. They freely cross over into parkland to gather wood and honey, herd cattle, raise potatoes or tobacco, and set traps for antelope and other game.

It would seem, then, that Rwanda cannot afford its national treasure, the mountain gorilla sanctuary. In fact, in 1969, the Rwandan government converted about a third of its Virunga parkland into farmland for the growing of pyrethrum, which was sold as a natural insecticide to North Americans and Europeans until the market collapsed. More recently the Rwandan Ministry of Agriculture considered removing a substantial portion of the remaining land to provide for cattle grazing.

In spite of its regular excision of parkland, in the 1970s the Rwandan government began seriously to deal with some of the threats to the park. Beginning in 1974, the government discouraged illegal cattle grazing in the park by fining owners the equivalent of ten dollars per cow, well above the average monthly income in Rwanda. One official, Paul Nkubili, terminated the market in gorilla heads and hands as tourist souvenirs by instigating severe penalties against buyers and sellers.

During the 1970s Rwanda's steadily shrinking sanctuary was also actively protected by Dian Fossey, an American woman who had been chosen by the famed anthropologist Louis Leakey to study gorillas in the style of an anthropologist, to watch them unobtrusively as they went about their ordinary lives.

Fossey first set up camp in the Zairean portion of the Virungas, but within months civil war broke out, and she was kidnapped by government troops. She escaped, went to Rwanda, and established, in a 10,000-foot-high meadow between two volcano peaks, her permanent campsite, a collection of corrugated metal cabins known as the Karisoke Research Centre. It took Fossey months of following mountain gorillas, often crawling on her hands and knees through mud, brush, and stinging nettles, sometimes reassuring the animals by mimicking such gestures as chest slapping and chewing on wild celery, before her presence lost its threat. After four years, finally, a young male gorilla she had named Peanuts actually reached out with his hand and touched hers. For her, that touch was a greeting of trust.

Fossey's seminal research transformed popular images and scientific knowledge of the gorilla. And for many years scientists and students hiked up the narrow, muddy trail to Karisoke. But Fossey saw her own research population decline — six of the eighty gorillas she studied were killed by poachers — and she recognized that the final few mountain gorillas of the Virungas were being extinguished by the illegal incursions of cattle and by the snares and traps of poachers. So she turned her attention from science to conservation. She began driving cattle out of the park, created antipoaching patrols, and concentrated on removing the dangerous snares. (In 1984 alone, her patrols destroyed more than 2,200 snares.) No longer the disinterested observer, she made many enemies. Once she shot thirty Watusi cattle at close range; another time she spray-painted obscenities on a cow. She sometimes pursued poachers back into their villages, once setting fire to a poacher's hut, another time seizing a poacher's young daughter, planning to exchange her for a captured baby gorilla. For that act Fossey was nearly prosecuted as a kidnapper.

Supposedly some Africans feared her as a white witch, a cunning and tyrannical Prospero of the mists, an image she may have capitalized upon in her efforts to control poaching in the park. And she took advantage of local superstitions by regularly seizing poachers' sacred amulets and *sumu* (magic) pouches. A few months before her death, Fossey tore away a poacher's sumu pouch that had been sewn into his shirt sleeve. The pouch, according to her later report, contained "bits of skin and vegetation, all looking like vacuum cleaner debris."

On the morning of December 27, 1985, Fossey was discovered dead on the floor of her cabin, her skull split open by a machete. Her cabin had been ransacked, with grass matting torn, drawers roughly opened, clothing and bedding strewn about. Nothing had been taken from her cabin, though — not liquor, her radio, cameras, guns, camping equipment, or $1,300 in cash — nothing but her passport and, according to one person, the sumu pouch she had earlier taken from a poacher. (The mystery of Fossey's murder has not been adequately solved or resolved. Wayne McGuire, an American colleague of hers, was tried, convicted, and sentenced to death *in absentia* by a Rwandan court for the murder; but he argues his own case, convincingly, in the February 1987 issue of *Discover* magazine.) Eventually she was buried near her cabin in the Virungas, alongside the graves she had dug for gorillas slain by poachers.

The 1983 publication of *Gorillas in the Mist* made Dian Fossey a celebrity in the United States. After her death, however, some American newspaper reporters preoccupied themselves with her forceful eccentricity and supposed misanthropy. A writer for the *Philadelphia Daily News* stupidly suggested, as the article was titled, that "Dian Fossey Asked for It" because she was "known to prefer gorillas to people" — as if one were not only strategically required to choose sides but morally obliged to select the correct side, or die. A writer for the *Boston Globe* passed along the irrelevant and uncharitable comments of a Boston-based academic: "Scientifically, you have to wonder just what she was doing the last 10 years." Yet, for all her genuine faults, it is ultimately to her credit that Dian Fossey neither lingered to engage in timid, cross-cultural analysis nor turned to advance on the air-conditioned battlefields of academia. "She gave herself to our animals," said Rwandan journalist Aloys Mundere. "Because of her, Rwanda and the park are known around the world."

While Fossey was carrying on her private battle to save the gorillas, in 1980 a few international conservation organizations created and funded the Mountain Gorilla Project to assist Rwanda in its conservation efforts. The Mountain Gorilla Project approached the problem of saving this great ape from multiple directions. Most immediately the gorillas were threatened by poaching within the park. Dian Fossey's antipoaching patrols protected only the area around Karisoke, about a tenth of the parkland; the Mountain Gorilla Project contributed funds to further Rwanda's antipoaching efforts over the remaining nine tenths.

But over the longer range the gorillas were threatened by the reality of Rwandan poverty. Somehow the sanctuary had to acquire an economic value at least equivalent to farmland.* So the Mountain Gorilla Project sponsored a plan to make tourism in Rwanda's park a genuine source of national income. And in 1980 two Americans who had been studying gorilla ecology and land resource management in the Virungas, Amy Vedder and Bill Weber, began carefully habituating three groups of wild gorillas to daily observation by tourists.

"When we first started habituating wild gorillas to tourists," Vedder told me, "the adult gorillas would scream or bark or hoot at us, and demonstrate. Sometimes an adult would charge to within a yard away from us — you hold your ground and act submissive and lower your head, and they stomp away. The little ones would often climb up where they could peek their heads out, and look. Then the group would move off. But little by little they gained trust and finally realized we were only going to watch." Today, tourists at the Parc National

*Some experts have argued that the forested park is crucial to Rwanda's agriculture. The park catches 10 percent of the rainfall in the country; its thick vegetation absorbs the sudden seasonal rains and gradually disperses the water to surrounding farmland. In addition, if the small park were converted to new farmland, it would provide only enough food for three months' worth of Rwanda's regular population increase. Yet such theoretical arguments, however true, tend to be ignored by people and governments facing immediate, dire economic problems.

des Volcans pay to be escorted on foot, in carefully supervised groups of a half dozen or less, through very rough terrain into areas where the habituated groups feed. The tourists are allowed to observe for an hour. Typically, they are able to sit within several yards of a family of wild mountain gorillas and watch the animals, often in full view, go about their daily lives. Tourists usually arrive in the middle of the day when the gorillas have finished their morning feeding, so they often see a placid group of animals, the adults resting or napping, the juveniles playing and wrestling with each other.

Finally, the Mountain Gorilla Project sought to generate within Rwanda a long-term commitment to preserving the Parc National des Volcans. It thus sponsored an extensive public education program about the park and its gorillas. Surprisingly, perhaps, many Rwandese had never heard of the mountain gorillas, and many more knew little about them. As part of the education project, Bill Weber distributed audiovisual educational materials to all the elementary schools in the nation and personally visited all the high schools. Soon the subjects of conservation and the mountain gorilla were formally integrated into the nation's high school curriculum. And with radio broadcasts and other sorts of communication, the education project attempted to reach the general populace. In addition, the project created a mobile education unit (a van with exhibits, slide and film projectors, and so forth) to travel into rural areas, particularly around the Virungas.

Some critics object that "gorilla tourism," even when it is minimal and carefully controlled, makes the noble apes less nobly wild and diminishes their protective wariness of humans. Nonetheless, the park's unique program for tourists has increased its revenues twelvefold. The recent influx of visitors has made tourism the nation's third greatest source of income. The park and its mountain gorillas have become economically important to Rwanda, and the government has recently shelved its previous plans to open some parkland to cattle grazing. The Mountain Gorilla Project, with its recognition of the social and political dimensions of a biological problem, provides a model of how conservation can succeed in the third world.

THE ARK OF THE VIRUNGAS: its near future will unfold in the context of poaching and protection, tourism and politics. But its long-term fate may be dictated by the more powerful forces at work in biological islands.

About two decades ago Robert H. MacArthur of the University of Pennsylvania, an expert on population dynamics, and Edward O.

Wilson, a Harvard entomologist who had just finished ten years studying ants on Pacific islands, began analyzing the extensive data on the numbers and distributions of plant and animal species on islands around the world. Biologists had long accepted a view that may seem obvious: that larger islands contain more species than smaller ones. But when MacArthur and Wilson looked closely at their own and others' data — on the distribution of beetles, amphibians, reptiles, and birds in the West Indies; ants in Melanesia; birds in the East Indies and the islands of the Gulf of Guinea; land vertebrates on islands of Lake Michigan; land plants on the Galápagos Islands — they found that the *ratio* of number of species to island size was impressively consistent. Consistently, they found that the number of species doubled for every tenfold increase in island size.

Islands are especially useful in population studies because they are self-contained ecological units, tolerably finite natural laboratories. But when MacArthur and Wilson considered the data on species living on pieces of land that were not islands, isolated neither by ocean nor by any other habitat gap, they found that the same species-to-area ratio held. That is, a tenfold increase in area meant double the number of species. The same basic ratio held. However, quite remarkably, they found that no matter what its size, an island almost invariably contained fewer species than a continental piece of land of the same size within a larger stretch of continuous habitat.

Consider the California Channel Islands — Ánacapa, Los Coronados, San Clemente, San Miguel, San Nicolas, Santa Barbara, Santa Catalina, Santa Cruz, and Santa Rosa — situated eight to sixty-one miles off the California coast between Santa Barbara and Los Angeles. Comparing each island with a same-sized piece of similar habitat on the mainland, we find that the islands all hold far fewer bird species. The 96 square miles of Santa Cruz Island, for instance, contain around thirty-seven bird species, whereas a 96-square-mile piece of comparable mainland contains about ninety bird species. On average the California islands possess about half the number of bird species found on comparable pieces of mainland habitat.

The species paucity of the California Channel Islands is true for islands around the world. By their very nature as discontinuous pieces of habitat, islands hold significantly fewer species than mainland that is part of a larger, continuous habitat. Wilson recently recalled that when he and MacArthur publicized this finding in 1967, "A dark cloud then fell over the conservation community." For, as the unprotected habitat around them is developed and destroyed, parks and reserves become islands — habitat islands, surrounded by seas of non-habitat. If the number of species on islands is inevitably lower than the number of species on same-sized, nonisolated mainland areas, then the creation and isolation (by human development of surrounding habitat) of a park or reserve should be followed by a regular series of extinctions, an inexorable decline in species diversity.

Conservationists had the example of Barro Colorado Island to ponder. During the construction of the Panama Canal, Gatun Lake was formed by rising water. What had previously been an elevated area of forest became in 1914, as the water rose, Barro Colorado Island. At that time the Smithsonian Institution established a biological research station on the island; since then it has more or less continuously monitored species changes and has patiently recorded the general decline in species diversity. The number of bird species on Barro Colorado has steadily declined from 208 at the time of the island's creation to 160 today. About two thirds of those local extinctions may have been caused by habitat changes, but the remaining third were probably the result of an *automatic* loss of species numbers associated with insularization — what might be called the *island effect.*

Having observed that islands hold fewer species per area than mainland habitat, MacArthur and Wilson tried to explain the pattern with what they called the theory of *island biogeography.*

They noted, first, that the number of species in any habitat, island or mainland, is stable — in equilibrium. Test case: in the mid-1960s two biologists enclosed four small mangrove islands in the lower Florida Keys with plastic sheeting erected across scaffolding. Conceptual art? More like conceptual science. The biologists hired a professional exterminator, who sprayed methyl bromide inside the plastic enclosure, killing all animal life on the four islands. The sheeting and scaffolding were then removed, and the two biologists monitored what happened next. Within six months the number of animal species on the islands rose to almost exactly the original number and then leveled off. It appeared that the four islands — like hotels with a fixed number of rooms — would hold only a certain number of species, an equilibrium.

Islands maintain an equilibrium of species (which is, as we have seen, lower than the equilibrium maintained on a same-sized piece of mainland). Interestingly enough, though, the species living on an island are not necessarily the same over time. The animal species that repopulated those four small mangrove islands in the Florida Keys were about the same *number* as before, but often not the same *kind.* And comparative censuses have shown that the number of bird species inhabiting each of nine California Channel Islands in 1968 has stayed very close to the number in 1917 — but about a third of the 1968 bird species were different from those of 1917. The loss of one species from the Channel Islands, the peregrine falcon, could be accounted for by human activity: pesticides had by 1968 eliminated this bird from most of North America. But the other bird species that disappeared from the islands between 1917 and 1968 did not disappear from other islands or the mainland. Their disappearance seems random. Likewise, a few of the replacement species can be accounted for by human intervention. Some game birds, mostly pheasants and quail, have lately been introduced onto some of the Channel Islands. But most of the replacement species seem to have appeared randomly, blithely taking over the spots left by the disappearing species. In short, beneath the fact of steady numbers lies the deeper reality of perpetual change. If islands contain an equilibrium of species numbers, it is yet a *dynamic equilibrium,* in which over time some species depart or die out (*local extinctions*) and an approximately equal number of other species arrive to fill the vacancies (*colonizations*). The idea of dynamic equilibrium, the dynamic interaction of local extinctions and colonizations, provides a key to understanding island biogeography and the fate of any isolated Ark.

Local extinctions. In any given spot of habitat, species come and go. As weather changes, climate alters, disease microbes arrive and dissipate, food and predators appear and disappear — as all of these events and more occur, species change. A patch of species located in Spot A today will, with some probability, not be there tomorrow: local extinction. The same pressures of adversity and change operate on a small island as on a larger island or even a continent — except that the larger area will carry, on average, more individuals of a species, and more individuals provide more of a buffer against adversity and change. To put it another way, a small island that has room for only ten rhinoceroses is going to lose the entire species, through the random slings and arrows of ordinary adversity, more quickly than a larger island or continent that holds 1,000 rhinoceroses. The smaller the piece of continuous habitat, the faster it loses species: the higher the rate of extinction.

Colonizations. In dynamic equilibrium, local extinctions of species are regularly replaced by colonizations. Guests checking out of the hotel are steadily replaced by new guests checking in. Members of the same or another species walk or fly, float or drift in. But when a piece of habitat is discontinuous, whether isolated by ocean or hostile habitat, colonization becomes more difficult. Many animals will not cross even a narrow gap of hostile or open habitat. Larger mammals may never colonize oceanic islands. (Their existence on some islands — rhinos on Java, for instance — is the artifact of a previous land bridge.) Smaller mammals and reptiles may colonize ocean islands slowly, traveling on drifting bits of vegetation or even occasionally on floating islands of matted vegetation, broken off and set adrift by a storm. Birds and insects will fly or be blown across ocean or hostile land habitat. Some will cross only distances of a few miles or so, but others may cross larger distances. And plant seeds may be carried by the wind or inside birds' bellies, finally to be dropped within nutrient-rich globules of excrement. However difficult or easy the colonization, though, the overall speed and efficiency of species colonization is obviously much higher when the two areas are connected by continuous habitat than when they are isolated by ocean or any other sort of habitat gap. The greater the isolation, the lower the rate of colonization.

In a stable environment, whether island or continent, local extinctions reach a level where they are stably replaced by colonizations. The result is an equilibrium, a stable number of species. But that equilibrium represents a complex balance, maintained under pressure by the opposing tendencies toward extinction and colonization. The lower rate of extinctions and higher rate of colonizations of a continent (or a piece of land within larger, continuous habitat) create pressure for a higher dynamic equilibrium — more species per area. The tendencies toward faster extinction and slower colonization on an island create pressure for a lower equilibrium — fewer species within the same area. The concept may yet seem abstract, but it is critically important to the future of the world's Arks. When a park or reserve is cut out of what was continuous forest and isolated by surrounding development, it becomes an island, and over time its dynamic equilibrium lowers. The number of species collapses to a lower level.

I LEFT THE ISLAND of the Virunga Volcanoes in a four-wheel-drive vehicle. I had a passenger, a young Canadian, and some luggage and supplies. It was dusk, the sky mottled with purple clouds, and then it was dark. I turned on my headlights and drove slowly along a very rough and rocky dirt road through several miles of farmland and mud houses. People were standing, sitting, and walking along this dark country road. Illuminated by my headlights, faces and figures appeared and then disappeared. People emerged from behind banana trees in small banana plantations, from cornfields, from the insides of small, candlelit houses. People ran in the dirt ditch alongside the road. Men stood, turned away from the road, urinating in the ditch. People held out cupped hands, both a begging and a hitchhiking gesture. I finally picked up two adolescent Rwandan girls, who began giggling, and made my Canadian passenger sit in the back, on top of the luggage and supplies. But once I had picked up the girls, several other people approached the car and demanded rides too. A young boy ran alongside my open car window and shouted: "Cinq francs, monsieur. J'ai faim." His expression feigned pathos — so I thought as I handed him some change.

Then my headlights illuminated on the edge of the road an old woman, dressed in traditional robes, holding a walking stick and balancing a large bundle on her head. She paused to let me pass. Her thin, high-cheeked face was tilted slightly, and her eyes seemed to look at me with a deep patience and vague sadness. Suddenly — and this was as close to an actual vision as I am capable of experiencing — I saw an image of my mother's face. That used to be her expression. The image dissolved. But I was tired and, suddenly moved both by the old woman's apparent sad weariness and by the memory of my mother's sometime sadness in life and her difficult death by cancer, I felt like crying.

Island biogeographic theory divorces its predictions about the fate of islands and other isolated areas from the potential impact of human intervention. Such a divorce is theoretically neat. But the fate of real Arks connects not only with the steady and somewhat predictable natural pressures at work on islands, but also with the explosive and unpredictable forces of a constantly expanding and increasingly needy world of people. The problem of hungry people surrounding an Ark that contains sleeping farmland and walking food finds no facile solution, as each year, in our brave new world, more people try to squeeze more of a living from the same land.

Seven

HEALTH

I WENT TO MADAGASCAR seeking lemurs, but I arrived thinking about malaria.

A former Peace Corps volunteer I met in Rwanda, who was chronically infected with malaria and regularly experienced relapses, told me he had learned to recognize the early symptoms in time to take the necessary emergency treatment before succumbing to delirium. He had trouble articulating what the strange predelirium stage felt like, but I could imagine it: a sense of heat and a tingling, otherworldly light-headedness — like a glass of champagne for breakfast.

An American scientist, an expert on lemurs, had advised me more particularly about malaria in Madagascar. She told me that a flu virus in Madagascar will give all the symptoms of malaria — sweating, fever, chills — plus a splitting headache. If you get *those* symptoms, relax: you merely have a case of the local flu and ought to take two aspirin. If you get the same symptoms minus the headache, however, you have malaria and ought to take the emergency treatment and find a doctor right away. Or was it the other way around? I couldn't for the life of me remember whether malaria or the innocent flu included a headache.

Every year, approximately one million people die from malaria, and perhaps another hundred million are chronically debilitated by the disease. Until the nineteenth century malaria was usually considered the work of supernatural creatures or, at least in post-Renaissance European lore, the unfortunate consequence of breathing bad air from fetid swamps. We now recognize that malaria is the work of two natural creatures: first, a microscopic one-celled animal, a parasitic plasmodium; and second, the female *Anopheles* mosquito, which, flying mostly at night and buzzing only faintly, can carry in its salivary glands many tiny plasmodia.

Upon biting a human, the mosquito transfers malarial plasmodia into the human bloodstream. The parasites quickly move on to the liver and begin breeding. Once established in and nourished by the liver, they venture back out into the blood and take over red blood cells, eventually rupturing them and producing the clinical picture of malaria. The human host, meanwhile, has become a reservoir of plasmodia and can pass the parasites on to other humans via *Anopheles* mosquitoes or more directly via blood transfusion.

Throughout my trip into the tropics, I killed all suspicious-looking mosquitoes and took the standard weekly dose of antimalarial medicine, which is supposed to provide prophylaxis against one of the four varieties of malarial plasmodia. (The other three are not so easily prophylaxed, I think.) I also carried a second type of medicine, with which I planned to dose myself as an emergency treatment should I start feeling any preliminary symptoms.

So, when I arrived in Madagascar one night, I arrived thinking about malaria and had already started drifting into the first stage of hypochondriasis by the time I emerged from the airport of the capital city, Antananarivo. Clear skies and the sight of Orion sprawled overhead provided some salubrious reassurance, however, and the city itself, by starlight, seen from the back seat of a taxi, presented a gentle, fairy-tale aspect of winding streets, small wooden houses with second-story balconies, and, toward the center of town, elegant French colonial buildings of brick and stucco.

By morning light, from the window of my hotel, Antananarivo appeared less a fairy-tale construction, more a city antique and heroic. The city swelled grandly against rolling hills; reached skyward with white church spires, a grand cathedral, a rectangular chateau cornered by four round towers; expanded outward with ornate wrought-iron balconies, fancy brickwork, and window shutters painted green and blue and white.

Under the midday blast of sun, and seen from the street, Antananarivo began to appear faded, mildly decayed, vaguely desperate. First the flower sellers, mostly women and young girls, thronging around the entrance to the hotel, all trying to sell flowers. They begged, bargained, followed, cajoled. A fine-featured adolescent girl, flowers in one hand, child in the other, small baby slung over her back, approached me with a desperate, pleading look. "Please, buy some flowers," she said several times in French. I didn't want flowers, but pressed some change into her hand and hastily walked into the post office.

I had already bought a postcard with a picture of a lemur on it. In the post office, I bought a stamp with a picture of a lemur on it and scribbled some words on the card to my daughter's kindergarten class. But while I was scribbling, a small boy about my daughter's age approached me and began begging (in French): "Give me. Give me. Give me. Give me. Give me." He didn't stop repeating that single phrase until I — reluctantly, because his whining style aroused irritation more than sympathy — gave him a coin.

I left the post office, walked through a public plaza occupied by soldiers swaggering in camouflage suits and berets, holding semiautomatic rifles, and descended a long series of stairs toward the city center. On either side of these stairs, sidewalk merchants were tending booths and tables full of glasses, both sunglasses and prescription eyeglasses. I counted them: exactly twenty-four sidewalk opticians' shops, each displaying nearly identical wares, the merchants either sitting behind their glasses, waiting for customers, or standing in front and polishing them up. In spite of all those willing merchants, though, I saw no customers. In fact, as I looked at the faces around me, I saw no one wearing eyeglasses. I was the only one.

The stairs eventually led into an alleyway lined with more stalls and a few garbage piles. An old man, barefoot, in rags and a straw hat, squatted over one garbage pile and carefully raked with his hands, back and forth, over a patch of broken peanut shells, looking, I thought, for a whole peanut.

The alleyway opened into a street, which soon expanded into the city center: three or four blocks of a wide, divided boulevard lined with shops and arched, covered walkways, terminating at a grand, European-style train station. Small tables, booths, and goods had been spread across the walkways, and great crowds of people stood around. Mostly, it seemed, they were trying to sell things to each other. They tried to sell peanuts, tennis shoes, flip-flops, meat cooked in fat, wooden boxes, carved pipes, leather hats, and so on, which were grouped (like the eyeglasses) in areas of specialization along the street — peanuts on this block, flip-flops on that — which made the entire boulevard into one gigantic department store, except that there seemed to be many more salespeople than customers. Young men moved in and out of the crowds with fistfuls of watches for sale — How much will you pay? — and old beggars sat crumpled in shady corners, holding out their hands. Professional child beggars with runny noses worked the crowds, looking out especially for the uncommon face of an obvious visitor to Madagascar. Finding the visitor, they followed as persistently as mosquitoes, stepped trippingly in the way, held their hands up, and whined over and over and over again, "One franc, if you please. One franc. One franc. One franc. One franc."

Begging is a standard profession in many urban parts of the tropics, and I had placed coins in hands before coming to Antananarivo, operating both out of natural inclination and on the presumption that the exchange is really a payment of social tax, a hand-to-hand, personalized alternative to the often thoroughly institutionalized charity of most developed countries. But in Antananarivo I began to notice the beggars more acutely.

Some seemed merely poor, desperate from poverty, including many young women carrying infants and children. But many were sick or crippled, and I was reminded of the whole tropical community of sick and crippled beggars I had seen before coming to Madagascar. I recalled sitting at an outdoor café table along the Copacabana beach in Rio de Janeiro, and being approached by an incontinent old man who could barely hold out his hand. (The Brazilians next to me histrionically pinched their noses and laughed at his incontinence as he wobbled away.) I recalled seeing in Dakar, Senegal, the first of the many four-legged beggars I was to notice throughout urban Africa: polio victims and others with withered legs, who, without wheelchairs or other prosthetics, walked the streets and sidewalks on all fours, often protecting their hands from rough pavement with worn slippers or torn flip-flops. In Dakar one legless boy specialized in zipping on a skateboard among the tables of outdoor restaurants, approaching people eating dinner with an irresistible smile and an outstretched hand. In Freetown, Sierra Leone, a young boy with elephantiasis came up to me, pointed to his foot, which looked like the base of a tree trunk, and wordlessly held out his hand. In Nairobi, Kenya, a woman with her nose missing, the skin of her face stretched brutally clean as if it were a chalk drawing someone had wiped with an eraser, held out her hand.

But in Antananarivo that morning, I saw a sight so appalling it troubles me still, and so distressing then that my only reaction was one of nearly complete avoidance: the Traffic Girl. She was a pretty little girl who couldn't have been older than my daughter, five or six years old, dressed in scraps of clothing, sitting on the curb of a heavily trafficked street just around the corner from the Centre Culturel Albert Camus. Her legs, probably crippled by polio, were useless, dragging matchsticks. She would wait for the corner traffic light to turn red and the traffic to stop, then crawl out into the street between cars and hold up her hand for money. With luck, the drivers saw her leave the curbside and crawl in front of and between their cars, where she disappeared from their line of vision. Thus — and I believe this was her particular strategy — people were forced to lean out of car windows to make sure she wasn't about to be crushed under a wheel and were greeted by her little face and outstretched hand. I suppressed the vision of my own daughter in her place and winced at the thought of her for days.

After lunch that first day, I began to feel feverish and lightheaded — preliminary symptoms of malaria — so I went to bed in my hotel room and prepared to dose myself with the emergency medicine before delirium overtook rationality. After the angle of the sun coming in my window declined and the heat of the day diminished, though, my malaria went away.

By late afternoon I felt well enough to sit at a sidewalk cafe and watch big clouds move in and overtake the sky, collide violently, and fill the sky with pewter and purple — an El Greco delirium, sweeping and swirling. Then it became dark. It started to rain, with explosions of thunder and blue-white flashes of lightning breaking open the black sky, and the street became a sheet of black glass broken by circles and speckles of white. Everyone on the street took cover under the roofs of buildings, and then the rain came in sheets. Suddenly a whole skyful of water seemed to be coming down, and all the white and blue lights of the city, and the green and red of a few neon signs, were muted behind rain and reflected feebly and waveringly on the streets. Cars, buses, and taxis whooshed across the watery streets with a sound like tearing cloth.

I had dinner then, went to bed, and listened for a long time to muffled thunder and a soothing background of rain — a pleasant sea-in-the-seashell noise, a soft agglomeration of drips and gurgles and splashes. Next morning I took a train east to consider the health of two of Madagascar's most endangered primate species: the biggest and the weirdest.

SOMETIME near the beginning of primate evolution, several ancestral primate species expanded over much of the forested part of the globe, into Africa, Asia, and the Americas. These early and primitive primates, the prosimians, were small, nocturnal, and highly oriented to the sense of smell. After about 20 million years of arboreal success, from this early primate line emerged a second major type, the simians. The simians, which evolved into today's monkeys and apes, competed well enough to displace most of the earlier prosimians.

A few prosimian species — lorises, galagos, and tarsiers — have endured in limited regions of Africa and Asia. They survived the emergence of the larger, diurnal monkeys and apes probably because they remained small and nocturnal. In other words, they never directly competed with the simians. The greatest number of prosimian forms endures on Madagascar, primarily because that island was separated by massive geologic forces from the African mainland about 50 million years ago, before the simians and several large predators evolved.* So, the prosimians of Madagascar — now fewer than thirty species — proliferated quite nicely and thoroughly until about 1,500 years ago, when the first human settlers arrived on the island: Pacific Islanders from the east and, later, Africans from the west.

Portuguese sailors were the first Europeans to sight Madagascar, in 1500. Captain William Keeling, commanding the English East India Company's *Dragon,* explored the mouth of the Onilahy River on the southwest coast in 1608, and wrote in his journal probably the earliest account of the island's prosimians (in this case, a ring-tailed lemur): "In the woods neere about the River, is a great store of beasts, as big as Munkies, ash-coloured, with a small head, long taile like a Fox, garled with white and blacke, the furre very fine." Of all the European explorers and traders, though, the French were most intrigued by Madagascar. Etienne de Flacourt, Directeur de la Compagnie Françoise de l'Orient & Commandant pour sa Majesté in Madagascar and adjacent islands, resided at the outpost of Fort Dauphin, to the southeast, between 1648 and 1655, and published in 1658 an extensive account of the island. He described "several kinds of monkey." Some had "a long muzzle like a fox" and "make so much noise in the trees, that if there are two of them it seems that there are a hundred." Another species was "gray, with a long muzzle, and a big velvety tail, like the tail of a fox, which all the other species have as well." He wrote of "yet another species of white monkey, with a tan cap, and which most often goes on its hind feet." Another white monkey, "whose tail is striped in black and white, they go around in groups of

*An alternative theory considers that Madagascar was separated from Africa much earlier, not by drifting but by a subsidence of land, which created the 500-mile-wide Mozambique Channel. According to this second theory, smaller African mammal types subsequently colonized the island beginning roughly 50 million years ago, by rafting across the channel on floating vegetation. In either case Madagascar's wildlife evolved apart from that of the African mainland for dozens of millions of years and never had to contend with large predators or competing simian forms.

30, 40, and 50." Another lemur he considered "a species of gray squirrel which they call *Tsitsihi,* which usually hides in tree holes and is neither beautiful nor good to tame."

Following this and other reports, Linnaeus in the mid-eighteenth century invented the name *lemur* (from the Latin *lemures,* which for the Romans were wandering spirits of the dead, with glowing eyes and eerie cries) and placed three prosimians in that group, one of which was from Madagascar. Eventually, however, the name *lemur* was assigned exclusively to prosimians of Madagascar and a few small neighboring islands.

Like nearly all other primate species, prosimians have grasping hands and feet. Most species today are quite small (and include the smallest primate genus, *Microcebus,* or mouse lemurs, which weigh about two ounces on average) — but some recently extinct forms in Madagascar were as large as apes, and a few still living species rival good-sized monkeys. Like many monkey species, most prosimians move through the trees and on the ground quadrupedally, and several are strong leapers. Some lemurs, however, hop kangaroo-style on the ground. Some transport themselves primarily by leaping vertically from one tree trunk to another — kicking off with long, powerful legs and landing upright.

At any rate, in terms of size and locomotion most prosimians are not entirely different from their monkey and ape cousins. The division is more fully defined by other characteristics. For example, most prosimians today are, like their ancestors, nocturnal, whereas only a single monkey species, the night monkey of South America, sleeps during the day and prowls at night. And although several lemurs of Madagascar have become diurnal, all but one of the prosimians retain an important anatomical mark of nocturnal ancestry, the *tapetum lucidum.* Most people are familiar with the external evidence of the tapetum in domestic cats, whose eyes become shiny reflectors at night. The function of this anatomical feature, a sort of shield that lies directly behind the eye's retina, is to reflect light that has already passed through the retina, thus reusing or amplifying the light and allowing the animal to see better at night.

Today's prosimians are the most smell-oriented of the primates. Most species retain a rather long and sensitive nose, and (except in the three tarsier species) the tip of the nose remains moist. As well, the small prosimian brain is more attuned to smell than to sight.

The sight-oriented monkeys and apes developed several features useful for visual signaling, including highly mobile lips and faces, but the prosimians retained the inflexible snout-noses and faces of the earliest primate ancestors. Although prosimians do communicate through visual and auditory means, they have continued to rely on the kinds of scent marking used by many other smell-oriented mammals, including today's domestic dogs. Prosimians communicate territorial possession, sexual availability, and other information by depositing spots of urine or feces on leaves or branches or by rubbing them with various scent glands. One species, the ring-tailed lemur of Madagascar, even engages in occasional stink fights during mating season. Two or more males of this species will sit back, rub the tips of their long, bushy, ring-patterned tails against potent scent glands on the chest and inner forearm while staring at each other and vocalizing. Thus prepared, the adversaries descend to all fours and point and shake their odorized tails at each other until one is disheartened enough to retreat.

Some prosimians are further distinguished from the monkeys and apes by protruding lower front teeth, the *dental comb.* This toothy protrusion serves both as a comb, for grooming self and others, and as a useful rake for harvesting gum from trees. Sticky? These primates keep their dental combs clean with a strange horny structure beneath the tongue, called the *sublingua.*

The biggest primate of Madagascar, indeed the biggest prosimian in the world, is a long-legged, twenty-two-pound tree dweller called the indri, which was given that name only two centuries ago by Pierre Sonnerat.

As France's Naturaliste Pensionnaire du Roi, Sonnerat explored several islands of the Indian Ocean between 1774 and 1781. In a multivolume account of his explorations, *Voyage aux Indes Orientales et à la Chine* (1782), he described and named several lemurs of Madagascar, including one he called *l'indri.* In the native Malagash language, *Indri* means simply "There it is!" So perhaps Sonnerat mistook his native guide's factual statement for a name.

With large yellow-brown eyes, large round ears, a stump of a tail, bare black muzzle, and thick black-and-white patterned fur, the indri vaguely resembles a tree-dwelling teddy bear. But its hands and feet are quite long — six times longer than broad. Its thumbs and big toes are unusually well developed. It sits and often moves upright. Although it will walk slowly and quadrupedally across long branches or boughs, the indri most commonly progresses through the forest by leaping. Using long, powerful hind legs, it leaps, still maintaining an upright posture, from trunk to trunk. On the ground it typically moves by hopping on its hind feet.

Resting motionless in a tree fork, an indri's black and white markings blend thoroughly with the patterns of light-colored lichen across a shadowed tree trunk or of dark tree limbs against a cloudy sky. It is as easy to hear as it is difficult to see, however, and people searching for the timid indri in its rugged terrain usually first locate groups by sound rather than sight.

According to some scientists, the indri has two major calls in its repertoire. One seems to be an alarm, emitted by isolated or disturbed individuals, which is quite loud but carries over a relatively short range: "repeated cries like blasts of a horn." The second call, more frequently used, consists of "barks, followed by long, modulated howls," which are uttered by all members of the group and which carry well over long distances. Communal howling, usually broadcast from the highest areas of the forest during morning hours, can continue for as long as thirty minutes. When one group has finished, a neighboring group will take up the chorus. The indri chorus is a type of intergroup communication, probably announcing territorial possession.

The indri social group is a nuclear family, a monogamous mated pair with their offspring, inhabiting a well-defined home territory. The mated couple bears a single little indri every two or three years. Upon birth, the infant hangs onto the fur of its mother's belly, remaining in that state of literal dependence during daily travels for the first two months and thereafter clinging to its mother's back.

The family group sleeps in trees at a height of thirty to one hundred feet. During sleep the family is dispersed, except that during its first year and occasionally thereafter the youngest offspring sleeps cradled within the limbs and curled body of its mother. The next youngest may sometimes sleep with its father. An hour or two after sunrise, members of the group awaken, stretch their limbs, and if food is near, they may pause for a quick breakfast. Then the family descends for a communal session of urination and defecation, typically in an area previously used for that purpose. Depending upon a number of factors, including weather and season, an indri group may at some point in a typical morning ascend to the treetops and announce its position and territory to other groups with a howling chorus. Also at some point during the day, the group will descend to the ground and briefly consume handfuls of loose dirt, which probably provides some essential minerals. But by far the greatest portion of the day is spent progressing through the trees, foraging for food. Indris carefully select the young leaves, buds, stalks, flowers, and fruit of about seventy species from all levels of the forest. The females and young usually take the best and highest feeding areas, while the males are relegated to inferior, lower regions. This situation seems to be maintained by the female, who will actively defend her superior feeding sites from the male's occasional incursions. The day concludes well before sunset with a brief session of grooming.

Indris used to be common and widely distributed in Madagascar. Not anymore. A naturally low birth rate, dependence on undisturbed forest, and loss of habitat have pushed this beautiful animal to the edge of extinction.

I sought indris by sitting on a train for several hours, traveling east from Antananarivo across Madagascar's central plateau, until I disembarked about halfway between my starting point and the east coast at a place called Andasibe.

Andasibe consisted of a village, a hotel, and a forest. The village was small and unprepossessing, the hotel much more impressive. Actually, the hotel was part of the train station: antique and wooden, two stories, with a large restaurant and bar on the ground floor that may have been a station waiting room a few decades ago. The restaurant's open arched windows looked out onto and across the tracks, and the breeze passing through the windows billowed translucent white curtains. The ceiling was supported by huge beams and columns, all wood, and the walls were covered with murals of Malagasy men sawing and chopping trees, Malagasy women trading food in a marketplace, and so on. A dog slept on the herringbone parquet floor in one corner, and a cat sat and cleaned its fur between books on a bookcase. Pleasantly melancholic French music played on a hidden radio somewhere, but once the train ground away down the tracks, I was the only person left in the restaurant. It was quiet enough then for me to be hypnotized by the ticking of an ancient round clock framed in a wooden cabinet over the restaurant bar. I was glad to be there, feeling healthy and at peace in the countryside, away from the diseased, crowded chaos of the city.

At least two American primatologists had told me, independently, that when I came to Andasibe, I should find a boy known as Little Joseph, who was a superb guide, a walking miracle at finding indris. So I asked at the hotel for Little Joseph. He wasn't available, but his older brother, Maurice, said he could help me find the lemurs. Maurice was young, probably in his late teens, with a not unusual Malagasy combination of African and South Pacific features, a handsome face, and a serious but direct and uncomplicated manner. His feet and legs were bare, and a long, faded blue jacket that looked like an old laboratory coat covered him down to his knees.

We walked along a road and then entered the Andasibe forest.

Maurice proceeded ahead, and I followed, looking around me at the forest. It was vined, damp, and mossy. The trees in the forest were modest — comparatively low, extending up on thin, gray, lichen-covered trunks into a shifting, dappled canopy of leaves only perhaps thirty to sixty feet above us. Giants did grow in this forest — not trees, but mainly giant bromeliads and ferns. The bromeliads looked like enormous pineapples, splashing outward with thick, long, serrated, spear-shaped leaves. The largest ferns sprouted lacy leaves big enough to serve as tablecloths. The ferns here came in all sizes, and we passed through whole fields of smaller ferns, emanating lacy fingers that divided and divided and divided again until they were lovely and spread, delicate and filamental, like soft green sheets of frost dissolving into lace, tracery, and byzantine curlicue. The forest was modest, but that very modesty, actually a feeling of delicacy and fragility, made it beautiful. Smaller things provoked one's greater attention: dragonflies, for instance, flitting with transparent wings and bright red velvet bodies; small berries growing close to the ground, shaped like tiny apples but colored a bright, almost garish blue; furry blueberries, dark blue, that according to Maurice were good to eat and an aid to digestion. I tried one, and my digestion felt pretty good.

We walked along a ridge in the forest, past crowds of those narrow, gray, lichen-splotched tree trunks, until Maurice stopped, raised his arms like a conductor hushing an orchestra, looked into the treetops around us, then told me to wait as he scampered down the hillside, disappearing behind gray trunks and green leaves. A quarter of an hour later he reappeared and beckoned me to follow him. I followed. We walked among the gray trunks, skirted many clumps of ground-level vegetation, and at last arrived at a spot where, looking up, I saw three large, mostly white-furred animals sitting in the crotches of a single tree.

They were indris, and they looked down at us. One was leaning partly back on a branch, rather like a reclining odalisque. The other two sat straight up, clinging to the upright trunk of the tree with hands and feet, resting their stumpy-tailed bottoms on very thin branches. They seemed inappropriately large, like big dogs sitting in small trees. But the first thing I noticed thoroughly was the size of their legs, which were long and obviously powerful; in the sitting position, with both feet together and clasping long, fingerlike toes right around the trunk, their knees extended up and loosely apart, the indris reminded me of crew oarsmen. They had small teddy bear heads, marked black and white, with furry, muzzled faces, and big round, black teddy bear ears. Their backs were dark, like shadows;

but their arms and legs were white or, rather, whitish-grayish, much the color of the tree trunks and branches they sat in, and, in fact, about the same thickness. They looked down at us from time to time, but mostly they remained still and relaxed, like people sitting comfortably in a living room.

Maurice said he was a naturalist, and he gave some samples of indri calls. I was interested, but the indris weren't. After watching those placid animals for half an hour, we decided to leave. We walked out of the forest to the road, and after agreeing on a time and place to meet again, we went separate ways.

I walked along the road for a mile or two toward the hotel and enjoyed seeing the forest from its edge. I came across more fern fields, flowing through the forest, rippling down hillsides. I came across bushes with great clusters of sweet-smelling cream-colored flowers; and later, along the road, I looked at intensely bright red flowers on long stalks. They shivered in the breeze like scarlet rags.

My walk was interrupted, however, by the appearance of a grizzled middle-aged man walking erratically along the road toward me, wandering from one side to the other, alternately glancing at me and looking into bushes at the side of the road. He was a shifty character, with all the style of a New York City pederast walking along a playground fence. When he came within speaking distance, he began spouting French words, which I understood as: "Snake? Snake? Snake? Crocodile? Crocodile? Crocodile? Orchid? Orchid?" He took a stick from along the road and scratched a price on his forearm. He would show me big snakes (he stretched a large circle with the joined fingers and thumbs of both hands, his eyes widened in imitation amazement, to express how big), as well as crocodiles and orchids, for that price. I explained several times that I wasn't interested and began walking again. But he turned around behind me and began following me, still erratically, still darting back and forth across the road and looking into bushes. I slowed down enough that he was forced to pass me, so he wandered in front of me and eventually came to a large bush beside the road. He reached into the bush and pulled out a plastic bag. He crossed the road again and pulled two more bags out of another bush. He crossed the road a third time, threw one of the bags into a bush there, and pulled out yet another. I was walking fast by then and had almost passed him when he grabbed my arm and insisted on showing me the contents of his bags. He had little yellow frogs for sale. Did I want them? He thrust a large, armored chameleon on a stick in front of me. It was healthy: what would I pay?

Next day I met Maurice at the wooden shack where he lives, and we went out again looking for indris. It had rained during the night, so the forest was dripping wet. As we walked, Maurice talked. He talked in a rough French and had the habit of repeating everything he said many times, just to make sure I understood. In Madagascar, he told me, life is very expensive. For Americans it isn't, but for the Malagasy it is. If you wish to take a job, you must pay your prospective employer much money. If you wish to go to school, you must pay your professor much money before you can take your exams. He wanted to be a naturalist, but he couldn't afford an education.

At last we found a group of four indris, a family of two adults and two young, sitting in three different trees. First we saw the father sitting on a small branch, his black-furred back to us, his white-furred bottom resting on the small branch. He was big, and, as I walked closer, I could see his smallish, white-furred scrotum hanging down. Like the indris I had seen the day before, he squatted on a small branch and clung to the tree trunk with long-digited hands and feet, knees up, reminding me now of a telephone lineman on a pole. He was eating leaves, reaching out with his left hand, pulling leaves to his mouth, chewing rapidly, and I could hear, softly, his chewing. He leaned out almost horizontally to grasp more leaves. I was below and behind him, and two or three times he swiveled his head to look right in back of him, to stare calmly at me with lemon-colored eyes surrounded by black fur. The swiveling, which involved an invisible twisting of only his neck, was astonishing.

Then I saw the mother sitting several trees away, eating leaves, and once in a while swiveling her head to get a good look at us; the offspring were sharing yet another tree. All four animals were very quiet, and they kept on eating.

Time passed, and then the father sprang up with his long, powerful legs, using them like coiled springs to leap up and out and land on another tree. It was a hop as much as a leap, rather like the motion of a pogo stick or the bounding you'd expect of an arboreal kangaroo. He kept on bounding from tree to tree, sometimes higher, sometimes lower. Each time he leapt, I listened to a small rainfall of water shaking off the tree he jumped from and then off the tree he landed on. He was moving fast now, making a lot of short, quick hops but some impressively longer ones too. I paced out the distance of one of his hops, and decided it was around fifteen to eighteen feet.

The rest of the family had begun bounding, too, shaking drops of water off trees as they sprang and landed, and gradually they all traveled to another spot in the forest, took up their telephone lineman's position, and began eating leaves. I was regularly impressed by the smallness and apparent fragility of the branches they sat on. Once the mother sat on a small branch that broke. She didn't seem put out by the accident, though, and still held on to the trunk with hands and feet. Then she moved on.

They sat and moved in trees as low as fifteen feet, as high as forty feet, and after some time, all four moved into a single small tree and climbed up to the very top, where there was a cache of young red leaves. The tree was so burdened by their weight that it swayed slowly and widely, back and forth. But they stayed there, silently eating leaves, and sometimes slowly looking at us with those very pale lemon eyes, set in black fur like yellow buttons. It must have been very peaceful and pleasant up there, the sky clear blue, the sun out, a slight breeze blowing.

Later in the morning I started hearing a distant chorus of cries, which had some of the wavering quality of air whistling out of a toy balloon, in a pitch that rose slightly at the end. Maurice said those were the cries of another indri group in the forest. The indris we watched began making kissing noises — loud smacks — alternating with noises like stifled burps. The high-pitched calls in the distance continued for some time, and then, very suddenly, the four members of our group lifted their muzzles and opened their mouths in the manner of howling dogs, and began blasting the air with an ear-shattering, high-pitched siren chorus. The sound was amazingly loud, the loudest noise I'd ever heard an animal make, with a wavering, high pitch — still like a balloon whistle, only an actual blast of sound. I suppressed the impulse to cover my ears while the chorus went on and on, but after several minutes, as suddenly as it had begun, the chorus stopped. And my still tingling ears were stimulated then only by the distant calls from two or three directions of two or three other groups.

Some time after that startling production, all four seemed to become sleepy. I watched one rest its chin on its knees. Others slowly blinked their eyes and then closed them: from lemon within black to black.

Maurice and I went into the forest several other times to look for other indri groups and to find some of the other lemurs of Andasibe. One night I waited for him in pitch dark along the road. It had just rained, so the road was lined on either side with puddled water. I stood on the edge of the road, listening to the whines of many mosquitoes. I thought briefly about *Anopheles* mosquitoes, how quiet and sneaky they are, and then thought that there would be fewer mosquitoes in the middle of the road, away from the puddles, so I stood in the middle of the road. I grew tired of standing, however, so I lay down in the road and looked up at the dimly moonlit clouds above. Traffic almost never came along that road, and I wasn't worried about being run over, but as I lay there and time passed, I began to hear from somewhere inside the forest a ghostly sound of voices, chanting and mumbling. At first the voices were very faint, but gradually they came closer and closer, until finally I thought I recognized the litany of a Catholic mass: a single voice mumbling something, followed by a group of voices repeating the same mumbles. A Catholic mass? Out here in the middle of a forest? At last I saw the dark forms of a family of five walking toward me on the road, and I finally realized that they were listening to the mass on a radio. I sat up. They passed in the darkness without a word.

Half an hour later Maurice appeared, barefoot and wearing his lab coat, and we walked into the forest looking for nocturnal lemurs. We found two mouse lemurs, which announced their presence with amber, roadside-reflector eyes. They looked indeed like mice, with pretty palamino fur and mouse-sized bodies, and they scurried fluidly up and down, over and under branches, confused by the beam of our flashlight and not clever enough to run into the darkness.

Another time, at dusk, we found three trees full of brown lemurs, perhaps four dozen of them. When we shone a light into the trees, I saw lots of small animals with long tails looking down at me. A few of them seemed to clear their throats and begin a call — a threat or alarm — that sounded like a crow's call: caw, caw, cawwwwww. The call spread in a fast contagion of caw-caw-caw-caw-caw that very quickly infected the lot of them, until we were inundated with the cawing of forty voices at once, almost but not quite synchronous, so that the final effect was a loud, rippling sound, as if the echo control had been turned all the way up.

Another night we went looking for bamboo lemurs. We walked through a dark forest spotted with phosphorescent fungi and dotted with wandering fireflies that never turned off their lights, and finally shone our flashlight into a clump of giant bamboo that rose before the moon like a huge dark tower. Two bright red eyes shone back at us from near the top of the tower. In the morning, just before dawn, we returned to the bamboo clump, hoping to see the body that held those eyes. The bamboos were huge, thick as legs, and clustered tightly enough to make a solid mass fifteen feet across at the base. Maurice approached the cluster and began knocking on the bamboo, just like someone knocking on a door. I thought of the children's story about a little fur child who lives inside a hollow tree with a door and curtained windows. The bamboo lemur wasn't home, however.

I ALSO WENT LOOKING for the weirdest prosimian of Madagascar, discovered for Westerners two centuries ago by Pierre Sonnerat, who returned to France with a pair of live animals of a type he called *le Aye-aye.* The name represented, he said, the sound of their alarm call. The animals died within a few weeks, so various European zoologists were able to examine them quite thoroughly. But Sonnerat's aye-ayes were so different from any known lemurs or other prosimians that for a time most zoologists considered them a type of squirrel or rodent.

Superficially, the aye-aye does resemble a large, dark brown squirrel. It is small, about three feet long from nose to tail tip, with a slender body and a full, bushy squirrel's tail. Its coat includes thick, short white fur beneath a layer of longer coarse fur, which is mostly dark brown except for a white tip. The full effect is of a coat of blackish-brown fur flecked, in places overcome, by white.

The aye-aye has large, round eyes surrounded by protective bony ridges and protruding, mobile, membraneous ears like a bat's. Unlike any other prosimian, it has only eighteen teeth — most have thirty-six — and it lacks the characteristic dental comb. Instead, two front incisors in the upper and lower jaws grow continuously and wear down continuously, like rodents' teeth. Indeed, this feature persuaded many zoologists to list it with the rodents until a nineteenth-century English scientist, Richard Owen, discovered that its immature milk teeth more resembled those of typical lemurs. Most peculiar, however, are the aye-aye's hands and feet. Typical of primates, its hands and feet grasp, but atypically the animal has claws on all digits except the big toes, which are protected on the ends by flattened nails. All the digits are quite long, but, very strangely, the middle fingers are twice as long as the others, and they are thin and spiny, even withered-looking. Those weird middle fingers terminate with a long hooked claw.

Unique in appearance, the aye-aye is also highly specialized in behavior and feeding. A somewhat solitary animal, it sleeps in a high nest during the day, curled up in a ball with its bushy tail wrapped around itself. (The nest, roughly spherical, constructed of leaves and green twigs in a high tree fork, is semipermanent and may last for years.) At dusk the aye-aye becomes active. It jumps between branches or climbs slowly, quadrupedally, and it can dangle upside down by its hind legs, leaving the hands free for eating and grooming. Often the aye-aye will walk slowly upside down beneath large horizontal branches. With its elongated, toothpicklike middle finger, the creature may comb its fur or scratch or clean various spots on its face and body. But primarily the middle finger serves as a probe and as an eating and drinking utensil. The aye-aye crawls across or beneath branches, head close to the branch, tapping with that middle finger, listening intently for a hollow sound or some other indication of burrowing beetle larvae. When it senses the location of beetle larvae, it furiously gnaws a hole in the branch, pokes in with the claw of its elongated middle finger and, if lucky, withdraws succulent mushy hunks of beetle grub. To drink, the aye-aye dips its middle finger into a liquid and rapidly draws the wet claw to its mouth. It also feeds on fruits, especially coconut, and other vegetation, but beetle larvae are its main source of protein.

Aye-ayes are usually silent except when gnawing or actively eating, but occasionally they will call to each other with a sound that has been compared to the scraping of metal or the rattling of a tambourine: rron-tsit. When frightened, they do not call *aye-aye* as Sonnerat indicated when he named them. (Perhaps Sonnerat, unfamiliar with the native language, mistakenly described the surprised exclamation of his guide upon sighting one.) Approached closely, they will hiss and threaten to attack.

So anomalous is this small, dark prosimian that zoologists now assign it a superfamily all its own, the Daubentonioidea. Thus, this one species has its own superfamily, family, and genus. All other prosimians of Madagascar belong to a second superfamily, the Lemuroidea. Apparently a much larger prosimian closely related to the aye-aye once inhabited southwestern Madagascar, but it is extinct.

The aye-aye itself, once widely distributed in eastern and perhaps western portions of the island, was considered extinct by the 1930s. When in 1957 aye-ayes were rediscovered in a small area of eastern rain forest, that habitat was quickly protected as the Mahambo Reserve by the Madagascar government. By 1963, however, the reserve appeared to contain no aye-ayes, and because the area had proven hard to protect, it was halved. In 1966 the government declared a special reserve especially for the species on a small island, Nosy Mangabe, located in the Bay of Antongil, in Madagascar's northeastern corner. Nosy Mangabe was uninhabited and virtually undisturbed because of its significance to local people; ancient Malagasy kings are buried there. With financial assistance from the International Union for the Conservation of Nature, mango and coconut trees were planted on the island as aye-aye food. And the following year French biologist Jean-Jacques Petter trapped nine aye-ayes (five males and four females) on the mainland and transported them, accompanied by considerable fanfare, to Nosy Mangabe. Soon, however, the political winds in Madagascar shifted, and Western scientists were no longer welcome. As a consequence, no one returned to Nosy Mangabe to see if the aye-ayes were surviving until 1975.

That year Elizabeth Bomford, an English wildlife photographer, won first prize in a photography contest. The prize was a round-trip ticket to anywhere in the world, and Bomfort chose to go to Nosy Mangabe. After an arduous trip in which all her photography equipment was destroyed, Bomford at last arrived on the island to discover, indeed, a living aye-aye. She returned to Nosy Mangabe a year later with more adequately packed equipment, accompanied by her family and the director of Madagascar's Antananarivo Zoo. After two weeks of searching, Bomford's group sighted glowing eyes in the beam of a flashlight: "Then two eyes peered down from a branch just over the path. We had scarcely stopped when another pair flashed for a second, and then disappeared. . . . Not at all perturbed by our intrusion, the aye-ayes stared right back at us. With a graceful, liquid movement, the boldest swung beneath a branch. The other used its skinny middle finger to examine the wood. Tapping like a blind man with a walking stick, it searched for signs of life within the bark."

To board the Ark of Nosy Mangabe, it is necessary first to be a scientist or someone on official business, and second to acquire a permit from Madagascar's Department of Waters and Forests in Antananarivo. I hadn't quite understood that the permit could only be issued from the office in the capital city and that I had to present myself there in all four dimensions. Thus I spent some time in the eastern port city of Tamatave, by night sleeping in a hotel Henry Miller would have loved, by day taking the ricksha line to the provincial Waters and Forests Department office and telephoning the American Embassy in Antananarivo, asking for help. The embassy enlisted the generous support of a British scientist in Madagascar, Martin Nicoll, and at last I was given enough stamped and signed paper to leave Tamatave and go to the far northeastern corner of Madagascar, to a small town called Maroansetra.

Maroansetra began as a few cinder-block buildings and several wood houses along a single paved road, and moved outward into wooden and then bamboo and palm houses along a grid of dirt roads and paths until it ended at the edge of rain forest, river, and the Bay of Antongil. It seemed that not many foreigners passed through Maroansetra — my appearance in the town market became spontaneous entertainment for crowds of people — but the town did have a hotel, the Coco Beach Hotel, which included a palm-thatched restaurant and several comfortable palm cabins. A few foreigners were bivouacked at the hotel, including an Australian seed collector.

This seed collector was a skinny man, about my age, with ropes of veins in his arms. He had a scrawny rodentine face, a puffy black beard, and long hair. He wore a leather Stetson hat, chin-strapped to his head; his shirt was an Indonesian affair, mixing blue, green, red, and black into patterns of paramecia and flaming birds' heads. An expert practitioner of body language, the seed collector emphasized whatever he said with dips and bobs and leans, with smiles and grimaces and sweeping gestures of the hands. "I'm into both fruits and palms," he said. "Not many people are into both fruits and palms." Madagascar, he told me, has two hundred endemic palm species. "Some of them are absolutely stunning! And the Malagasy don't even know what they have!"

He had come to Maroansetra following the footsteps of the great palm seed collector Darian. Darian is the real thing — he has collected seeds hanging from a rope suspended from a helicopter. A year previously Darian had discovered a new palm, called the *Marojeja darianii,* which grew only in a tiny patch of rain forest just twenty kilometers west of Maroansetra, but he wasn't able to get good seeds, and as a result only two or three of these palms were alive outside of Madagascar. The seed collector figured the area containing this palm would be cut down or burned over soon, and he wanted to get some good seeds while the forest was still healthy and intact. "You cut down a forest, the lemurs can always go to another forest. The trees can't walk away." He had hired a guide, and the guide insisted they hire a small fishing canoe, a pirogue, to go upriver. But the weather wasn't promising, and no canoes were available. "I've got a curly one!" he said, shaking his head, distressed by his lack of progress. "It's only twenty K from here. Well, what's the bloody problem? In Queensland, where I live, twenty K is nothing. You walk. You just have to watch out for the bloody crocodiles." The last I saw of the Australian seed collector, he was running after his guide, who rode a bicycle. They thought they had found a boat.

Meanwhile I had purchased some supplies, registered with the local Waters and Forests Department office, and acquired a guide for the trip to Nosy Mangabe. My guide's name was Demondiny (like many Malagasy, he had only one name), and he chose to bring along another person, a sulky local teenager named Albert, whose function on the trip remained unclear. Early one morning someone else took us out in a speedboat to the deserted island and left us there on the beach.

We hauled our supplies along a path through forest and flowers to a couple of small wooden cabins: one a cook house, the other a sleeping cabin. The island seemed beautiful to me, close to paradise, but Demondiny and Albert appeared unimpressed. They sat in the shadows of the sleeping cabin, opened a bottle of wine, and began drinking. I tasted the wine — it was potent — and realized I had most of the day to enjoy the island by myself, to explore the forest, walk on the beach, and swim in the Indian Ocean.

The ocean felt exactly body temperature, a slippery extra skin, smooth as milk, clear as glass with a green tint. Looking out toward the horizon, I could see the forested hills and mountains of Madagascar, and between that great island and the little one I stood on, the ocean softly rolled, reflecting with glistening distortion the bright colors of sky and cloud, gleaming pale blue and dimpled white. Out to sea every once in a while, I could see the black triangle of a dorsal fin, then the rolling back of one, then two dolphins, and occasionally one leapt entirely into the air. They were eating fish, I supposed, while the ocean rolling my way curled in and over softly, collapsed into foam, and slid far up the sandy beach, then slid out again with a long-drawn hissing kiss.

The beach was a crescent of yellow quartz sand, and thousands of baby crabs the size of spiders, the color of the sand, zoomed and zithered sideways across it, periscope eyes up. Many small lizards made short decisions on the beach, scurried and stopped, scurried and stopped; bigger brown-striped lizards chuttered away from the sand into the grass whenever I approached. An electric-blue bird kept appearing out of the forest at the edge of the beach; and once, looking into the grass above the sand, I saw a large chameleon, salmon-colored with a black diamond pattern along its back, its eyes like eggshells with holes at the ends that pointed this way and that. The chameleon rolled its eggshell eyes and wavered, shifting its weight cautiously, two legs down and two legs up, alternating legs, and moving forward, back, forward, back, forward again. It suddenly climbed a stalk of grass and swallowed an insect. It turned darker, so that the pattern on its back became a woodcut print, black ink on salmon paper, then two-stepped away into the deeper grass.

The island itself was a croissant-shaped series of steep, densely forested hills, and the forest came right down and hung thickly over the beach. From the beach you could look right inside the rain forest yet be cooled by an ocean breeze.

Tree trunks and limbs hanging over the beach were so thoroughly splotched by lichen it was hard to know what color the bark itself was:

white, gray, brown, bone, ivory, bluish, pinkish, or brownish cream. The trees looked like painters' dropcloths. Vines climbed them from bottom to top and in many places spread forth broad displays of flowers. Some of the flowers came in clusters of tiny yellow and orange vases. Some reminded me of blue morning glories, ethereal blue oblong two-petaled cups, finely veined, turning into pure white and a yellowish ivory white on the inside base of the cup, where a second set of much smaller petals, folded, repeated the coloring of the larger set: labia major and minor. Others were small, bright red, and shaped like starfish, with small yellow centers. The trees leaned over the beach and dropped strange green fruits and sweet-smelling ivory petals onto the sand. One tree held a whole gaggle of brown lemurs, long-tailed, short-limbed, brown-furred. The males had white cowls of fur. Whenever I approached that tree — I called it the Lemur Tree — they would toss themselves into a denser tree away from the beach and stay in the shade of the leaves there, looking at me with cocked heads, making nervous little grunts.

When I walked away from the beach and up a trail into the forest, the sea breeze instantly disappeared, and within a few seconds I was sweating. The forest seemed like other rain forests I had seen: buttressed trees, some of them huge, amid giant bone boulders; hidden insects and frogs perpetually in noisy motion within a quietly dynamic construction of height and space, shadow and light; high up, a filtering and speckling of light, light straining into green; down low, dead tree trunks recumbent and rotten with whole gardens of new life pouring forth.

I was astonished by the sheer diversity of vegetation there. Indeed, an American botanist I met in Madagascar told me that Nosy Mangabe probably contained in its two square miles as many different plant species as exist in a typical state in the United States, about 600 to 700 species. I concentrated on leaf shapes: spoons, spatulas, and shovels; stilettos, spears, and swords; ribbony leaves, serrated leaves, deeply veined leaves; leaves shaped like dinosaurs' three-toed footprints, like five- and seven-toed footprints; leaves like pangolin scales; heart-shaped leaves; small leaves that whorled upward like a circular staircase, narrow leaves that spread densely out and down like a fountain. Many palm and fern leaves appeared in the form of great, green feathers. *Why is a leaf like a feather?* I briefly wondered, and decided that the shape nicely, with great strength, resolves a conflict between surface area and weight.

Meanwhile, back at the ranch, smoke was pouring out the door and window of the cook house. Demondiny and Albert, moving sluggishly, slurred of speech, had begun cooking dinner. We ate it, a small mountain of rice topped by a canful of hash, with a flat, undivided palm leaf serving as both communal plate and tablecloth. After dinner, Albert threw our garbage out the window, and we sat for a long while and waited for dark. Demondiny smoked cigarettes and talked with me in rough French, while Albert loudly sucked his teeth.

At last it was nearly dark, so we went down to the beach. On that island only two prosimians are nocturnal and have reflective eyes, mouse lemurs and aye-ayes, so our search began with a hunting for eyes. The technique is this: you hold a strong-beamed flashlight on top of your head or at least close to your face, so that when the light does strike a pair of mirrored eyes, the reflection will return directly to your eyes. We walked back and forth along the beach twice and along some trails in the forest, stroking branches and trunks and the insides of trees with our tube of light. Once we stood beneath a sleeping bird with an orange breast, and twice we saw glowing eyes, but they were very small and fast-moving in a confused way — mouse lemur eyes. Another time we saw briefly a pair of eyes half hidden, way back in the underbrush. "Perhaps," said Demondiny. "Perhaps. Perhaps." But when we tried to approach the eyes, they disappeared.

Then it started to rain, and soon it was pouring rain, so we returned to the cabin and prepared to sleep. The rain came pounding in waves on the tin roof and later turned into a dripping and gurgling that lasted all night and into the morning.

Half a century ago conservationists transported individuals from two endangered lemur types to Nosy Mangabe. One was a subspecies of brown lemur, whose grunting descendants I had seen in the Lemur Tree on the beach. The other was a subspecies of ruffed lemur, and on the second day I climbed to the top of the island looking for descendants of that marooned group. I found one sitting high in a tree, very furry, black and white, with a long hanging tail. It looked over its shoulder at me and made a strangling sort of call, an awk-awk-awk-cawk-cawk-caw-caw, like an inhaled, hoarse dog's bark. Then it purred a throaty purr, and made the call again. It jumped from one tree to the next two times, then stayed where it was, its big furry tail hanging down, and looked at me and called.

The second night we went out looking for aye-ayes again. At last, beyond the far end of the beach and up a trail into the forest, we shone our light into a tree and saw, at the end of a deep tunnel of vegetation, two large, bright circles. The circles slowly flattened out and disappeared — a slow blink — and opened again. They were large and didn't move very much. "Aye-aye," Demondiny said, and I thought so too. It started to rain and then to pour, however, so that was all I ever saw of the weirdest and one of the rarest of the world's prosimians, on the Ark of Nosy Mangabe.

NEARLY 1,000 MILES LONG, about 360 miles wide at its widest, and around 230,000 square miles in area — slightly smaller than Texas — Madagascar is the fourth largest island in the world. It lies like a great long ship off the eastern coast of Africa, separated from that continent by the Mozambique Channel. Topographically the island is dominated by a long central plateau of rolling hills and volcanically formed mountains. To the east the central plateau terminates abruptly with a rugged, eroded escarpment and a low coastal strip. To the west the central plateau descends more slowly, merging with a gradually sloping sedimentary basin.

Madagascar straddles the Tropic of Capricorn, so the climate is generally tropical; more particularly, though, Madagascar includes several kinds of environment within a comparatively wide climatic variation. Winds from the Indian Ocean carry considerable moisture onto the eastern edge of the island, depositing about 120 inches of rain annually onto the luxuriant eastern rain forests. The high, barren central plateau is colder and seasonal in climate, while the western side includes grassland and temperate forests, with deciduous trees and evergreens. To the south the climate is arid, and the land includes scrub and broad areas of spiny, desert-adapted plants.

Madagascar's long isolation from the African mainland has resulted in an isolated evolutionary history, and the island today beckons as a biological lost continent, a drippingly rich Atlantis of endemism. Of the island's many plant species, four-fifths are endemic, that is, native to and occurring only on the island. Of the island's tree species, more than nine tenths are endemic. Madagascar hosts 10,000 flowering plant species, most of them endemic; nearly all the island's 1,000 orchid species are found only there. Over half of the island's 250 bird types occur only there; almost half of the world's chameleon species live and twist their eyes in different directions only on the island. For the purposes of my journey, though, Madagascar beckoned as a dream theater of primate evolution, possessing as it does the great bulk of our planet's lingering primitive primates, the prosimians.

The prosimians, however, along with Madagascar's many, many other endemic plant and animal species, are now being displaced by an explosively growing human population. In 1900 the island was home to about 2.5 million people. By 1950 that number had doubled, and by 1987 it had doubled again, to 10.6 million. The population is projected to reach almost 16 million by the year 2000, and about 26 million by the year 2020 — a tenfold increase in little more than a century. Many Malagasy live on the edge of malnutrition, and their per capita income, now about $250 per year, continues to decline. The effect of that rapidly growing human population and declining personal income varies from one region to another, but essentially the human population of Madagascar grows at the expense of the island's unique flora and fauna. Madagascar's more than two dozen primate species are now highly threatened both by hunting and by deforestation.

To gain some insight into the effects of hunting on Madagascar one need look back only several hundred years — an instant of time in the biological history of that island. Archaeological studies tell us that the first human settlers arrived around 500 A.D. By 1100 most of Madagascar's recent larger animals, alive when the first humans appeared, were gone, including a dwarf hippopotamus, a type of aardvark, a carnivorous cat, two species of land tortoises, and between six and a dozen species of *ratites* — large flightless birds somewhat resembling ostriches. The largest of the ratites, the elephant bird, stood nine to ten feet tall, weighed around a thousand pounds, and laid twenty-pound eggs. It was the largest bird ever to have existed. Today we are left with samples of ratite bones, scattered carpets of broken shells, and a few intact but empty eggs. During that same early period at least fourteen prosimian species were blotted out. Most were large and ground-dwelling. Some were very large — human-sized, in fact.

Most of these animals were apparently hunted to extinction, and hunting continues to threaten the remaining lemurs. For many species hunting is a relatively minor but chronic problem, but for a few it is a major threat. The government of Madagascar is fully aware of the importance of its endangered prosimians, and all are entirely protected from hunting by law. However, the law is ineffectual. In some places people are not even aware of the law, and often it is selectively enforced. The penalties include large fines and even jail terms, but the severity of such penalties may actually contribute to a reluctant and uneven enforcement. In any event, tribal Malagasy continue to hunt lemurs for food — not usually out of necessity, but as a delicacy. Sometimes they cut down trees containing nests of the small species, then seize the animals from their nests. In other instances they capture and hunt lemurs with traps, snares, sharpened sticks, slings and stones, and blowguns. The one ground dweller, the ring-tailed lemur, is often hunted with the efficient assistance of dogs.

Guns are common enough throughout Madagascar, although gun

permits and ammunition are expensive and hard to obtain. Tribal Malagasy sometimes possess guns, but many are in poor condition and often are used only for ceremonial purposes. Gun-toting Europeans and Malagasy town dwellers who hunt for sport, on the other hand, could easily become a more serious threat. One sport hunter was recently pleased to kill twelve of the highly endangered Verreaux's sifaka in a single afternoon.

Some native tribes have traditionally believed that some of the prosimian species, including the indri, embody the souls of their ancestors. The Tanala people living in parts of the eastern rain forests maintain a taboo against killing indris, as do the Betsimisarakas. But the power of traditional beliefs and taboos is declining, and hunting seems to be an increasing threat to that highly endangered prosimian. As for the aye-aye, according to a traditional belief of the Betsimisarakas, its entrance into a village portends death. Thus many villagers will kill this extraordinary and extraordinarily rare prosimian on sight. In addition aye-ayes enter cultivated areas and eat such crops as mangoes, coconuts, and litchi fruits, so they are sometimes killed as crop pests.

The real Deluge in Madagascar, however, is a Deluge of deforestation. Slash-and-burn farming, fuelwood harvesting, commercial logging, as well as clearing for commercial farming and ranching and road building: Madagascar may have been almost entirely covered by mature forest before the first human settlers arrived, but today the great central plateau is denuded; large areas of deciduous forest to the west have been replaced by grassland; the mature rain forests of the east have been reduced by half since 1950 and now are fragmented or replaced by impoverished secondary growth. Altogether, less than a fifth of Madagascar's original forest and arid zone vegetation cover remains, and that is being destroyed very rapidly. As the forests go, so go the lemurs.

ONE DAY I ordered a bowl of soup at the Andasibe train station hotel. It turned out to be enough soup to fill three or four ordinary bowls, and it tasted bad. To take away the taste of the soup, I ordered a Coke and some solid food and was served a chicken breast (at least I thought it was chicken breast) the color of steak, sliding in grease; a heap of string beans cooked in thick oil; a heap of small, pale, greasy potatoes, cooked once upon a time but cold now; and a plastic basket of crunchy-stale bread chunks. I was hungry and tried some of everything, then made the mistake of allowing dessert to appear, which was bananas, soaked in oil and sugar, and set on fire. After that I ordered another Coke, hoping the carbonation of the drink would react with the grease in my stomach and provoke a few salutary borborygmi. It didn't. I felt overfed in an underfed country: not good. I felt plump in the midst of lean: not good. I felt sick.

Two French families on vacation — the only tourists I saw my entire time there — also happened to be in the restaurant, and as I pondered the future of my digestion, they began talking to me. Dressed for Disneyland, with very bright clothes and clean white tennis shoes, they had stopped in Andasibe, hoping perhaps to see a few lemurs before driving back to Antananarivo. What was I doing there? When I said I was looking at indris and working on a book about endangered primates, one of the men said, "You must have seen lots of the human primates on your trip." "They're not hard to find," I replied. He said, "I think they're the most important to protect, no? This is a sanctuary, is it not?"

It took a few seconds to puzzle through his meaning, but after I had, my dyspepsia moved from stomach to brain, where it registered as anger. He was suggesting — "This is a sanctuary, is it not?" — that somehow the indris were all taken care of, snug in their little fur family homes, when in fact Andasibe is barely protected and trivially small, a besieged forest slightly larger than three square miles. It must share with Madagascar's approximately three dozen other parks and reserves a total annual government budget of $1,000.

He had, moreover, played poorly an old tune I had already heard too often: the idea that humans and wildlife are enemies, opposites, that you save one at the expense of the other — the idea, to chase it right down to its absurd conclusion, that the Traffic Girl was crippled and crawling in the streets of Antananarivo because too many lemurs were still healthy and hopping in the trees of Andasibe.

I recalled the words of another person in another place, who said, during a larger conversation expressing her self-preoccupation and limited sympathies, her yawning lack of curiosity about the world, "I think humans are the real endangered species on earth," and then smiled to let me know she had just said something clever. But if humans are an endangered species, in any reasonable sense of the term, we are unique in the floral and faunal world for at least two reasons. First, we are the only species endangered by its own activity. Second, we are the only endangered species so astonishingly prolific and numerous, supposedly endangered and yet dominating most environments of the planet.

Perhaps humans ultimately are threatened with extinction, but if so, we will be the last of the primates to go. Just as the faces and bodies of primates in the forest sometimes mirror the human face and body, so the ecology of primates reflects the human ecology — and the fate of the nonhuman primates in this world may foreshadow the fate of humans. Without realizing it, we may be conducting the largest medical experiment of all, with nonhuman primates as research subjects, on the consequences of depleted natural environments. It is certain, in any case, that in Madagascar the health of primate environments declines in concert with the health of human environments, for they are inextricably interconnected. In the words of Joseph Randrianasolo, minister of livestock, fisheries, and forests in Madagascar: *"Tsy misy ala, tsy misy rano, tsy misy vary!"* If there is no more forest, then no more water, and no more rice!

In a larger sense the health of the world's tropical forests could sustain the health of humankind, since the tropical forest genetic reservoirs contain a great many naturally occurring pharmaceuticals.

The scientific search for natural pharmaceuticals now concentrates on a particular class of plant-produced compounds known as the *alkaloids*, which proliferate within tropical forests. Most plant biochemistry is involved in the rather standard business of processing proteins, fats, and carbohydrates. But some plants also produce more unusual, often complex compounds, which biochemists label the secondary plant products. Many of these products that happen to contain one or more nitrogen atoms in their molecular structure — the so-called alkaloids — have proven to be pharmacologically active.

Take curare, for instance. Curare has long been used by many Amazonian Indian tribes as a hunting and warfare poison. Darts and arrows are dipped into the prepared curare brew (which may contain two or three dozen other ingredients, including snake venom, poisonous ants, and some vegetable substances to provide stickiness) before being fired. Through even a minor scratch in the skin, curare can enter the bloodstream and cause a painless but progressive paralysis, followed, as the muscles controlling breathing become paralyzed, by death.

The possible medical uses of curare became apparent not long after Charles Waterton returned to England from South America with a supply of the poison in 1812 and began experimenting. Injecting the drug into animals as small as a chicken and as large as an ox, Waterton watched them all die, without any apparent pain, in five to twenty-five minutes. Waterton's most interesting test occurred when he injected curare into a donkey. The donkey soon collapsed and lost consciousness, but Waterton cut a hole in its windpipe and forced breathing artificially with a bellows. That procedure revived the animal, and it survived. The experiment suggested that asphyxiation was the cause of death, and curare's potential as an extraordinary muscle relaxant soon became apparent.

Curare was first given to patients who exhibited muscle rigidity or spasms: people suffering from acute tetanus or lockjaw, for example, as well as victims of strychnine poisoning, epilepsy, infantile paralysis, nervous tics, and so on. More recently the poison has been used in the treatment of spastic paralysis, multiple sclerosis, hereditary Saint Vitus' dance, and Parkinson's disease. It has also worked well in reducing the incidence of spine and skull fractures and broken arms and legs caused by muscle spasms during electroshock treatment of psychiatric patients. And it is effective as a muscle relaxant during certain types of abdominal, rectal, and eye surgery, and tonsillectomies.

Another traditional tropical forest poison, tetrodotoxin, squeezed out of a few sorts of Central American frog species as well as some tropical marine fish species, has been found to possess 160,000 times the ability of cocaine (another tropical forest alkaloid) to block nerve impulses. Tetrodotoxin has great potential use as an anesthetic, and in fact it is used in Japan as a local anesthetic and general painkiller for patients with terminal cancer and neurogenic leprosy. Tetrodotoxin may be even more important as an experimental drug for the study of nerve impulse transmission. And from West African forests a third arrow poison, ouabain, is now used to stimulate the heart during medical emergencies. The nineteenth-century British explorer David Livingstone is credited with the Western discovery of that particular alkaloid.

Beyond the few traditional alkaloid poisons, scientists are looking at a very large number of alkaloid-based medicines that healers in the tropics have prescribed for years, often for centuries. Amoebic dysentery, a severe inflammation of the intestines caused by infestation of one particular amoeba, has long been a killer. Yet Amazonian Indians have long known that roots of the ipecac plant contained a cure in the form of an alkaloid we call emetine. The earliest European explorers brought back ipecac roots, enabling Louis XIV's personal physician to cure the French king's amoebic dysentery in 1682; today the illness is still treated with synthetic emetine, along with antibiotics.

Reserpine is the name given to one alkaloid compound of dozens that have been isolated from roots of the Indian snakeroot plant, *Rauwolfia serpentina.* Species of the *Rauwolfia* genus grow in tropical regions throughout the world, and native peoples in many widely separated areas have traditionally used these plants to treat snake-bite and other forms of poisoning, to tranquilize the anxious or even the actively insane, and as an emetic and purgative. In Liberia and Mexico, *Rauwolfia* extracts are recommended for eye and skin diseases. African and South American healers have also prescribed the plant for fevers.

Thirty centuries of common use as a near panacea in India might have alerted Western physicians early on to this compound's potency. But most Western physicians and scientists have been rather arrogantly contemptuous of folk remedies; only since 1949, when Dr. Rustom Jal Vakil of Bombay reported his own well-controlled research on *Rauwolfia* in the *British Heart Journal,* did non-Indian physicians at last take notice. One of them, Dr. Robert M. Wilkins, director of the Hypertension Clinic at the Massachusetts Memorial Hospital, experimented with *Rauwolfia* extract on fifty hypertension patients and found that it lowered blood pressure dramatically and functioned as a powerful tranquilizer without the side effects, such as drowsiness and stupor, associated with the sedative then commonly used. Soon after Dr. Wilkins's enthusiastic report, psychiatrists also began experimenting with *Rauwolfia* extracts. By 1953 many psychiatric journal articles were reporting its great effectiveness in the calming of psychiatric patients, and it was soon commonly prescribed for patients suffering from schizophrenia, mania, and chronic alcoholism.

The alkaloid reserpine was finally isolated from *Rauwolfia* by chemists working for Ciba in Switzerland, who found it to be 10,000 times more potent than the raw plant. Today synthetic and naturally extracted reserpine supports a retail industry valued at a quarter of a billion dollars per year; it has undoubtedly saved millions of people from strokes, heart failure, and kidney failure associated with hypertension.

Yet another compound from the tropical forests has had a major impact on Western medicine: quinine, an alkaloid isolated from the bark of the beautiful flowering cinchona tree that grows wild in the South American Andes. The first indication of the bark's usefulness in treating malaria was provided to Europeans in 1633 by a Jesuit missionary living in Ecuador, Father Calancha. A decade later, another Jesuit brought samples of cinchona bark to Rome, which was then not only the center of Catholicism but one of the great centers of malaria. From Rome the "Jesuit bark" cure was soon dispersed throughout Europe. Cardinal John de Lugo used the bark to treat the young Louis XIV for malaria, but since the treatment was associated with Catholicism and traded mainly by Jesuits, many Protestants refused to consider it. England's Oliver Cromwell, for instance, thought the Jesuit bark superstitious nonsense and died of malaria in 1658.

By 1820 two French pharmacists had isolated an important alkaloid of the cinchona bark, which they called quinine. That was useful, for among other things the raw bark was bitter and tended to produce gastric reactions. At that time Bolivia, Peru, Colombia, and Ecuador, enjoyed a monopoly on the critical substance, and both the English and the Dutch began attempting to smuggle out of South America cinchona seeds and seedlings. Finally, around midcentury, an Andean Indian by the name of Manuel Incra Mamaní passed on some cinchona seeds to a British trader, Charles Ledger, who sent them to his brother in England. The British government foolishly turned down an offer for the seeds; but when the Dutch finally tried them out in Java, they discovered that the species, named *cinchona ledgeriana* in honor of Ledger, produced a much higher concentration of quinine than any other known species. Soon, the Dutch were controlling nine tenths of the world quinine trade.

Quinine enriched the Dutch colonial empire, but the importance of this particular medicine became most apparent during World War II. At the start of the war, the world's major stocks of processed quinine lay in Amsterdam, and the Germans took it. The world's major supply of raw quinine grew in the Dutch plantations of Java and American-controlled plantations in the Philippines. (The Americans had earlier stolen some *Cinchona ledgeriana* seeds and seedlings from Java.) But in 1942 the Japanese moved into Southeast Asia and quickly took control of those plantations.

The last American B-17 Flying Fortress to depart from the Philippines in 1942 carried a supply of cinchona seeds and seedlings, which were germinated and nurtured in the U.S. Department of Agriculture laboratories at Glen Dale, Maryland; ultimately 4 million seedlings from this supply were sent out to various plantations in Latin America and elsewhere. Yet those young plantations would not produce quinine-laden bark for several years. In desperation, Americans journeyed to the Andes to gather bark from wild cinchona trees. They returned with 12.5 million pounds of bark: not enough. Botanists were put to work screening another 14,000 naturally occurring compounds that might have potential for control of malaria. At the same time scientists were using the chemical structure of natural quinine as a blueprint in developing synthetic antimalarial drugs. Finally, a series of synthetic antimalarial compounds eased the crisis and carried the United States and its allies through the rest of the war.

*

Like quinine, most naturally occurring medicines, isolated and analyzed in the laboratory, can be synthesized. And in some cases, synthesizing drugs from blueprints provided by natural compounds is sensible and cost-effective. Aspirin, once derived from European willow leaves, is today manufactured more economically by synthesis. Should we now throw away the willow leaves? Emetine, traditionally extracted from ipecac, today is synthesized and sold in drugstores as a powerful treatment for amoebic dysentery. Can we now throw away the ipecac plant?

Most natural compounds can be synthesized. Yet of the seventy-six prescription drugs sold in the United States in 1973 that were based on the chemistry of higher plants, only seven were purely synthetic. The remainder, over 90 percent, included direct extractions from plants. The chemical structures of most natural drugs are very complex, and simple extraction is usually less expensive than synthesis. In the case of pharmaceuticals, moreover, synthetics are stable, precisely defined barriers against disease, whereas disease-causing organisms are inherently changeable. Like crop-damaging insects, disease organisms can mutate around chemical barriers. Thus, even if synthetic versions of an important drug exist, the high genetic variability of wild-growing pharmaceutical plants remains a critical safeguard against the variability of disease organisms.

For instance, quinine. By the 1960s, as the United States began to occupy itself with a major war in Southeast Asia, some strains of the malarial parasite in that part of the world had developed significant resistance to the usual synthetic versions of quinine. One strain had expanded its resistance to synthetics by a factor of 4,000, but had increased its resistance to natural quinine only twofold. Wild-growing cinchona trees seemed once again important weapons for the war against an old, protean enemy.

Two more superstar alkaloids have lately been found in the tropical forests, both of them produced by a small, pretty flowering plant growing only in Madagascar. The rosy periwinkle yields altogether some sixty different alkaloidal compounds and was once screened by the U.S. National Cancer Institute as a possible source for anticancer drugs. The National Cancer Institute's analysis, however, failed to turn up any useful material, and the plant would have remained on the discard heap forever had not the claims of traditional healers led to further reviews at the University of Western Ontario and at the Eli Lilly company of Indianapolis. Further reviews eventually isolated two important alkaloids, called vincristine and vinblastine, which now provide potent treatments for Hodgkin's disease and other malignant lymphomas; for breast, cervical, and testicular cancers; for choriocarcinoma; and for some cancerous tumors. Thanks in part to these two drugs, patients with Hodgkin's disease now have a 58 percent chance of surviving ten years after treatment, as opposed to the 2 percent chance they used to have. Additionally, vincristine has helped alter the survival rate of acute childhood leukemia from one in five to four in five. Vincristine and vinblastine provide the world pharmaceutical industry with a $90 million yearly income.

How many other powerful anticancer drugs are yet to be discovered in the tropical forests? Latin American forests alone contain roughly 90,000 different plants, of which only 10,000 have been examined for possible use as anticancer compounds. The world should have the opportunity to consider the remaining 80,000. The search for anticancer compounds continues, and at least two other tropical plants already show excellent potential. The Indian mandrake provides a substance which, when modified slightly, yields two promising possible anticancer agents named VM-26 and VP-16. The *Brucea* genus of African plants, yielding the alkaloid bruceantin, is also considered very promising by cancer researchers.

Almost as exciting as the search for anticancer alkaloids is the search for natural birth control compounds. We know they are there. A nut from the greenheart tree of Amazonia, for example, is already commonly used by forest-dwelling people of Guyana as an oral contraceptive. Worldwide, about 4,000 plant species have been identified as possible sources for contraceptive drugs; and the tropical plant genus *Dioscorea* has already helped provide contraception for hundreds of millions of women around the world. With some minor chemical manipulations, some *Dioscorea* species yield an alkaloid known as diosgenin, which until recently was crucial to the manufacture of birth control pills. These plants also contain cortisone and hydrocortisone for the treatment of rheumatoid arthritis, rheumatic fever, Addison's disease, and certain allergies and skin ailments. Diosgenin is still used in the manufacture of 95 percent of all steroid drugs sold today; and steroid drugs, as of 1973, were included in one seventh of all prescriptions filled in United States pharmacies. In 1977 the wholesale diosgenin market was valued at $10 to $25 million, based mostly on two *Dioscorea* species taken from Mexican forests, both of them wild climbing vines called Mexican yams.

Indians of northwestern Amazonia use at least 1,300 plant species for various medical purposes. Forest-dwelling peoples of Southeast Asia use around 6,500 different plants for the treatment of many diseases, including malaria, syphilis, and stomach ulcers. In India 2,500 of the 18,000 flowering plants provide medicines. Most of these traditional medicines of the tropical forests have yet to be analyzed or tested by scientists, but clearly they represent some huge value for the millions of people in the tropics who already rely on them, and some vast untapped medical treasure for the rest of us. Considering that half the world's commercially marketed pharmaceuticals are based on plants, as are about one fourth of the pharmaceuticals sold in the United States, it should not be surprising that today medicines based on tropical plants and animals amount to a $20 billion a year business, worldwide. By what proportion could intelligent exploitation of tropical forest plants improve world medicine in both tropical and temperate locales? What is going down the drain as the world's tropical forests disappear before we know more about what they contain? What is the simple dollar value that may be lost? What is the health value we might lose?

IT HAD BEEN so constantly wet on Nosy Mangabe and in Maroansetra that washing clothes seemed futile — nothing would dry. When I eventually returned to Antananarivo I was carrying several extra ounces of drying mud and sweat, so I took a room in a moderately priced hotel, luxurious by Madagascar standards, where I could have a hot bath in private and wash my clothes.

The hotel was named after a finance minister of France's Louis XIV, but after I passed through the desperate crowd of beggars and flower sellers outside the entrance, I began to think of it as the Hotel Marie Antoinette.

I had eaten almost nothing all day, so when the hotel restaurant opened at seven that evening, I was the first to take a table. A recorded Scott Joplin tune, played tremulously on a trombone, tumbled gently through the air. A soft blue carpet covered the floor. The walls were paneled with antique wood and red velvet and hung with tasteful French landscape miniatures. Each table was covered with a red cloth and decorated with a single red rose in a silver vase; the roses had been finely sprayed with water to look fresh and dew-dropped. I sat in an antique-looking red velvet chair and felt distinctly out of place in my rough clothes and still muddy boots. On my table were a folded red napkin, three glasses, and more silverware than I knew what to do with: four forks, three knives, and two spoons. The sweet melancholy of Scott Joplin proved irresistible, though, and I soon began to feel homesick.

I was also hungry, so when a waiter appeared, dressed in red velvet, I ordered steak. Only it was steak in a fashion I didn't understand, but thought would be solid and well cooked: steak tartare.

I was even hungrier when the red-velveted waiter finally appeared with a large silver platter supporting two plates and a bowl. The bowl contained something that looked like raw egg; one of the plates was empty, and the other contained three piles, one of them raw, chopped meat, the other two finely chopped onions and some other chopped vegetation. Two waiters dressed in white took up positions behind the red waiter and occasionally removed and deposited plates and glasses, while he, with great drama and deftness, holding two spoons in one hand, mixed the raw meat with everything else, and finally spread it all out onto the clean plate. It looked exactly like a huge, sloppy, raw hamburger patty. Some cooked potatoes and a sprig of green were added to the plate before it was ostentatiously placed in front of me. Wine was poured into my glass. Was there anything else I wished? Of course, I wished for my hamburger to be cooked. But hunger and fatigue, the muddiness of my shoes, the roughness of my clothes, the inadequacy of my French, the elegance of the surroundings, the aplomb of the three waiters, and the redundant silverware all conspired to intimidate me. I said nothing, ate that huge, sloppy, raw hamburger, every bit, and even took dessert. The bill was presented in a finely carved wooden box, and I paid it and left.

Early the next morning I checked out of the Hotel Marie Antoinette, walked through the crowds of beggars and flower sellers, and climbed into a taxi. Flower sellers called through the windows as we drove away. The taxi descended a winding street into the center of town, and then I realized that the taxi had turned onto the Traffic Girl's street. When we stopped at a red light, the Traffic Girl left her spot on the curb and crawled right out in front of us. I couldn't see her face or the little hand I knew was stretched up, but the taxi driver leaned out his window and yelled something at her in Malagash. The light changed to green, and we turned onto the town's main boulevard. There had been a riot in Antananarivo while I was gone, and now, as we drove along the boulevard, I noticed that many plate glass windows on this street were shattered and boarded up. At last the taxi arrived at my destination, Hotel Terminus, right next to the train terminus.

Hotel Terminus was cheap yet tolerably clean, but when I lay down on the orange-covered bed, looked up at the dripping pink walls and cracked white ceiling, and realized I was suddenly feeling terribly sick, I began to think of it as the Hotel Terminal.

Perhaps the problem had grown out of yesterday's raw hamburger. For whatever reason, my insides wanted out while my outsides resisted with shivers, trembles, and sweat. Later that day I remembered that I had in my pack some medicine for just such an occasion, so I took it, and my insides became cement. The next day I felt well enough to go to the city zoo and look at lemurs, but two days later the cement began disintegrating, and I felt even worse, much worse, than before. I was incapacitated, weak, in pain, barely able to move: down and out, out and down. I had, in short, lots of time to think, and, among other things, I thought I had discovered an important difference between lemurs and people. Lemurs, precarious in the trees, are either healthy or dead, while humans, supported on the ground by a social contract, have a third option: they can be sick for a long time. In any case, the internal explosions continued for about a week, until my body finally healed itself.

On the plane out of Madagascar, I sat next to a tall, distinguished-looking American economist, who was conducting a World Bank study comparing the recent economic history of Madagascar with that of another African country. He told me that when he had been in Madagascar three years previously, he hadn't seen any beggars. But Madagascar's per capita income has been declining by about 2 percent each year for the last fifteen years. "Even more disturbing," he said, "there has been a gradual increase in the inequities of distribution. And you're getting that from the horse's mouth, since I'm the only person who has so far calculated distribution, and I haven't published it yet." I mentioned that the decline in per capita income closely paralleled the rise in capitas: population growth. But he dismissed the issue of runaway population growth and pointed instead to poor planning: a $100 million fertilizer factory based on an outmoded technology requiring raw materials that cost more than the final product brings on the open market, a tannery built near an existing tannery that was not working to capacity, a flour mill built too far away from the wheat, and so on.

He too was interested in primates. He had once written a paper comparing the economics of chimpanzees with the economics of baboons. "Primates are a marvelous mirror; they're us," he said with some enthusiasm. And we agreed that in a healthy economy, the fate of humans and animals is intertwined. But in the desperation engendered by an unhealthy economy, he thought, humans and animals become enemies. He wondered if there might come a time when the people of Antananarivo would scale the walls of the city zoo and eat the lemurs. "In the downward spiral," he pronounced, "animals are at the bottom."

Eight

WATER

IT WAS JULY, summer monsoon season, when I arrived in Bombay, and I was not surprised to walk into a night of falling water and old men crouched against walls. I had imagined that July would be the wettest month of the year, but as I traveled south and east from Bombay, July became a cruel month. The monsoon had failed, I was told, and by the time I reached the southeastern city of Madras there was no rain at all, only sun and unbearable heat. Madras had begun rationing water and now turned on its spigots only every other day.

I continued south across a great arid plain dotted with clay villages and stone shrines, and eventually came to the holy city of Madurai, in the state of Tamil Nadu. Madurai, a city of a million people, built around a massive, ancient Hindu temple, remains one of the great southern pilgrimage sites for devout Hindus. And I, weak from the heat, entered the crowded city, drank tea and soft drinks, and presented myself at the Tamil Nadu Hotel, hoping for a cool room and a good night's sleep.

The man at the desk had one bad eye, a pearl in a withered socket, and his good eye never moved in my direction. I spoke to him, but he knew no English, so finally I cornered the hotel manager and asked for a room with air conditioning. The manager's fingernails and toenails were painted pink, and he loosely waggled his head *no* when he meant *yes,* and spoke a demotic English I could barely understand, but gradually I made my wishes known. I wanted a room with air conditioning, since I was not used to the heat and needed to sleep. They had no rooms with air conditioning, he told me, and anyway, an air-conditioned room was twice as expensive as a regular room, the rupee equivalent of two dollars instead of one. If they didn't have such a room, I told him, then I would have to find another hotel. But, yes, they did have a room with air conditioning after all, and it would be available soon, if I would just sit down and wait a few minutes.

I sat and waited, and then was shown a room that could be made cool if you shuttered all the openings and turned the machine up full blast. I had dinner and, when the evening turned violet, went to bed. Not to sleep, however, because right outside my shuttered window an engine roared and a power hammer knocked metal against metal. I unshuttered my window and looked out. In the hotel's back yard several turbaned men stood around a giant diesel engine and air compressor. The engine and compressor, mounted on the bed of a truck, drove a vertical shaft that screwed a round toothed bit into the earth, seeking water. The engine revved and compressed enough air to hammer, while with each hammered turn and drive, a ratchet clacked into place. The evening's noise seemed as oppressively loud as the day's sun had been hot, so I gave up trying to sleep for the moment and opened a book, waiting for the water drillers to finish. I finished the book first, though, and since it was getting late, I went downstairs to the hotel desk and asked the manager with the pink fingernails whether the drilling would stop soon. He waggled his head *no,* which meant *yes,* and assured me they would be done by ten o'clock that evening — in half an hour. "Thank you," I said. "Denko," he replied. The machine finally fell silent at three in the morning, when I finally crawled under a sheet and went to sleep. I slept until the engine roared up and the hammering began again, three hours later.

Over 1,500 languages and dialects are spoken in India. Sixteen languages are formally recognized by the constitution, including the official Hindi and English, but in Tamil Nadu almost everyone spoke Tamil as their native language, and I met very few who appeared to speak more than a few words of English. There was also a paucity of English language signs, so finding my way south from Madurai became a puzzle. The solution ultimately was to stand in the middle of a crowded bus terminal and repeat the name of the city I was bound for: Tirunelveli. "Tirunelveli? Tirunelveli?" I asked a group of people. No one understood, until I repeated it several more times. Finally, my mysterious pronunciation sorted itself out in someone's mind: "Ah, Tirunelveli!"

The bus had no glass in the windows, and the driver, sitting in a wicker chair beneath the picture of a strange Hindu deity, was a pock-faced, lead-footed demon, who would rumble up behind a pedestrian, a bicyclist, or a slow-moving bullock cart, then madly crush the rubber

Andrew L. Young

MURIQUI OF SOUTHEASTERN BRAZIL

I watched a muriqui reach from the branch of one tree to the long, hanging branch of another. In a single easy motion, he transferred his weight, holding on with an arm, two legs, and his tail.

Russell A. Mittermeier

GOLDEN LION TAMARIN OF SOUTHEASTERN BRAZIL

They looked like the miraculous offspring of a squirrel and a bird, scrambling here and there, then suddenly casting off in superb leaps from one branch to another in near-flight.

Georgina Dasilva

WESTERN BLACK-AND-WHITE COLOBUS OF WEST AFRICA

When at last I was able to get a good look, I was impressed by their tails, which hang like thick white ropes with thicker white tassels at the ends.

Robert P Carr/Bruce Coleman Inc

EASTERN BLACK-AND-WHITE COLOBUS OF EAST AFRICA

As I watched, my fantasy engine slipped into high gear, and I began to see them as a tribe of glum clergymen dressed in clerical black and white.

Opposite: SOUTHERN BEARDED SAKI OF THE BRAZILIAN AMAZON

I heard a high-pitched whistle, then another, and another. I looked up to see a dark, thick-tailed monkey the size of a domestic cat, about twenty feet above me.

Russell A. Mittermeier

Frans Lanting

INDRI OF MADAGASCAR

The first thing I noticed was the size of their legs, which were long and powerful; with their knees extended up and loosely apart, the indris reminded me of crew oarsmen.

Frans Lanting

AYE-AYE OF MADAGASCAR

We shone our light into a tree and saw two large, bright circles. The circles flattened and disappeared—a slow blink—and opened again. "Aye-aye," Demondiny said.

Opposite: MOUNTAIN GORILLAS OF CENTRAL AFRICA

A female gorilla sat facing us, potbellied and dull-eyed, apparently not interested in us. And then from behind her appeared a toddler gorilla with a yellow root in its mouth.

Peter G. Veit

Gerry Ellis/Ellis Wildlife Collection

LION-TAILED MACAQUE OF SOUTHERN INDIA

Their faces seemed distinctive, individual, and I spent a long time considering one solemn gray face that looked like an old man's face.

© Richard Tenaza

YOUNG BILOU OF THE MENTAWAI ISLANDS

At last I saw clearly a small black gibbon, illuminated from behind by a clear morning sky, hurling itself through the trees hand over hand, swift and straight.

Opposite: ADULT MALE ORANGUTAN OF SUMATRA

He seemed very big, peering out at the feeding animals from his shadowy vantage point: a mournful voyeur, an aging playboy spying on a high school prom.

Wardene Weisser/Bruce Coleman Inc.

Roderic B. Mast, World Wildlife Fund

COTTON-TOP TAMARIN OF COLOMBIA

Their faces seemed remarkably expressive, particularly given that plume of pure white hair, a pompadour rising at the forehead, reminiscent of the dramatic hairdo of composer Franz Liszt.

Wildlife Education

DOUC LANGUR OF INDOCHINA

I noticed once again the yellow, leathery quality of their faces, those impossibly large almond eyes, the bright patches and bands of color on their fur.

bulb of his ooga horn and swerve into the center of the road, whether or not a vehicle was coming in the opposite direction. Pedestrians, bicyclists, bullock carts, even cars and trucks, from both directions, were forced to scamper, scurry, or drive onto the rubble at the edge of the road while our shadowed steel leviathan roared on its way.

We roared south into a land beaten by the sun, hazy in the distance, shadow-crossed by moving clouds the color of sand. It was a shriveled land, where rivers were arroyo dreams and bridges ran over dust, where rectangular rice paddies were broken into crazed chips of pale clay. It was a hard-earth land, where houses and villages were built of the material they stood on — rock and clay, sand and dust. Houses, indeed whole villages, rose and fell, crumbled and were extended even as we roared along. Here a rock wall slid halfway down toward the rock it was made from; a clay house disintegrated into clay rubble on top of clay; a brick house dissolved into brick dust. There a white-haired old man, half wrapped in a red cloth, broke granite boulders with a small hammer and created heaps of gravel for construction; a woman patted wet clay on top of dry, reinforcing a clay fence; three brick makers stood beside a pyramid of brick dust and stacked new bricks into neat new piles.

In villages and outside of them, dark and ancient stone effigies slouched within shrines. Sculpted clay bodies and body parts, animal and human, remarkable avatars, had been conglomerated into small and large temples, painted garishly in hot colors outside, hallowing cool mysteries inside. The shrines and temples seemed ancient, a hundred or a thousand years old, yet they looked hardly older than the houses and villages. Past extended, present attenuated: the turbaned peasant piling mud chips into a wood wheelbarrow, the old woman wrapped in liquescent color and carrying a copper water jar under her arm, the bullock cart driver urging motion into moronic beasts, appeared then, to my heated and impressionable mind, flimsy shadows of a fevered present toiling before a tyranny of the past.

We roared past a few crooked trees whose roots clutched the hot earth, while weary people sat in the trees' crooked shadows. And in the distance, as we traveled south, palm trees emerged, sometimes in rows or clusters, swaying tall and slender with tufts of green on top, like long-handled brushes, the tufts rippled and whipped by a harsh, dry wind. But the sun was hot as iron, and the air tasted of dust.

We came at last to the southern city of Tirunelveli, and I dashed away from the sun and heat and took a room in a hotel.

Yet there was water in Tirunelveli. The city straddled an old brown river. That night there was even an expectation of rain. The air turned briefly humid and heavy, rolled through open windows like cream, and I heard from a distance the reverberation of thunder. In the morning I looked out my window to see a hazy sky and circling birds, and then I walked along a city street to the bridge over the river. Looking east from the bridge, I saw crowds of people walking into the water, soaping themselves and the clothes they wore, washing and combing and braiding their hair, or smacking cloth onto flat rocks, soaping it, rinsing, and smacking again. Beyond the people a few cattle also walked into the water, which flowed slowly around their legs. Looking west, I saw more people bathing and washing, and four columned stone buildings, tumbledown Parthenons, situated at the flat and rocky river's edge, providing, I thought, sanctuary from the sun for people who had come to the water. The river itself flowed slowly toward me, wrinkled brown, and glittered diamonds from the rising sun, while far beyond, above the river and its arid plain, far into the distance toward the river's source, a haze in the sky smeared itself across high, lumpy hills and made them smoky purple.

Those were the Ashambu Hills, southern foothills of the Western Ghats, and I had come south to enter the forests there and find India's rarest and most endangered monkey, the lion-tailed macaque.

I knew I would recognize the lion-tailed macaque by its sleek black coat and full, nearly white ruff and mane surrounding a bare black face. I knew also I would recognize it by its tail: for all adult males and some adult females, the short tail terminates in a leonine tuft, in honor of which this animal is named.

The lion-tail is a medium-sized member of the macaque genus. An average male of the species measures almost twenty-two inches in body length and weighs seventeen or eighteen pounds; females are smaller. The genus originated in the Mediterranean region and spread into Southeast Asia about three million years ago, eventually diversifying into the twelve to seventeen species we recognize today. Macaques of all species live in relatively large social groups or *troops,* which include adult males and females, juveniles, and infants. Females form the permanent core of the group, while males are likely to leave at least once in their lives, either to join another group or to live alone for some period. All macaque troops are organized in dominance hierarchies, largely determined by the physical power and aggressiveness of individuals, and defining the individual's access to life's basics: food, space, and sex.

In some macaque species the hierarchy of dominance is established somewhat peaceably during the juvenile period, when the young animals test each other's strength and nerve in play wrestling. In other species the struggle for dominance is more extended and serious. Male toque macaques of Sri Lanka, for example, begin competing with each other and with adults at around four years of age, even though they do not reach full adult size until the age of seven. The competition may take the form of both physical and psychological warfare, and the weaker and smaller males survive only by acting submissive, lurking on the periphery of the group, eating leftovers, and generally avoiding any serious encounter with larger males. Female toque macaques are severely harassed by older males and females even when they are very young and begin to acquire a position of status and power only at adolescence. As a result of this fierce dominance competition, around 90 percent of the male and perhaps 85 percent of the female toque macaques never live to adulthood — the competition continuously selects for survival only the most powerful and aggressive individuals. Barbary macaques of North Africa and Gibraltar also struggle to define their position in a dominance hierarchy, but the adults, both male and female, are innately protective of their young. The young are not severely harassed, while subordinate adult males sometimes defend themselves with the neat trick of taking and presenting an infant to the dominant male in a gesture of appeasement.

All macaques are opportunistic omnivores. Various species feed on fruits, flowers, insects, eggs, crabs, tree frogs, and so on. They are generally adaptable in other ways. Many species move and forage well both on the ground and in the trees, and most will travel considerable distances on the ground. Various macaque species inhabit open, lightly wooded land, secondary forest, forest periphery, cliffs, and rocky coastal areas. In India the rhesus and bonnet macaques have survived the recent destruction of forests by learning to live in and near human settlements; for them, garden and garbage can provide. Lion-tailed macaques, however, depend almost entirely upon undisturbed forest, although on occasion they move into lightly logged areas. They are the only truly arboreal macaques, probably spending no more than one percent of their time on the ground.

Yukimaru Sugiyama of Kyoto University carried out the first serious field study of lion-tailed macaques in the early 1960s in a rugged, forested hill region of southern India's Kerala State. The area was, he recalls, "frequently disturbed by wild elephants, snakes and leeches." Over a total of about four months, Sugiyama followed the progress of a few solitary males and two troops, of sixteen and twenty-two animals.

In both of these troops the adult male-female ratio was roughly one to three. Both troops were apparently led by large adult males. When a troop traveled on the ground, one adult male always traveled well ahead of the rest. When traveling fast, adult males took positions both at the head and rear of the troop.

Both troops foraged in ranges just under a square mile, and their ranges overlapped. (A later, longer-term study found that lion-tails, following seasonal variations in food supply, will cover up to two square miles or more of optimal habitat in a full year.) When both groups fed in the overlapping area, which contained several excellent food trees, the adult males of each group broadcast threats with a "whooping display." Although Sugiyama observed no direct fighting between the two groups, one appeared to be dominant: the other group would always retreat when threatened. He found the whooping threat to be similar enough to human vocalizations that he was able to provoke a male lion-tailed macaque by imitating the call while hiding behind a tree. Sugiyama distinguished nine other calls, including a contented "muttering or murmuring call" that individuals produced when they were feeding peacefully or resting or dispersed. Sugiyama was able to imitate that sound as well. For nearly an hour he called, and individual animals responded in turn.

Lion-tailed macaques eat fruits, flowers, and leaves from every level of the forest. They also pluck out insects and larvae, lizards, tree frogs, and fungi from beneath bark or from inside rotten logs. During rainy periods they drink water from pockets in tree forks or lick water droplets off leaves; in dry periods they will sometimes descend to the ground and drink directly from streams or rivers.

Like most other primates, lion-tailed macaques bear only a single offspring at a time. The young reach sexual maturity at about five years for females, eight for males — and it is likely that each adult female bears only two or three offspring in her lifetime. Such limited reproduction, once sufficient, now combines with the pressures of illegal hunting and continuing habitat destruction to place this handsome, unique primate on the verge of extinction. It is gone from the northernmost portion of its original range and now lives only in fragmented islands of forest in the Western Ghats of southern India — within the state of Karnataka to the north and straddling the border of Kerala and Tamil Nadu states to the south.

Sugiyama believed that at the time of his study, in the early 1960s, around 1,000 of these animals remained. A thorough study in 1975 covering the entire range placed the surviving population at 405 individuals altogether — a disastrously low number. Since then the population may have increased in some portions of well-protected habitat. That increase, along with increasingly accurate surveys in Kerala and Tamil Nadu, led to a general consensus, at the 1982 Lion-tailed Macaque Symposium in Baltimore, that 915 to 2,000 of the animals remained.* A single catastrophe, such as drought or disease, could push the lion-tailed macaque beyond return.

I had come to India with written permission from Tamil Nadu State's chief conservator of forests to enter the Kalakad Sanctuary, the Ark of the lion-tailed macaque, in the Ashambu Hills. But it still took me a full day in Tirunelveli to meet with the wildlife warden for Kalakad, get his formal permission, arrange for a guide, and buy food and supplies.

English on signs and tongues had all but disappeared by then, so I was grateful, the next morning, to be introduced to my guide, who told me his name was Anbalagon. Anbalagon was very slender. He had a narrow face, a long jaw, and a big mustache; his black hair was cut short around the sides, but it swept across his forehead in a full wave. He wore khaki pants, a khaki shirt with shoulder epaulets, and yellow flip-flops. He carried himself stiffly, in the manner of a newly recruited soldier, and had very little to say — he knew only a few words of English. But he seemed to know more or less what I was seeking.

Anbalagon directed me onto a bus, where we sat and waited. After a while he decided we were on the wrong bus and led me onto another, where we waited some more. As this bus slowly filled up with passengers and their possessions, beggars appeared. A white-haired old man with a full, yellow-white beard walked beneath my window and greeted me solemnly, hands together at the forehead in a praying salute. He wanted a coin. A little girl climbed onto the bus, dragging along her tiny brother, and both walked from the front of the bus to the back. They wanted coins. A young girl, perhaps seven or eight years old,

*Those numbers were mostly based upon a presumption that habitats to the north, in Karnataka State, were so fragmented that few or no viable troops of lion-tails remained there. Such may not be the case. In 1983 and 1984 K. Ullas Karanth surveyed forested regions in Karnataka (by quizzing local residents about monkey sightings and examining the quality of habitat for himself) and totaled sightings of 133 different troops. Very optimistically adjusting that number upward by about half to cover troops that might not have been counted, Karanth reported in 1985 that possibly 200 lion-tailed troops, perhaps a total of 3,000 individual animals, still lived in Karnataka, although nearly everywhere the species had diminished or disappeared locally within the last ten to fifteen years. In any event, the total number of individuals representing this species is still dangerously low and probably still declining.

climbed onto the bus and presented herself as, I believe, an image of religious significance. She wore a sequined gold crown on her head, and her face and upper body were painted white, overpainted with red lines and circles, while her lower body was covered by an elaborate orange skirt, sequined, feathered, mirrored, and belled. In one hand she carried an ornate gold spear decorated with feathers and bells, and in the other hand she held a fancy golden cup. But most grotesquely, a small golden arrow hung vertically before her mouth. One end of the arrow was drawn tightly against her chin, the other end against her upper lip, while the middle of the arrow pierced the tip of her tongue, forcing the tongue out and mouth open. Her tongue moved in and out softly, an involuntary tic, as if she were permanently trying to swallow, causing the arrow to twitch softly; she paused before each person on the bus, twitched the arrow, and thrust out her cup. She wanted coins.

Soon after, the driver climbed aboard, and we were off. We now traveled west, not south, and the land seemed to become even hotter and thirstier. We crossed flatlands that had become desert, cracked rubble supporting a few cactus plants, and we passed massive rocks, wrinkled and dark like gargantuan fossil turds. Two or three hours later we came to a small, hot village, boarded a second bus, and sat body to body with far more people than there were seats, plus complicated bags of possessions, bound chickens, and large sacks of rice. The minute this bus pulled out of the village, everyone on it, as far as I could determine, went to sleep. Eyes closed and heads began to nod loosely. We were sitting inside a steel oven: suspended animation seemed a reasonable response. But then the road began winding and climbing — moving into the Ashambu Hills now — and I kept my eyes open to see the desert disappear. First there was grass, mangy and anemic. As we climbed, the grass turned yellow, and leafless trees appeared. Then trees began to sprout a few green leaves, and as the bus slowly ground up and twisted into hairpin turns, and as the sleeping passengers wavered like dreaming swimmers back and forth with each new twist, real forests suddenly appeared on either side of the road. We crossed real rivers with clean water, passed through more forests, until the forests fell back before hills and hills of bright green, tangy-smelling bushes: tea.

The road gradually leveled, and finally we passed into an open green meadow and stopped before a stone house. Everyone got off the bus and stretched before the miracle of cool air and a perfect meadow that rolled far into the distance until it slid into trees and then stopped before solid forest. Streams cut through the meadow and here and there lingered in grassy pools and reed-edged ponds. Some cows were eating the grass of the meadow, and, remarkably, they weren't white, dull-eyed, arthritic zebus with crescent horns, but small-horned brown-and-white cows, perky and fat, with bells tied around their necks. In the distance three women walked along a path in our direction, wearing bright blue, green, and red saris. As a breeze billowed the fine cloth and raised it dreamily, I was stunned by their far-off beauty.

All the bus passengers entered the stone house, which was a rest stop on the bus route, and stood and drank water from a communal cup in a communal pail, or sat on teetering benches at flimsy wooden tables and ordered tea and simple lunches. People stared at me, since I was the only obvious foreigner, but no one spoke English. I sat at a table, too, and politely made faces and gestured for tea and a couple of soft drinks. All the passengers except Anbalagon and I climbed back on the bus, and the bus went away. I looked across the meadow once more and thought how fine and green it was, almost like a golf course.

It was a golf course, actually. The British had built it a few decades ago, and now it was maintained for any golf-minded Indian managers of the Manimuttan Estate, a tea plantation leased from Tamil Nadu State for ninety-nine years by the Bombay to Burma Trading Company, Limited.

We walked across the golf course and eventually were taken in and given dinner by a field officer of the tea plantation, who lived with his wife in a modest but solid and pleasant stone house. After dinner I was kindly encouraged to take a hot bath in a casket-sized, moss-lined bathtub, and then encouraged, with what seemed the style and language of biblical hospitality, to lie down: "Come and take rest, sir." After permission had been sent to our field-officer host by the manager of the plantation, Anbalagon and I were taken across the green to a clubhouse for golfers. The clubhouse contained a table, a chair, a hard bed, and a soft couch. Anbalagon took the couch, and I took the bed, a wooden platform without mattress. It was built for short people, that bed, but the headboard and footboard were made of widely gapped slats, so I could stretch out full length by sticking my feet through the gaps in the footboard. Before he went to sleep, Anbalagon cracked his knuckles and whispered a long conversation with himself in Tamil.

The Manimuttan Estate abuts the northern boundary of the Kalakad Wildlife Sanctuary, and the golf course, where I stayed for a couple of days, is situated at the northeastern corner of the sanctuary.

I was told that a troop of lion-tailed macaques could be found in the sanctuary forest not far from the golf course, and my first morning there Anbalagon and I were joined by three tea workers who thought they knew where the macaques were. The tea workers were dressed in shirts and shorts, flip-flops on their feet, and they all carried sharp, curved knives. One of them wore a shirt with pictures of army tanks on it and a turban on his head.

The tea workers took us down along the bus road and then off on a trail into the forest. The trail descended steeply and crookedly, and soon we were enveloped by trees and ferns and flowers, the sound of insects churning in percussive waves, and also the rushing white noise of water coursing down a gorge below us. Once we stopped and looked into some trees along a hillside rising above us, where I thought I saw a furry body in motion. A few times we saw Malabar giant squirrels, which are reddish brown with big furry tails, slipping around tree trunks. We heard a distant hooting, soft, almost a cooing, which was the call of India's only other black monkey, the Nilgiri langur.

We stood beneath a large beehive hanging from a branch forty feet above us, and paused to consider it. One of the men pointed to the hive and said, "Sweet." I made climbing gestures with my hands: perhaps we should climb the tree and get some honey. Everyone laughed. "No! No!" one of the workers said. "Injection! Injection!" And he began injecting his arm with his finger and then wiggled all his fingers to indicate swarming bees.

The trail kept descending, sometimes very steeply, and finally we came to the bottom of the gorge, where two rivulets arrived from different directions. One flowed smoothly down a smooth stone slide, the other tossed and broke into a white falls. They joined and mingled in a large pool before continuing as a single narrow river through rocks. The water in the pool was rust-colored, deep in places and as dark as black olives, while ripples moving away from the falls reflected the green of trees and white of a hazy sky. In the water at the shallow end of the pool, we found an entire skeleton, almost intact, including skull and large horns, of a samba, the largest deer of India. It must have died or been killed recently, since oil still spread thinly into the water above its bones. But the flesh had been picked clean, so that now only tiny, white shreds clung to the bones and wavered like small worms in the flowing water. I stood at the pool and put the screws to my imagination: a big samba buck, tossing nervously, standing in the water to drink; a bigger tiger, shivering with tension, slouching toward the water to eat. But there are only three tigers in all of Kalakad, and the organized bones in the water gave little evidence of a violent struggle — the skull was a few yards distant from the still articulated spine, ribs, and leg bones — so probably the recent past had not been as dramatic as I imagined.

We took a trail that ascended along the side of a steep hill. I thought we were on our way out of the forest and was disappointed that we hadn't seen the lion-tailed macaque. Anbalagon added to my disappointment by saying several times with a tone of finality, "No monkey."

But then a large, green fruit, half open and white on the inside — a breadfruit — fell onto the trail right in front of us and began rolling our way. Above the spot where the breadfruit fell, a large, leafy branch bridged the trail, and walking across that bridge were two glossy black monkeys with beautiful white manes and tufted tails: lion-tailed macaques.

They didn't scatter or leap when they saw us, as I expected, but merely walked, four-footed, tails up, into other branches and other trees on the downhill side of the trail. We quickly and quietly walked up to where the fruit had fallen and the monkeys had crossed, and I saw several more white-maned black monkeys on the uphill side.

Amazingly, they were not alarmed by our presence and went about their business, eating breadfruit and leaves, almost as if we weren't there. They didn't cry out or call, but rather sat still in the trees or moved very quietly, only making small grunts every once in a while. One quietly climbed up a tree trunk, hand over hand, foot over foot, and then sat on a branch and looked over his shoulder at us. He scratched his side casually with a hand. Another walked quadrupedally on the ground toward us, looked up, and then climbed up a tree and disappeared into other trees. About twenty feet up in the trees sat another, a mother with her baby. I could see her bare, gray face, framed by a white mane that rose up high on her forehead, parted in the middle, and scrolled away to either side, and I could see the baby's tiny gray face near her stomach. The baby had a leaf in its mouth.

Roughly thirty feet up, another lion-tail sat on a branch with both feet outstretched, holding onto the branch with its feet while it hunched over a round breadfruit half its size. This hunching monkey held the breadfruit with both hands and quietly ate it corn-on-the-cob style, reaching in with mouth and teeth and jerking back to pull out mouthfuls. Breadfruit and pieces of breadfruit were all over the ground there, and so I picked up a piece and tried it. It was very pulpy and bland, with a tinge of bitterness.

Other lion-tails sat in the trees around us, some eating, some not, and gradually, one by one, they walked across the leafy bridge over the trail, entered the trees on the downhill side of the trail, and began disappearing into the lower forest. They weren't running, leaping, diving monkeys, merely walking monkeys. And in profile, as they walked, they looked very much like lions, with leonine manes and tails, large chests, and a strong, slow, four-footed walk.

Three or four remained in the trees at the downhill edge of the trail for some time, and because the forest dropped steeply from there, they were easy to see against a background of white sky. Two sat on one branch and slowly ate tender new leaves. Their branch rose and fell slowly in a breeze. A third seemed merely to be looking over the valley, eating leaves peacefully by himself, and slowly rocking in the breeze. I took out my binoculars and examined their furless faces carefully. The faces seemed distinctive, individual, and I spent a long time considering one solemn gray face that looked like an old man's face.

Toward evening that day, in another part of the forest near the tea plantation, we found another troop of lion-tails. These, sitting in much larger, higher trees, grunted and squeaked. This time I noticed that even their manes were individually distinctive. They slowly fled from us and followed each other along regular branch highways forty to eighty feet above the ground.

Lion-tailed macaques have been hunted for centuries by some tribal groups in southern India, both for their meat and for their sleek black pelt. Other indigenous groups traditionally have regarded the lion-tailed macaque as sacred, and refrain from hunting it. Even in the middle of the nineteenth century in India, most wild animals were commonly hunted, but human populations then were small enough and the forests extensive enough that most wild fauna, including the lion-tailed macaque, could maintain their numbers.* However, a twelve-year survey of wildlife in peninsular India, begun in the early 1960s, concluded that tribal groups "indulging in regular orgies of hunting" with bows and arrows, spears, nets, traps and snares, as well as professional meat hunters and amateur poachers ("many of them high-placed in status") using guns, were reducing the populations of nearly all the wild animals of southern India. Although the India Wildlife Protection Act of 1972 prohibited killing or capture of the lion-tailed macaque except for scientific or educational purposes, sub-

*India's population has more than doubled since independence, after World War II, and today's population of 785 million is expected to reach the 1 billion mark by this century's end.

sequent reports indicate that hunting has continued in several parts of its habitat. Hunting has eliminated the lion-tailed macaque from much of its former range and is probably a major cause of its journey toward extinction.

The Nilgiri langur, sometimes known as the black leaf monkey, shares the lion-tailed macaque's habitat and happens to be the only other black monkey of India. Although the Nilgiri langur, also severely threatened, is protected by law, it is often hunted for its pelt and meat. The meat, mixed with herbs, is sold commercially in southern India as a tonic called *karium kurangu rasayanam* (black monkey medicine). The lion-tailed macaque is sometimes mistaken for the Nilgiri langur and shot by hunters who would otherwise leave it. In other instances the lion-tail has been recognized, known to be a legally protected and highly endangered species — and shot anyway.

The legal protection given particular animal species in this century, moreover, has not commonly included protection of their habitat. Although some animals were specifically protected as early as the 1920s, at that time forests throughout India were being logged and clear-cut, and many areas were planted with such imported species as Australian wattle, eucalyptus, casuarina, cashew, and rubber; elsewhere privately owned tea and coffee plantations replaced large sections of forest. People increasingly settled in and around forests and contributed to deforestation with the gathering of firewood and clearing for agriculture and cattle grazing.

Unfortunately, that steady forest destruction was probably exacerbated by a single historical event. After India achieved independence in 1950, the central government gave its authority over provincial forests to the individual states. As a result, no strong national policy preserves the forests, while at the state level shifting political alliances have promoted short-sighted and parochial, but irreversible, policies. In some instances state governments have established new protected areas and otherwise displayed concern for their vanishing forests and wildlife, but generally the Indian states since independence have contributed to an overall loss. They have given away forest areas to private individuals or groups (in Bihar and Tamil Nadu, for instance), located major developments in or near some of the nation's richest forests (such as those at Moyar, Parambikulam, and Ramganga), and have perpetually increased the grazing and collecting rights of villagers living near supposedly protected areas.

By the 1960s, naturalist and photographer M. Krishnan, conducting a major survey of peninsular Indian flora and fauna, found very few forests not severely affected by the human presence. Official records indicated almost no change over the previous fifty years, but according to Krishnan, the forests of the plains "have disappeared entirely from many parts of south India and in places I have actually watched their disappearance." Some plains forests of central and northern peninsular India remained, "but everywhere they have deteriorated by human exploitation." Even the less accessible forests of the hills were cleared in places, opened up, and "deeply penetrated by human enterprise: in many areas they have degenerated."

Lion-tailed macaques are highly dependent upon their natural habitat — the seasonally wet tropical forests of the Western Ghats, sometimes known as *shola*. But much of the shola has lately been fragmented and diminished by roads, railroads, dams, large plantations, logging, subsistence farming, grazing, firewood gathering, and other sorts of destruction associated with human enterprise. Since lion-tailed macaques will neither use nor migrate across nonforested areas, the fragmentation of the Western Ghats shola has also fragmented remaining lion-tail populations. The species' current population of a few hundred to a few thousand animals in Karnataka, Kerala, and Tamil Nadu states is now splintered into genetically isolated subpopulations living within steadily declining forest islands.

By the 1970s, only one fragment of habitat, in the Ashambu Hills region, remained with enough intact forest to provide a reliable Ark for lion-tailed macaques. In 1976 the Tamil Nadu State government, which controls most of this region, set aside a small part of the northern forests and a major portion of the southern forests in the Ashambu Hills, altogether around eighty-five square miles of land, as the Kalakad Wildlife Sanctuary. Today, the superb Kalakad Sanctuary is established and protected not only by law but, in the words of a Western scientist who recently visited the area, by "local sentiment and strong-willed forestry officers who want to see this land preserved."*

*Nonetheless, pressures to exploit the region continue. A powerful political lobby in Tamil Nadu has recently pressed for construction of a large dam inside Kalakad to block the Pachayar River for irrigation purposes. The dam would flood a significant area of forest and require the temporary settlement of 1,000 workers and their families within the sanctuary. Although it was rejected in 1979, that proposal is currently being reconsidered. The portion of the Ashambu Hills region outside of Tamil Nadu (to the west, within Kerala State) may never be well protected. Kerala has shown no interest in preserving the area, and the fact that India's central government favors this preservation makes it even less likely, since Kerala has traditionally been antagonistic to central Indian authority.

After our visit to the lion-tailed macaques near the golf course, I spent a day and a half walking across the Kalakad Ark, from top to bottom, along with Anbalagon and two of the tea workers. We started at the golf course one morning, under a sky of white swirled over blue, and entered the high forest. Until now I hadn't noticed how dry it was. About ten feet of water falls annually on the forest, mostly during India's two monsoon seasons, but none had fallen lately. The forest was rich and green and full of life, yet dry. You could recognize the dryness not only by observing the absence of humidity in the air and the lack of drips and drops on leaves and stems, but by looking at the ground, which was almost sandy and dry enough that you could sit on it without getting your clothes damp. I was thus reminded that this was not tropical rain forest, but tropical moist forest — seasonal forest — and I was walking through it during an abnormal drought.

It was a high forest, though, beautiful, full of vines and ferns, noisy with insects and occasional birds. Unlike the rain forests I had walked through elsewhere, here no ornate gardens sprouted high in the trees. I noticed a few buttressed trees, but the buttresses were minor. Nor, it seemed to me, was this quite a closed canopy forest; a strong breeze boldly wandered right in and shook the leaves.

I kept hearing a random loud whistling, like a man's whistle, but irregular and almost tuneless. I often heard the distant coo-hooting of Nilgiri langurs. All along the trail for much of our walk, I saw brown and hairy pineapples — elephant dung — and when the trail made a sharp turn, I sometimes saw scrapings of dry mud across the trees, where elephants had made the turn but with a tighter fit than we required. I wondered how tight the fit would become if we met elephants on the trail.

We descended into a gorge and came to a stream and a pool swelling amid an avalanche of boulders. There we rested and ate a little. I carried my own food, nuts and crackers and so on, and I wasn't very hungry, but Anbalagon and the tea workers had brought cooked rice and some kind of meal-cakes wrapped in newspaper, which they thrust at me insistently. "Eating. Please. Please." "No, thank you. I'm not really hungry."

The glen was surrounded by high trees, including three or four giants, but open to the sky enough that the sun shone through. There were vines everywhere, on the trees and even falling into gardens on the rocks in the glen; scattered violet wildflowers grew among rock-top moss and fern and vine gardens. A black and white butterfly flickered around. A dragonfly with red and black wingtips stitched the air. Many circular cobwebs arched with the breeze. The pool swelled with a glass surface, and beneath the glass, sunlight and shade stroked an olive, silky silt bottom. Heart-shaped yellow and green and brown leaves slowly floated on top of the water; skimming waterbugs rowed on the smooth surface and made little ripple-shadows on the bottom. Beneath the surface small, salmon-colored crabs slowly moved from shadow into light, then chased pieces of rice dropped into the water, while small fish wavered from light to shadow, dragging their small wavering shadows along beneath them.

Meanwhile Anbalagon and the two tea workers were becoming more and more concerned about my appetite. "Eating. Please. Please." "No, I'm not hungry. Full. Full." But they insisted and insisted, and when that didn't work, they insisted some more, treating me with a strange mixture of deep obsequiousness and stern severity, as if I were an idiot of royal blood. "Eating. One. One. One." They wanted me to have at least one meal-cake, which they forced into my hand. I ate it at last.

During the next couple of hours we passed through a good deal more forest, and then the trail emerged onto a high rock ledge. There everything dropped one or two thousand feet before us, opening up a vista of rock face and plummeting forest and forested hills that rose and fell beyond until they exhausted themselves and gave way to yellow arid flatlands and a haze in the far distance. I imagined I could see the brown river that flowed into Tirunelveli. A strong wind blew toward us and rose against the cliff we stood on; three or four brown, white-headed eagles were gliding, wheeling, and swooping across the rising air against the cliff.

We sat down on the ledge, drank water from my canteen, and rested. The hill nearest our cliff included an open field with bare boulders, a few trees, and many clumps of olive-colored dry grass; as we rested, we noticed in that far field two elephants eating grass. They seemed almost indistinguishable from the grass and boulders at first: gray like the boulders, but with an olive tint, a reflection of the grass perhaps. But they moved slowly, their ears slowly flapping, their tails swishing, and with their trunks they plucked clumps of grass and dirt out of the ground, dashed the clumps back and forth to remove the dirt, then curled their trunks and brought the grass to their mouths. They were a mother and son, I judged, since the smaller of the two had tusks.

Turning back into the forest, we walked for another hour or two, passing through a plantation of tall cardamom bushes growing beneath still-standing forest trees, until we came to a shrine and a river. The shrine consisted of four black sandstone deities, two wrapped in old, blood-red cloth, two wrapped in dirty white cloth, with daubs of red dye on their foreheads. A change box was suspended by a wire from a tree over the shrine, and I dropped a couple of coins in the box and looked at the deities. Then I took off my boots and dipped my feet in the river. The walk had put blisters on my feet, and the cold water of the river extinguished the irritation like a wonderful medicine. I had a small pineapple in my pack, which I cut into four pieces for the four of us.

We crossed the river and continued briefly in forest before we came at last to the sanctuary's rest station, built ten years previously, where we stayed for the night. The rest station was called Sengaltheri, and considering that I was the only visitor in the entire sanctuary, which does not cater to tourists, it was impressively luxurious. It included a long visitors' building, with a couple of bedrooms and some real beds, a semifunctional bathroom with running water, a large dining room, and so on. There was even a cook, who met us and said he was ready to prepare dinner.

The cook and my three companions started up a very lively conversation in Tamil, and then they caught a rat. They tied a string around its neck and began teasing it, jerking it up and down like a yoyo. Since I couldn't understand Tamil, and since I was bored with the yoyoing rat, tired, and didn't really want dinner anyway, I merely ate another pineapple from my pack, plus a can of nuts. I went into one of the bedrooms, placed my belongings on a wooden table there, closed the door, and went to bed.

It was dark then and I had fallen asleep, when suddenly Anbalagon burst into my room, placed a lighted lantern on the table, and said, "Tea, suh?" "No, thank you," I said. So he left, then returned with a pot of tea, poured some into a cup, and handed it to me. "Thank you."

I drank some tea and went back to sleep. An hour later Anbalagon and one of the tea workers burst into the room again. Anbalagon turned up the now fading lantern and said in a firm, loud voice: "Suh! Suh! Eating!" And he placed on the table a big plate with a heap of rice and a pile of something dark and lumpy on top of it. It looked like chicken stew. He also put three smaller plates on the table, one with more chicken stew, one with chutney, the other with an onion and tomato salad.

"Chicken! Eating!" I turned over in the bed and pulled the blanket over my head. "Suh! Eating! Suh! Chicken!" Anbalagon brought the lantern over and pulled the blanket away from my face. "All right," I said. I started to get out of bed, and then Anbalagon gasped. My nakedness was exposed, and he quickly grabbed the blanket and held it in front of me while I put on my pants. The tea worker left. I sat at the table and began eating the rice, while Anbalagon stood over me and watched. I started drinking from my canteen, which was near the table, but he grabbed the canteen out of my hands and poured water into a cup. He handed me the cup. "Drinking, suh!" I drank, and ate some of the concoction on top of the rice — powerful spices burned my tongue, my mouth, my throat. I started back on the rice, and Anbalagon pointed at the chicken. "Chicken, suh!" he commanded. "Yes, I know," I said, but I was happy eating the rice. "Chicken, suh!" he repeated, and at last he took a spoon, dipped it in the chicken mixture, and tried to poke the spoonful into my mouth. I pushed it away. "No, thank you." Then he pointed to the onion and tomato salad. "Onions, suh." "Yes, I know."

Anbalagon left after that, and I ate more of the dinner and went back to bed. An hour later he came into the room once more, accompanied by one of the tea workers. They had a quiet conversation in Tamil, during which I feigned sleep, and then they began rummaging through some of my things on the table. I had some toilet paper, which they examined and discussed. The tea worker picked up a box of matches I had left on the table and began lighting matches. They became very quiet for a while, and then I heard rustling noises near my clothes, so I sat up to see the two of them looking over my small tape player. They seemed a little taken aback by my sudden wakefulness, but they held the player out with a *Can we try it?* gesture, and after I vaguely assented, they took turns putting the earplugs over their ears and talking excitedly about the results. At last the tea worker left the room, and Anbalagon settled down on the bed next to mine and listened to the player. I could smell whiskey on his breath, and I thought I sensed a pathetic desperation in his heart as he clutched my little box and listened to music until the batteries wore out.

Sengaltheri was situated on top of a hill in a beautiful spot. The next morning I got up at dawn and went out to a promontory near the house, where I could see a scene that made me think of Yosemite: granite domes, steep forested mountainsides, and the long plummeting white ribbon of a falls. The falls appeared from the top of a cliff, tossed out white, broke over rocks, and split in two, disappeared behind more rocks and emerged as a single white ribbon again, fell into a deep olive pool, splashed out of the pool and slid down solid rock, splattered, reformed, slid in a sluice to the left, fell into another pool and out another sluice, and broke into a series of ribbon strands that entered a final deep and wide pool, then fell once again and disappeared, out of my vision.

We prepared to leave, but before we left, Anbalagon asked for a gift. He admired my socks and wanted a pair. I told him I had no clean socks; that was all right, he indicated, dirty ones could be washed. The tea worker who had been lighting my matches the night before asked for the box of matches; the other tea worker wanted nothing. Then we left.

Sengaltheri was actually near the bottom of the forest in Kalakad, and before long, as we walked down a winding road from the rest house, the steep forested hills turned into low, rolling foothills, with sparse low trees, palm bushes, and clumps of rough, olive-colored grass. It was quiet except for the breathing of wind, the occasional twittering of birds, and the distant hooting of Nilgiri langurs.

As we descended, the grass turned yellow, golden in the distance, and wavered like running water before the wind. We saw, then, on the other side of a small valley, a red-leafed tree with six Nilgiri langurs sitting in it, eating leaves. They were black with heads of short, thick, yellowish fur, and they had long black tails that hung straight down. They sat stooped and reached out to grab leaves. One stood up on his hind legs, reached high, grasped a branch and pulled it down to pick leaves. One leapt over to a nearby tree with three or four easy hop-leaps.

We continued down the winding road, and the grass turned white; the sparse trees became barren, leafless. We paused before a shrine consisting of a herd of miniature sandstone elephants and a jaguar, all facing the same direction, on parade. We walked past trees that contained big crawling congregations of giant centipedes. At last the road flattened, and we were returned to the hot shriveled land, scorched desert, cracked earth supporting only dry scrub and cactus. The suddenness of the transformation amazed me — it was as if we had walked from Oregon through California and into Mexico in two or three hours — and I was struck by the power of the relationship between altitude, trees, and water. In a small, hot village just outside the sanctuary I said good-bye to my companions, paid them for their help, and stepped onto a bus for Tirunelveli.

NOT LONG AGO, a British journalist named Henry Fairlie wrote: "Nothing of what I wear or eat or drink — except the occasional haunch of venison sent to me by a hunter, whose love of nature of course exceeds my own — comes to me from mother nature's larder. Even the conservationists are sustained not by nature but man's cultivation of it."

A bold statement that. And since Mr. Fairlie lives in an industrialized nation where indeed almost all foods come from a handful of plants and animals snatched from mother nature's larder so many generations ago we've forgotten, where indeed most energy is squeezed from the fossils of crumbs that dropped off mother nature's table so long ago it's hardly worth considering, perhaps he is right — assuming that he never eats seafood, that he lives in a house containing no wood, that he never rides in a car or bus rolling along with the help of natural-synthetic rubber compounds, that he is content to watch his food sources disappear year by year as diseases and pests override the limited genetic resistances of domestic cultivars, that he is willing to pay increasingly higher prices for his tea and coffee, that he habitually refuses to buy at least one fourth of the medicines offered by his local pharmacy, that he will be content with limited treatments for cancer and high blood pressure and several other major diseases, that he never rides on a wooden pleasure boat or sits in a wicker chair or sleeps atop a wooden bed, that he eschews most spices and refuses to smell many perfumes, that he — come to think of it, that he never breathes air, drinks water, or consumes or otherwise relies on anything else that requires air or water. For most of the rest of us, though, mother nature's larder as it exists on the Ultimate Ark has a more direct value.

The Ultimate Ark is the earth itself. But for the great majority of the primates and for two fifths of all other living species, the most critical planks of that Ark are to be found within the green and intricate structures of the tropical forests. In other words, preserving the Ultimate Ark of the primates means saving the tropical forests. I care about the primates and wish to see them survive this millennium and live in the next because I think they are beautiful and whole and worthy in themselves. Mine is essentially, then, an aesthetic and moral response — a poetic response. At the same time, I recognize that for some people the poetic impulse is touched only obscurely or furtively. For people who are hungry, poetry is meaningless. For societies that need or believe they need vast and quick economic expansion, poetry is the enemy. Thus, for those who are hungry or otherwise desperate, for those who see more beauty on the face of a dollar bill than in the timeless and moving faces of nature, and for those who think practically or find themselves struggling in a practical world, it may be appropriate to consider the economic beauty of a tropical forest left standing and intact.

In earlier chapters I have suggested some of the value of tropical forests as genetic reservoirs: rich sources of future medicines, foods, and so on. In a later chapter, I will suggest some of the value of tropical forests as a source of industrial woods, pulpwood, and fuelwood. With intelligent husbandry, still viable secondary forests and tree plantations on already cleared land could provide enough timber to satisfy the world demand for industrial wood, pulpwood, and fuelwood indefinitely: a potential income of $10 billion each year, year after year, for the tropical forest nations. But the best of the still untouched virgin tropical forests, with their immense diversity and fragility, would be destroyed by even very selective forms of traditional exploitation. Although exploitation of primary forests is now making and could in the future make a few wealthy people and corporations more wealthy for a brief period, the primary tropical forests can yield an immensely greater value for humanity at large if they are sustained and protected as genetic reservoirs and, perhaps most significantly, as ecological buffers.

Ecological buffers? All forests are buffers, modifying or averaging out the elemental extremes of sun, wind, rain. But some of the elements in the wet tropics are more extreme, and the buffering services of tropical forests are more extremely important. Tropical forests buffer themselves, of course, modifying the extremes of sun and rain and wind occurring above the protective canopy, and the importance of this internal buffering should not be minimized.

But the buffering services of a tropical forest can extend far beyond the forest itself. Consider the matter of patter: rain. About sixteen feet of rain falls over some parts of Borneo every year, or five times the rain falling onto New York City. A single rainstorm in Borneo can send down as much as two inches of water in half an hour, or forty times the typical New York shower. But when a storm pours such heavy rains onto a Bornean rain forest, the initial force of the wind and falling water is scattered by the canopy and the higher branches and leaves. By the time that water has descended to the ground through the forest's several levels of vegetation, it has been broken and percolated into mist and drizzle dribbling down leaves, stems, and trunks. Much of the water has already been absorbed by the plants. The rest soaks into the porous earth and is further restrained and absorbed by an arabesque system of roots. In short, the entire forest acts like a gigantic sponge, buffering and averaging out the extremes of periodic or seasonal rain.

A few attempts have been made to measure the potency of that sponge. In Borneo a virgin rain forest will soak up more than a third of the water from a rainstorm. A logged forest in the same area will absorb only one fifth of the water, and a tree plantation will absorb even less, about one eighth. The sponge effect not only lessens water runoff during rainy periods but, for those parts of the tropics where seasonal rains alternate with dry periods, it increases the water supply during the weeks or months of drought. At the end of Ivory Coast's dry season, for instance, rivers flowing through intact forests provide three to five times as much water as rivers flowing through coffee plantations.

Most official attempts to place a value on tropical forests tend to ignore this and other environmental services, perhaps because the services have always been performed in the past and therefore seem permanent — a discountable background value. A forest's environmental value becomes most powerfully apparent only when the forest is gone and the value is lost. And now we have many examples of that lost value. During the last three decades commercial loggers in Nepal and India have reduced by about half the forests growing around the watershed of the Ganges River. Downstream from that watershed in Bangladesh and India, one tenth of the world's population (half a billion people, expected to double during the next three decades) lives and depends on the somewhat regular flow of the Ganges. With the deforestation of the Ganges watershed, however, the area in India affected by annual flooding increased from 60 million acres in 1950 to 100 million acres by 1980. Flooding has become increasingly severe, and with the rising waterflows the annual cost of the damage in India has risen from an average of $120 million during the 1950s to around $1 billion today. Simultaneously the amount of water running in the Ganges during the dry season has declined in recent years by nearly 20 percent. India now spends $100 million each year on flood control measures, but only a trivial amount of that money goes to the literal source of the problem, logged-over forests upstream.

How much were those forests worth?

As the runoff from seasonal rains becomes more pronounced, so does the erosion of topsoil. When all the trees are removed from an area, the effects are extreme, but even partial loss of forest greatly increases the loss of soil. Six billion tons of Nepalese soil are now going down the Ganges River each year, suspended in water sixteen times muddier than the Mississippi, carrying five million tons of critical soil nutrients that might be sold for $1 billion if India and Bangladesh were able to draw them out of the water. Unfortunately, India and Bangladesh can't do that, and much of that floating Nepalese topsoil is settling onto the bottom of the Ganges, raising the riverbed in places by several inches per year, while some of the rest of it is silting up the ports of Calcutta and Dacca and creating in the Bay of Bengal a submerged sandbar not much smaller than West Virginia. When that sandbar finally breaks the surface of the water, it will be an island created mainly from reassembled particles of Nepal, but both India and Bangladesh would like to claim and name it. Meanwhile, Nepal's rice and corn crops have declined by a fifth and a third, respectively, in the last half decade.

Erosion of topsoils as a result of deforestation, accounting for some major proportion of the twenty-three billion tons of rich soils eroded from cropland around the world each year, creates a hard-to-calculate deficit in agricultural potential that our children will pay for. But we are already seeing measurable lost values as that eroding topsoil moves downstream and accumulates in canals and reservoirs. Three decades ago 85 percent of the watershed around the Panama Canal was protected by virgin rain forest. Since then, around 100,000 farmers and ranchers have moved into the region and burned and cleared over a third of the forests. As a result, siltation into the canal has increased rapidly and can be expected nearly to double between 1980 and 2000, effectively closing the canal to many larger ships. A $17 million project to reforest and otherwise protect the Panama Canal watershed is now under way.

Elsewhere siltation as a direct consequence of deforestation is reducing the capacities of many third world hydroelectric and irrigation projects. Destruction of forests around the Mangla Reservoir in Pakistan, for example, has increased the sedimentation rates in the reservoir enough to decrease the dam's productive life from the originally projected one hundred years to fifty years. Increased sedimentation as a result of forest destruction above the Ambuklao Dam in the Philippines has shortened that dam's productive life from an expected sixty years to thirty-two years, a loss valued at $25 million. Erosion caused by deforestation in the watershed above the Anchicaya Dam in Colombia had reduced that dam's reservoir capacity by one quarter a mere two years after it was completed. In Ecuador the destruction of half the forests within the watershed of the Poza Honda Dam is expected to result in a useless dam after only twenty-five years of operation, instead of the fifty years originally expected. The loss: $30 million.

How much were those forests worth?

The evidence is clear, and nearly everyone now recognizes the importance of tropical forests in regulating waterflow and inhibiting erosion. The value of tropical forests to agriculture in the third world is immense, and it has been drastically underestimated, as has the value of regular supplies of clean water to public health. However, the most important buffering services of tropical forests may have a much broader — and therefore less obvious — effect on related ecosystems. Some people still dismiss those services as science fiction. "There are plenty of valid scientific reasons to protect the environment," a Brazilian agronomist has stated, "and I am sure that politicians and decision-makers, as well as many other concerned individuals, will be more receptive to *scientific facts* rather than to mere prognostications." But before we so glibly dismiss *mere prognostications,* we might remember that prognostications suggest when a hurricane is coming, whereas facts only tell what damage a hurricane has done. It would be worth our while to look at some mere prognostications about the value of tropical forests as ecological buffers at the climatic level.

In complex ways that are not entirely understood, tropical forests serve a feedback function within the usual climatic cycles of evaporation and rainfall. We might imagine that the rain falling over Amazonia eventually works its way down to the ground, trickles into swamps and streams, rushes into rivers, and at last exits into the ocean, where it is gradually levitated back into the air by evaporation and finally returned to and deposited in the forest as rain. But that is only one aspect of the complex Amazonian water cycle. A good deal — perhaps half — of the region's seven-plus feet of annual rainfall does not drain into the streams, rivers, and ocean at all; instead, it is absorbed by the vegetation and soils and returned directly to the atmosphere both by evaporation and by a combination of evaporation and transpiration ("breathing out" of air and water through the pores or stomata of leaves), a process known as *evapotranspiration.* Mere prognostication suggests that destroying some major part of the Amazonian vegetation may alter the evapotranspiration cycle severely enough to produce a climatic drying out that will affect not only Amazonia but areas to the south, into the agricultural heartland of Brazil.

Forests also participate in cycles of solar energy transference. When energy from the sun strikes the surface of the earth, some portion of it is absorbed and some portion is reflected. Some of the absorbed energy powers photosynthesis, some of it warms water and produces evaporation, and some of it directly warms the lower levels of air. Some of the reflected energy heats up the atmosphere, but most of it penetrates the atmosphere and is lost in space. Since forests present a darker surface than most other ecological systems, they absorb more solar energy and reflect less. (We are all aware of this essential principle from everyday experience: wearing dark clothes in the summer makes you warmer than wearing light-colored clothes.) The ratio of reflected to absorbed energy can be measured precisely, and the measurement is sometimes described as *surface albedo.* Without becoming excessively technical, it can be said that calculations based on thousands of measurements over rain forests, recently deforested areas, and grasslands in Nigeria and Thailand show clear, distinctive changes in albedo values from one ecosystem to another. Altering ecosystems, in short, seems to alter the ways solar energy participates in local climates — changing the local patterns of air movement, cloud formation, and rainfall.

The effects of changing patterns of evapotranspiration and surface albedo are still under debate and in any event may alter climate only locally, but it is quite clear that tropical forests serve as a buffering agent in the earth's now imbalanced carbon exchange, which will probably contribute to climatic change at the global level: the *greenhouse effect.*

Without extraordinary protection, the bodily liquids of a person stepping onto the surface of Venus would immediately boil. Anyone stepping onto the surface of Mars would freeze by nightfall. The fact that the earth's temperatures are comparatively mild and hospitable to life has mostly to do with the atmosphere, that blanket of gases surrounding and insulating our planet.

The insulation provided by those atmospheric gases defines the quality and quantity both of solar energy striking the earth's surface and of solar energy radiated back into space. During the last two centuries, however, human activity has been releasing large amounts of certain gases into the air, gradually altering our atmosphere's chemical composition. Those chemical alterations, acting much like the windows of a greenhouse, still allow substantial solar energy to reach the earth's surface but have begun to reduce the amount of energy, largely heat, radiating back into space. Thus the theory of the greenhouse effect predicts climatic warming.

The most important chemical window in our global greenhouse is made of carbon dioxide, a byproduct of the combustion of fossil fuels and wood. Before the Industrial Revolution, carbon dioxide was present in the atmosphere at a concentration of about 290 parts per million. Largely as a result of the burning of fossil fuels, the concentration of atmospheric carbon dioxide has been increasing regularly, and the rate of increase itself has increased. By 1958 the concentration was 315 parts per million. The concentration had reached 349 parts per million by 1988; by the year 2030, mere prognostication tells us that carbon dioxide concentrations may be about double the preindustrial levels. Most of that "new" carbon dioxide in the atmosphere comes from the exhaust of cars and factories in population centers of the industrialized world, but it disperses so completely that increased atmospheric concentrations have been registered around the globe. The oceans are absorbing much of the additional carbon dioxide released by combustion, and if all cars and factories stopped burning natural gas, coal, and gasoline today, the oceans would eventually absorb 85 percent of the extra carbon dioxide. The process would take several centuries, though, because full absorption requires an extensive mixing between deep and surface layers of the ocean.

Global warming as a consequence of the greenhouse effect is still thought by some people to be mere prognostication. But during the last century the earth's average temperature has risen by about one degree Fahrenheit, and half that rise has occurred since the mid-1960s. According to Dr. James Hansen, director of NASA's Goddard Institute for Space Studies: "Global warming has begun." What can we expect? Some scientists predict an increase in the earth's average temperatures of three to eight degrees Fahrenheit by the 2030s, a climatic change comparable to the far more gradual changes after the last ice age, when temperatures were only ten degrees below today's average.

For anyone living in Boston or Siberia, a warming trend might seem like welcome news. But the warming trend will radically alter patterns of weather that humans have relied on since the beginnings of settled agriculture. Warmer oceans will increase the power and frequency of tropical hurricanes. And because the warming will occur much more intensely at the poles than in the tropics, a seemingly minor average increase will drastically alter prevailing wind and rainfall patterns. For instance, an average temperature increase of only two degrees Fahrenheit might result in higher evaporation and less rainfall in the American Midwest, reducing the American corn crop by one tenth. An average temperature increase of seven degrees Fahrenheit would give the American Midwest an arid climate, while the more northern areas of Canada, where the soils are not rich enough to support large-scale agriculture, would become warmer and much wetter. Altogether, the climatic shift might benefit agriculture in Mexico, some parts of the Middle East, northern areas of the Soviet Union, and China, but would likely disinherit today's high productivity regions in the United States, the Mediterranean, and the southern Soviet Union.

Most seriously, however, global warming will begin to melt the polar icecaps and thereby raise sea levels around the world. The United

States Environmental Protection Agency prognosticates that the global warming trend already under way will cause sea levels worldwide to rise some two to twelve feet by the year 2100. After 2100, if the melting continues, present sea levels will rise hundreds of feet before the ice caps will be entirely gone. Clearly, that would be bad news — but even small changes might be unwelcome. A mere three-foot rise, which may well occur by the year 2030, would inundate large areas of farmland in the lowlands of the third world and elsewhere. Salt water would invade many irrigation and drinking water systems, while cities built on river deltas, including New Orleans and Cairo, would have to contend with major flooding. A twelve-foot rise in sea levels, an event our grandchildren may have to face, would flood much of Florida, Bangladesh, the Netherlands, and other low-lying regions, while further increases in sea level would eventually inundate such major coastal cities as New York, Washington, Boston, Miami, Los Angeles, Montreal, Rio de Janeiro, London, Glasgow, Stockholm, Tokyo, Osaka, Calcutta, Shanghai, Jakarta, and Lagos.

Green plants absorb carbon dioxide during photosynthesis and store the carbon. Like the oceans, then, the world's forests can function as a sink, or a buffer, to decrease the amount of carbon dioxide moving into the atmosphere. Burning the forests, on the other hand, has exactly the reverse effect. The earth's green plants store some 800 billion tons of carbon, of which 40 to 50 percent is contained in the trees and other vegetation of the tropical forests. Burning forests transforms that carbon into carbon dioxide — and, given today's rates of tropical forest burning for slash-and-burn farming and domestic fuel, it may be that tropical forest destruction is adding another fifth or two fifths or even more to the total amount of new carbon dioxide entering the atmosphere.

How much are those forests worth?

I TRAVELED NORTH, taking the bus back from Tirunelveli to Madurai, where I stayed for a time to rest and visit the famous temple there.

One hot day I took a bicycle taxi to the temple and stood amazed by its sheer size. It covered a large city block and, walled like a fortress, reached into the crowded streets with five huge entrances. The entrances billowed and thrust ornately upward into five great towers, which were decorated entirely with layers and layers of painted clay figurines and which rose four-sided into slightly concave pyramids before terminating at their tops with elaborate and be-figured barrels.

I was instructed to take off my shoes before going into the temple's main entrance, and so I walked barefoot through a stone passage lined with ancient, imposing stone statues and came to a large room that contained many small shops selling devotional pictures of gods and many other items. In one part of this room stood a large bull elephant, harnessed and belled, next to his trainer. The elephant raised the tip of his trunk to me and opened it so that I could see pink inside and two dark nostril holes. I held out a coin, and the tip of the trunk closed around the coin. The elephant swung his trunk back and forth, then gave the coin to his master, who thanked me with a ceremonious bow.

From there I walked along dark corridors and into great, cave-like rooms filled with strange figures — dragons, monsters, balloon-breasted beauties, poets, prophets, and deities composed of human and animal parts, all carved with astonishing fineness out of solid stone. To be sure, the multitudinous figures, the monsters and deities, express holy logic to any reverent Hindu; but I was a pathetic foreigner, seeing only dimly through ignorant and unsanctified eyes, and to me the dark stone shapes all around seemed like an Aztec dream or a hashish nightmare, awesome and mildly frightening.

I walked through corridors lined with dozens and dozens of steel-barred cages containing dark Hindu deities of all sorts. People came before the cages, lit candles at their thresholds, bowed and prayed and kissed the stone thresholds. One devout man stood before a cage but faced away from it and looked over his shoulder with a mirror, so as not to subject his eyes to the shock of direct sight. One part of the temple, set aside for the powers of astrology, included stone representatives of the sun, moon, and all the planets; people wishing to appeal to those powers lit candles and prayed and walked in circles.

But the temple was so big and so complicated that I got lost and spent an hour wandering around in circles myself. By sheer accident I entered the holiest part of the temple, open, as I later discovered, to worshiping Hindus only, and I arrived at a great stone image of Siva himself in all his multilimbed, destructive-creative glory, lit by golden sunlight shining through a golden-glassed opening in the roof. There was a vat of ashes, and people put ashes on their foreheads.

A young man approached me and asked if I would like a tour of the temple. Since I was lost and impossibly ignorant of the significance of these strange surroundings, I agreed. And so I learned a little about the temple. "Hindus from all over India," my guide said, "come here to get their divine bless." He added, "To avoid proudness, people are putting ash on our forehead." And people say *Om,* he told me, because it sanctifies. "It will increase the spiritual feeling. It will decrease the sexual feeling."

Part of the temple had been built in the tenth century, and the rest had been constructed in various phases during the ten centuries following. It was a twin temple, my guide informed me, with two golden domes. The larger signifies a dedication to Siva in his avatar as Lord Sundareswarar, and the smaller golden dome signifies a dedication to Sri Meenakshi, Lord Sundareswarar's beautiful consort. The temple includes 1,555 sculptures, but since Siva alone has 1,008 forms, with 1,008 names, there is no lack of subjects to sculpt.

People sometimes wonder where all the stone came from to build the temple. According to tradition, a long time ago some devils sent a giant elephant, three kilometers long, two high, and one kilometer wide, to kill all the people. So Siva came down in the form of a hunter and turned the elephant to stone, thereby saving the people and incidentally providing enough stone to make the temple.

The temple was very dark inside, but then one of the corridors opened into light, and we came out blinking into the brightest sunlight. We walked onto a great, rectangular, colonnaded porch wrapped around a deep rectangular depression: an empty rectangular pool with stone steps leading down to the bottom on all four sides. That was the sacred Golden Lotus Tank, and we stood at the edge of the tank while aged Brahman priests sat cross-legged behind us, consulting with worshipers.

Looking up from the Golden Lotus Tank, we saw the highest of the entrance gate towers rising above us. My guide told me that the towers used to be open to the public, and you could climb to the top, but people kept committing suicide. "The local university students, they got some failure in the exams, so they jumped. People with a lot of problems, unable to solve by human beings, jumped because in front of a god they are dying. This way you will not be a sinner." It is said that a saint stricken with leprosy once jumped off a tower, and on his way down a god caught him, cured his leprosy, and turned him into a great poet. Anyway, because of suicides, lately the towers have been closed to the public.

We looked down into the drained pool, reflecting the sun's piercing light brightly on dry stone, and my guide told me about the Golden Lotus Tank. Legend has it that Indra, god of rain and thunder, first bathed there, cleansing his sins, and worshiped Lord Siva with petals plucked from golden lotus flowers. The sacred tank sits above a powerful underground spring, and for many centuries that spring naturally filled the tank to a depth of fifteen feet. "The spring was very powerful, and every day water was unceasing," my guide said. But during religious festivals when people walked into the pools, many people died. Children fell in by accident and drowned, and many unhappy people committed suicide by drowning themselves. Thus in 1965 the authorities decided to cement over the spring. It took six months to cap the spring, but now people at last have control over the holy waters.

My guide pointed out the plumbing, which I hadn't noticed before. Two large steel pipes led into the rectangular tank, one to drain water out, and the other to pump water in to fill the Golden Lotus Tank for special religious holidays. Indeed, a religious holiday was approaching, and I saw then a thin puddle of water at the bottom of the tank and a twisting rope of water pouring in from one of the steel pipes. We watched for a still moment as the lotus water rose quietly, quietly, and its surface glittered out of the sun's heart of light.

Nine

CHILDREN

MY SEARCH for the bilou — the gibbon of the Mentawai Islands — began late one night in a small, family-run snack, beer, and soft drink establishment on the waterfront of the city of Padang, on the west coast of Sumatra.

The walls and ceiling of the place were rough boards mostly covered by sheets of plain brown paper. Gray lizards slowly moved hither and thither on the ceiling and walls. Several men sat at tables, drinking and talking and smoking clove-laced cigarettes, and waited for their boat, a cargo boat to Siberut, the northernmost of the four Mentawai Islands. The owners of the place, a Chinese family, seemed slow and fatigued by the lateness of the hour; through a doorway in one wall I could see a sleepy, candle-lit domestic tableau: a man and woman sitting at the edge of a big bed, peeling paper off tin cans, and four children asleep in the bed, curled and sprawled in various positions.

I was sitting at a table, rehydrating myself after a hot day in Padang, waiting for the boat to Siberut like everyone else. Next to me sat a twenty-three-year-old Indonesian named Agustus Nadeak, whom I had hired as a guide and interpreter. Short and wiry, energetic and extroverted, his triangular face topped by curly black hair and bottomed by a peach-fuzz mustache and two or three chin hairs, Agustus helped pass the time that evening by explaining the Indonesian perspective on things. It's not allowed to criticize the government, he told me. Indonesians cut off the tails of their dogs and cats with the belief that the rest of the animal will grow better. I must be careful not to smoke marijuana. He would like to go to Europe to learn how Europeans think, to take it back to his country, where people are not so educated and still learning how to think. He maybe would visit the

United States but knew that would be impossible but if I found a job any job maybe I could tell him. Indonesians don't want money, which they don't have, but happiness in the family. He comes from a family of twelve. His parents are separated. In turn, I tried to tell Agustus a little about my experiences in Madagascar. "Is Madagascar a part of Europe?" he wanted to know.

We waited and waited some more. "Everybody is waiting to get boring," commented Agustus. At last, after midnight, we heard the sound of a motor starting. "Yo! Yo! Yo! Yo! Yo! Yo! Yo!" someone shouted. "Let's go!" Agustus said, so I picked up my pack and, along with everyone else, left the shop, turned a corner, walked in darkness across a precarious scattering of boards, and climbed aboard the boat.

The boat slipped out of the harbor across black obsidian, under a black dome dripping with stars. Orion had disappeared from the sky months earlier, so I concentrated on the Big Dipper. The sky soon turned hazy, though, and the stars frosted over and then faded until they disappeared altogether.

The boat was loaded to the brim. Barrels of diesel fuel were jammed all along the walkways of the deck. Some of the crew and passengers quietly moved into sleeping spots inside the cabin, while the rest of us settled into all the available flat spaces on deck. I lay down on top of some boards over the cargo hold and prepared to sleep: pack under my head, shoes to my side, blanket to keep me warm. And for a while I preoccupied myself with thoughts of my children on the other side of the globe. I missed my little boy, who had turned three in March, and who, according to telephone conversations with my wife, had been waking up in the night and calling for me. I missed my daughter, six years old the previous month, whose cheery, cheeky face I could see before me.

I woke at dawn and saw a massive, pink-tinted burst of cloud in front of us at the horizon. Behind us a yellow sun cracked open the sky and poured gold into the sea. The sea turned from a rolling blue and silver mirror into a ruffled, scalloped field of gray and white, with fleeting glints of gold, but when I looked directly down into the water, it was deepest blue, almost purple.

The cook squatted at the galley box, heating water for tea and a pot of rice, washing squares of tofu and small red peppers. People were talking cheerful wake-up talk. "What's call speak English: big fish and jumping?" Agustus asked, making a leaping motion with his hand, and I looked over to see a dozen dolphins swimming beside the boat, rising, rolling, sinking, sometimes leaping fully into the air for a moment.

Midway through the morning the low forested outline of Siberut appeared before us, gray at first, then pale blue, then green, as the boat slowly rolled and chugged toward it.

Gibbons, the smaller or "lesser" apes, are elegant and agile, furry yet bare-faced, long-armed and tailless primates. Gibbon embryos look almost exactly like human embryos, and the adult face and body, aside from the very long arms, exhibit an interesting human resemblance. They walk and run upright like humans, on their hind legs. But with long fingers and small thumbs and very long, powerful arms, they are also admirably suited for their most characteristic pattern of movement, known as *brachiation*: hanging, swinging, tossing themselves with unrivaled speed from branch to branch through the forest. They are the acrobats of the primate world. "Preeminently qualified for arboreal habits, and displaying among the branches amazing activity," wrote a nineteenth-century zoologist, "the Gibbons are not so awkward or embarrassed on a level surface as might be imagined. . . . It is, however, in the trees that they are seen to most advantage; there, free and unembarrassed, they appear almost to fly from bough to bough, and assume in their gambols every imaginable attitude; hanging by their long arms, they swing themselves forward with admirable facility, seizing, in their rapid launch, the branch at which they aimed: they throw themselves from a higher to a lower perch with consummate address, and again ascend to the loftiest with bird-like rapidity."

Today gibbons inhabit tropical forests throughout much of southern and southeastern Asia, as far north as southern China, as far west as eastern India, as far south as the Indonesian archipelago. The six to nine species (depending upon one's system of classification) of this genus are partly distinguishable by their fur color and markings, although a full identification of species includes major characteristics of behavior and anatomy. The largest gibbon, the siamang, stands about three and a half feet tall, with an arm spread of six feet. All other species are smaller, faster, and more agile.

The bilou, the Mentawai Islands gibbon, was discovered by Western scientists in the winter of 1902 and 1903, when W. L. Abbott and C. B. Kloss explored several small islands in the Indian Ocean southwest of Sumatra, gathering samples of representative animal life for the Smithsonian Museum in Washington, D.C. All told, Abbott and Kloss acquired specimens of around three hundred species, including over thirty previously undescribed types. The newly discovered species included five primates, one of which seemed to be a "dwarf siamang" from the Mentawai Islands. Abbott and Kloss brought back the remains of eighteen "dwarf siamangs": skins and skulls, and four infants small enough to preserve whole in jars of alcohol. One adult male weighed thirteen and a half pounds, which was about half the weight of adult male siamangs collected in Sumatra. Like the siamang and unlike all other gibbons, this new species was entirely black; it seemed to resemble the siamang along several basic anatomical dimensions as well. At Abbott's request, the new species was christened with Kloss's family name, so it became *Symphalangus klossii,* or Kloss's siamang. Later studies of the animal's physiology revealed that it is not so closely related to the siamang — not really a dwarf siamang at all. Now Westerners often call it Kloss's gibbon or the Mentawai Islands gibbon, while Mentawai Islanders continue to call it by the name they have always used: bilou.

The four Mentawai Islands comprise about 2,700 square miles of land, which is little more than half the size of Connecticut and by far the smallest area in the world harboring a unique primate species. In fact, those four small islands are home for at least four unique primate species and subspecies: the bilou, the Mentawai langur, the pig-tailed langur, and the Mentawai pig-tailed macaque. But aside from the occasional collectors who came, saw, and shot, no scientist attempted to study any of the islands' remarkable fauna until Richard Tenaza and W. J. Hamilton III spent a week there in 1970 observing the bilou. Tenaza returned in 1971 to begin an eighteen-month study of that gibbon.

We docked in the harbor of a town called Muara Siberut, which is the largest cluster of houses on Siberut, indeed on all four of the Mentawai Islands, and which serves as Indonesia's administrative center for southern Siberut. Muara Siberut consisted of perhaps 150 houses, most of them bark-walled, thatch-roofed, raised above ground level on short vertical posts. Ground level? Actually, the town was kept above swamp level by flattened coral heaps, and each house stood in the middle of its own rectangular coral lot. The lots were divided from each other and from the rectangular system of coral roads by a matrix of coral-lined gulleys and small canals. The houses, in other words, were moated, and to enter most you crossed a wooden bridge. The moats also served as the town's major sewage, garbage, and trash disposal system, flushed regularly into swamp and sea by rain.

Business necessitated staying in town overnight, so Agustus and I crossed a wooden bridge to the house of a woodcarver, where we dropped our packs and took tea. The woodcarver had grown up in Sumatra, but had moved to Siberut when he inherited this old house from his grandparents. Now he made a very meager living chiseling elegant designs into doors and beds, as well as hosting the occasional visitor in his spare bedroom. The house was larger than most in town, with a small central living room that opened into four side rooms, two on either side. The bark walls dividing the rooms were all ceiling height, but there was no ceiling, only the rise of the high-pitched, palm-thatched roof, so birds flying into the house could easily move from room to room and perch atop the walls. Above the interior doors hung the only decorative artwork in the house: one stylized picture of Virgin and Child; one plaster relief of Virgin and Child; and one picture of Jesus holding forth a red heart that looked like an apple, only it was torn and dripping blood, circled by a wreath of thorns, topped by two furling clouds out of which sprang a cross, surrounded by light rays. The roof had large holes in it, through which sunbeams and rain penetrated, and the floor was so cracked and knotholed that the woodcarver could dispose of his cigarettes by dropping them through the floor.

We had to buy supplies for our trip into bilou territory: food, and cigarettes and tobacco enough for all friends and hosts. We had to register with the local police, who were pleasant and efficient and took my passport as a standard procedure. We had to acquire a permit to visit the island's interior from the district commissioner, whose cherubic-faced assistant promised prompt attention but became the world's slowest typist (quickly plummeting from five words per minute to one, to one letter per minute) until we gave him a tip for his typing. Then, the next day, he became the world's slowest delivery boy until we gave him a much nicer tip for his delivery of the papers. We also had to hire a speedboat and driver.

The owner and driver of the speedboat was named Antonius Napitupulu, and, according to Agustus, he was a priest. (I later learned that he was a *badja gereja,* a respected man who can perform funeral services and Sunday services in the local Catholic church when no priest is available.) I met Antonius late that first afternoon on the woodcarver's front porch and liked him instantly. He was sitting in a wicker chair, and on his lap sat one of his daughters, a small, pretty girl of whom he was obviously very fond. I thought Antonius had a very kind and gentle manner, and I saw him as a good father. He was a little older than I, in his mid-forties, perhaps. He was Indonesian, the son of Sumatran parents who had come to Siberut as missionaries. His hair was dark and long, almost shoulder-length, curled at the ends. A few hairs made up his mustache and even fewer made a beard. When he smiled, I saw one gray tooth in front with the other teeth jumbled around it, twisted boards in a badly made fence. When we told him about the required bribe to the district commissioner's assistant, Antonius smiled, shook his head, and said to me, "I not like this."

That evening Agustus and I had dinner at Antonius's house, a small, three-room affair with palm-thatched roof and dark, dirty bark walls. The middle room of the house, lit dimly by a kerosene lantern, served as a living and dining room. Through an open doorway to one side of this room, the kitchen glowed gloomily from a few embers still stirring in a wood fire; through a doorway closed by a curtain on the other side of the room, I could hear the six children softly singing themselves to sleep. Antonius's wife, a health worker, had taken the boat back to Padang with a case of difficult childbirth, so this was a bachelor dinner — me, Agustus, and Antonius, eating what was to be the standard breakfast, lunch, and dinner for my entire time on the island: rice, noodles, and fish. During dinner two tailless cats wandered in and out of the house, looking for scraps, and after dinner one of the children came out of the bedroom and said she couldn't sleep, so Antonius stroked her head and then sent her back to bed.

All gibbons are monogamous and live in strict family groups composed of a permanently mated adult male and adult female and their young offspring. Gibbon females bear a single young about every two years. The infant is born with its eyes open, but completely bare of fur. It is thus absolutely dependent on its mother for warmth as well as nourishment. Within a day after birth, the infant is able to cling unsupported to the fur on its mother's abdomen, yet it remains quite dependent for about two years. (With siamangs, but not the other gibbon species, the father takes over much of the child rearing during the second year and often carries the young animal while the family is moving.)

The entire gibbon family, typically two to five individuals, moves and forages together and sleeps together in a preferred sleeping tree. At least with bilous, the mother usually leads the group as it travels single file across established routes through the trees, anywhere from twelve to one hundred feet or more off the ground. Upon encountering a potential threat, such as a human, the entire group will flee except the father, who places himself between the threat and his family, at times approaching and displaying — for example, by circling in branches above the threat. Even when seriously threatened, though, a gibbon group will not enter territory occupied by other gibbons.

Gibbon pairs apparently mate only during periods of a few months every few years, when the female is ready to become pregnant. Thus sexual attraction, or at least sexual action, may not be the most significant tie that binds the monogamous couple. Instead it has been suggested that gibbon monogamy is maintained by intrasexual antipathy. Adult males are intolerant of other adult males but not of females. Males, but not females, trespassing into another family's territory will usually be challenged with threats and physical violence by the resident male. Similarly, adult female gibbons are intolerant of other full-grown females but seem to be undisturbed by intruding adult males.

So monogamy, in the case of gibbons, is inseparable from territoriality. And all gibbon families stake out very well-defined forest territories. Obviously each family territory must contain enough food (predominantly fruit, such as figs, but also flowers, buds, young leaves, and occasionally bird's eggs, small birds, and insects) to provide for the group year round. A lar gibbon family group may occupy 125 to 150 acres. Bilous occupy the smallest territories of any species in the genus, averaging a mere 16 to 17 acres in size. Good bilou habitat in the Mentawai Islands is fully occupied, divided into a matrix of adjacent, precisely defined family territories, with only several dozen feet of demilitarized zone, mutual overlap, on the peripheries.

Gibbons reach sexual maturity at six to eight years. As the young male approaches adolescence and maturity, he is increasingly threatened by the father's innate intolerance of other adult males. Thus the maturing male begins sleeping farther away from his family and is eventually driven out of the family territory altogether. Similarly, a maturing female gibbon becomes increasingly peripheral to her family group because of increasing conflicts with her mother. Given the density of bilous, at least in many areas of their Mentawai Islands habitat, you might wonder where the newly mature males and females find new territory. If all territory is already divided up among the existing gibbon families, all of whom are intolerant of adult strangers, it would seem that there is no room for expansion. And in fact there hardly is. Like all other animals in undisturbed forest, gibbon populations tend to reach a stasis, a stable and maximum size.

A maturing female gibbon may form a mated pair with a single adult male in adjacent territory. Maturing male gibbons driven out of their family territories may become nomadic "floaters" in search of newly available habitat. The nomadic male may aggressively displace an already established older male or may gain a territory by mating with a widowed female. Among bilous, at least, new territory may also be forcibly seized by a maturing animal's family, and then passed over to the young adult. In his field study during the early 1970s, Richard Tenaza observed a family (which he labeled Group 10) with an adolescent male. The testy young male had already begun sleeping well away from the rest of his family and foraging at a distance of a hundred to a thousand feet away, while the other family members habitually stayed within about thirty feet of each other. One day Tenaza noted that Group 10 began expanding its territory, eventually adding 60 percent more land, by moving into the territories of two other adjacent groups.

Part of that expansion resulted in a major battle with one adjacent group, Group 7. The father and maturing son of Group 10 attacked the adult male of Group 7 in what had been Group 7's territory, while the adult females and other members of both families watched at a safe distance, uttering "soft hooting and howling vocalizations." Twice the male of Group 7 retreated to the ground, once driven there by the adolescent of Group 10, who chased him down a tree and then struck him. On the ground the two adults fought viciously, until finally Group 7 retreated, yielding for good a major part of its territory.

The adolescent male began sleeping even farther away from his family, spending almost all his time in the newly acquired portion of territory, while the rest of Group 10 began spending most of its daylight time as far away from him as possible, at the opposite side of their old territory. The young male also began singing in the predawn "male chorus" for what appeared to be the first time. Probably the young male was announcing his new possession of territory and his readiness to acquire a mate.

Late the next morning, having at last acquired our permit, we loaded food and possessions into Antonius's speedboat and began our journey upriver into the rain forest interior of Siberut. The speedboat was actually a large, old dugout canoe with an Evinrude motor clamped onto a flat board at its stern. My doubts about the stability of this vessel increased as Agustus stood up in front and awkwardly helped paddle us out of the garbage-laden main canal of town, but then diminished after Agustus sat down and I turned to observe Antonius the priest, cigarette in mouth, dented hat on head, smile softly and maneuver our vessel past floating logs and up-reaching snag fingers, up into the milky-brown, wrinkling river.

"Someone was washing in the river and he already died. And the crocodile took him away," Agustus said. "Last month many people was died here, in the river, and one children. Maybe there is here, in the river, crocodile." But the only evidence of a river creature we saw the entire trip was the whipping white line of a swimming snake.

I was surprised, though, by the amount of traffic on the river. All the boats were dugout canoes, and all except ours were powered by arms and hands holding finely carved paddles, pointed at the ends so that they looked like wooden swords. Almost all the dugout canoes were superbly fashioned from single pieces of wood, curved elegantly to finely pointed bows and sterns, perfect wooden peapods. The smallest were single-person affairs, narrow, hardly longer than a bicycle and seeming to possess all the intimate maneuverability of such a vehicle. The largest held entire families: father at the stern, mother at the bow, two or three or four children peering out from the middle, with perhaps a scrawny dog or two among the children, sometimes also a grandmother and grandfather wearing conical straw hats, once a mother nursing her baby under a pink umbrella. Sometimes near the front of a canoe a small smoking fire would be heating lunch. Some of the canoes carried cargo, often under a palm-leaf cover. What cargo I was able to see looked like banana bunches and long bundles of rattan, the raw material for wicker furniture. Antonius seemed to know many of the people we met and passed, and he called out to almost everyone. They would pause in their paddling, look up, and call back with a few friendly words. I began to notice more and more people, especially bare-chested men, covered with elaborate tattoos.

After two or three hours, we reached a fork in the river and turned up the left tine. Suddenly we were moving in a narrower channel against steeper, faster, more complicated water, which sometimes churned and boiled. Trees and vines hung over the edges of the river now, and dipped into the water. Logs floated quickly down, slowly spinning, and snags emerged from the river before us. Our progress became much slower, and then the sky clouded over, and it started to rain.

Agustus stripped down to his shorts, covered his clothes and our packs and supplies with a tarp, and hunched down low and began shivering. I covered myself with a poncho. Antonius pulled his dented hat down lower on his head. Soon it was raining so hard I could barely see the forest on either side of the river. At last we came to a place where several dugout canoes were tied to a tree at the bottom of a steep muddy bank, and we tethered our canoe there and stepped out. A notched log had been thrown down the muddy bank, the notches serving as steps in a ladder, which we climbed.

The people who met us at the top of the bank spoke in the Mentawai language. I had learned to say a greeting in Mentawai, so I said it several times, and we unloaded the canoe and waited at a wooden shack while Antonius hauled up the Evinrude and the gas tank and placed them inside the shack. From there we walked for half a mile along a muddy trail into the forest, until the trail turned into a walkway of single logs placed end to end.

We came to a village in the forest, a cluster of small houses on stilts situated on either side of the walkway. The village was called Rok-Dok. People inside the houses, lightly dressed, often heavily tattooed, looked out and shouted greetings and laughed every time I slipped on the wet logs. The houses were all palm-thatched, most with walls made of stick uprights and bark, some made of splintered bamboo. After passing about two dozen houses, we reached the far end of the village, where stood a more solid-looking dwelling, also palm-thatched but built of sawn lumber rather than sticks or bamboo.

This was the house of the village king, according to Agustus. (Mentawai society does not include hereditary leaders. This "king," I later learned, was a *kepala desa,* an elected head of the village.) The king's house included a large, L-shaped front porch lined by a bench with a back that served also as an outer railing; the entrance to the porch was two picket gates swung open. About a dozen people were sitting and standing on the porch, half of them adults, half children, and after we took off our footwear and climbed the porch stairs, the king came forward and greeted us with a warm smile and handshake. The king was a young man, in his late twenties, I think, bare-chested and barefoot, wearing brown shorts. He was powerfully built and about my height and weight (which made him the biggest person of the village). His hair was short, a crew cut, and his face was widened by heavy cheekbones and lengthened by a strong jaw. He had a disarmingly casual and open manner and a broad smile with brown teeth. His name was Karlo Saddeau, and he had been educated at a mission school away from the village, which helped explain why he had no tattoos and spoke good Indonesian and even a few words of English.

Karlo introduced me to his father and then his mother. His father, curly-haired, wrinkle-faced, thin and slightly stooped, was slow-moving and very quiet. He wore only a tightly wound maroon loincloth and was decorated top to bottom with a symmetrical and stylized system of blue tattoo lines. Two lines began at either ear, curled across and down his cheeks and throat, and merged into a series of broad scallops across his chest; lines at either shoulder branched into graceful rope ladder patterns down both arms; a woven glove of lines began at his wrists, swirled into curly stars on the backs of his hands, followed the sinews onto the fingers, and then swirled again into stars down his fingers; a grid of mostly lateral lines descended the front of his legs from groin to knee, looking rather like venetian blinds, and the highest lateral lines of this grid curved back into quarter circles across his buttocks. Karlo's mother, soft-fleshed, with a face as wrinkled as a dry apple but a young woman's beautiful long black hair, was likewise covered with stylized tattoos, although fewer were visible since she wore a long flower-print skirt and a tattered sleeveless T-shirt. Still, I could see that the tattoos on the fronts of both her shoulders expanded as leaf-cluster designs, and the tips of the leaves curled over her shoulders around to her back. Her feet were tattooed down to the toes, and the lines rose halfway up her calves, which they encircled in a woven pattern that made me think of rolled-down stocking tops.

We were introduced to the rest of the adults on the porch, some of whom were tattooed with the same patterns. We passed out cigarettes to all the adults; they finished the ones they had been smoking and lit the new ones.

Antonius and Agustus sat down with Karlo and explained that I had come into the forest looking for bilous. Karlo thought that seemed reasonable and said he would be happy to show us where they lived. He thought we should go out that evening, stay in the forest overnight, and find them at dawn. However, since the weather continued to alternate between heavy rain and dreary drizzle, that idea fell by the wayside, and we spent the rest of the day and evening socializing with the family.

Aside from smoking, the main form of entertainment seemed to be talking. And after sitting through that evening and others, listening to endless lively conversations and laughter that passed freely among all members of the family, young and old, male and female, I began to believe that this was a happy family. Sometimes, in some places, the line between squalid poverty and great simplicity is thin; this family lived well in great simplicity, I thought. I watched Karlo handle his children rather like a gentle bear. They would climb on and over him, and he would pat them and talk to them and sometimes play with them. He seemed a good father.

The children were delighted to have visitors. Agustus began teaching them a few English words and phrases, and one of the boys spent a good deal of time that first night saying, "Soogar cane. Soogar cane. Soogar cane," which described what was growing to one side of the front porch. Another of the boys had learned to say, very slowly and solemnly, "How do you do? What is your name?" And, since his first name sounded exactly like mine, we amused each other by exchanging questions and answers on that subject.

Adults and children from other houses in the village freely wandered onto Karlo's porch, and at first I had a hard time telling who was attached to whom, as up to ten children played on the porch day and evening. I had been reluctant to ask personal questions for much the same reason I never took photographs: I aspired to be a guest, not a voyeur. But on my third day there I finally resolved to ask Karlo how many of the children were his. To introduce the question, I first showed him three photographs I had been carrying around in my pack, one of my wife, two of my children sitting in a tree. Karlo admired the photos and passed them around to several members of his family. Then I asked him how many children he had, and he said six, but one had died so there were only five now. His wife, a pretty young woman who, shyly, never spoke to me and seldom looked in my direction, was nurturing another inside her womb.

The rear of the house opened into a garden of trees and plants, which may have covered an acre or two — including a fish pond, streams, sluices, a pig pen, and a palm-thatched shed shading four or five wicker baskets containing chickens — before it merged into forest. One day I asked Karlo what he grew in his garden. His answer, which required some animated discussion and translation to complete: bananas, coconuts, durian, sago, rice, sugarcane, taro, peanuts, peppers, tobacco, chickens, pigs, and children.

The children on the porch had only one toy among them, a small top fashioned from a nut penetrated by a thin wooden spindle. A hole had been cut in the side of the nut, and if the top was spun fast enough, it would whistle. The first day I was there, one of the little girls also played with a pet crab, hardly bigger than a spider, that she kept tethered to a fiber strand. She used it to tease an even littler boy, naked except for three or four bead necklaces around his neck, who when teased enough would sit down on the porch and bawl. A few days later the little girl's pet crab was gone, replaced by a small, bluish white butterfly tethered to a strand of fiber. The paucity of toys and regular pets made me remember what my mother once told me about her childhood. She was born on a homestead in northern Montana and had, at the age of five, watched her mother die. Her father, who had too many other children to care for, then left her with a childless Swedish couple living on a homestead many miles distant. She grew up in the middle of the prairie with a few toys made of wood and nails, and only dogs and cats as companions. And yet these children had the love of a whole family, and they entertained each other. What did they need of more toys?

On one side of the porch was a table, and on the table was a worn chess set. The children often played chess with each other, and that first evening I sat and watched Karlo play several games with them. He always won. Karlo invited me to play a game, and so I did. After I had easily taken several of his pieces, I began to wonder if it would be properly diplomatic to lose a piece or two or three, or even the game, but as I was wondering that, I found my queen caught in a trap. The best I could do was trade her for a bishop and a rook, which I did, but after that loss the contest disintegrated into a rout, and I lost. Karlo laughed. I laughed. We played another game, which I lost. And another, which I also lost. I realized I was competing against an excellent player, who always managed to have his power pieces crushingly aligned against my king.

There was also a guitar on the porch, and Agustus amused himself and others by playing it and singing with scratchy earnestness. He said his favorite kind of American music was "country music," and he knew a few inexact lines from one American song: John Denver's "Take Me Home, Country Roads." We also made up a song about the bilou, which consisted of repeating the name several times to the tune of a song my children sing, "Peanut Butter."

All adult gibbons sing. The songs and the nature of singing differ from one species to another, but they also differ enough within species that individuals and couples can be recognized by their songs. The singing of a single adult gibbon may proclaim the search for a mate. But most singing is done by a mated pair, proclaiming the possession of territory.

Lar gibbon couples sing in a duet, described by biologist Michael Kavanagh in the following manner: "A few, brief, melancholy whoops from either the male or female begin the song, and then the female takes over alone with a 'great call' consisting of a few small piping sounds followed by a crescendo of steeply rising notes, perhaps six or ten, that get higher and louder until they reach a peak and are followed by two plaintive little wails that die away, and the male chips in with more of his brief, melancholy calls. For the next few minutes, the male and, to a lesser extent, the female may hoot gently, before beginning their stylized song again." Bilou couples, though — and they may be unique among gibbons in this regard — sing separately. The male begins singing while still comfortably ensconced in the family sleeping tree, four hours before dawn. Although he may continue after dawn, usually his mate will take up the after-dawn chorus. While the males tend to sing in a relaxed manner, in sleeping trees well away from neighboring males, the females often approach other females on the peripheries of their territory, within sight of each other, and display vigorously during their vocal performances.

I went to sleep listening to a whole racket of frogs and insects threshing away in the night and woke up at dawn to the sounds of roosters and, in the distance, a whooping sound, rising and falling, the song of the bilou. It soon began to rain, but later in the morning the sky cleared, so we began our search for the bilou. Four of us: Karlo the king leading the way, then me, then Agustus the interpreter, with Antonius the priest at the rear.

Karlo was wearing a tractor cap on his head, brown shorts, and a yellow and white cotton shirt. His feet were bare. A smoking cigarette was poked into his mouth. On his left wrist he wore a watch, and in his left hand he carried a machete. His right hand was free, but over his right shoulder he had slung a rope that supported a quiver full of arrows and, settled across the quiver, a bow. The quiver was a thick tube of bamboo closed at one end, plugged at the other with a smaller piece of bamboo, nicely decorated with strips of monkey fur, bark, and woven and dyed strands of fiber. The bow was as long as he was tall, dark brown with an oily sheen, carved with decorative marks at either end, strung with an oiled fibrous material that looked like a single strand of vine. I examined the arrows inside the quiver only later, but I'll describe them now: shafts made of hollow reeds, points of wood carved into the shape of two thin cones, one penetrating the other, so that when the leading cone entered skin and flesh it would break off at the weak spot, the tip of the second cone. The arrow points had been dipped in poison which, I was told, would cause death within about ten minutes.

Thus went Karlo. And as we left the village and entered the forest, I was pleased to observe that he walked steadily but slowly. The trail quickly turned to ankle-deep mud, which Karlo, barefooted, walked right through. Glancing behind me, I noticed that Agustus was wearing tennis shoes, and when the trail turned to mud he tried to keep them clean by scrambling to one side and then the other. I was wearing leather boots, and at first I tried likewise to keep them from getting too muddy by scrambling onto the edges of the trail.

Soon, though, the mud trail descended into a knee-deep river. The stones in this river were all covered with a thin, nearly invisible layer of slime. They were slippery, but not for Karlo, who never missed a step, as he kept up his regular, steady pace, smoking that cigarette, not once looking back to see the three of us behind him disco dancing across rocks. Antonius, at the rear of the line, seemed the best prepared, since he was wearing knee-high rubber boots, but when the river became more than knee deep, his boots filled up with water. However, when we finally walked out onto the stones on the far side of the river, and my pants and boots were soaked and squishy, I watched Antonius merely pause for a second, cocking one leg behind him, then the other, to let the water flow out of each boot.

We soon entered the river again, and this time, as the water became thigh deep and then waist deep, we were not only sliding on slippery rocks but also pushing against a powerful, churning current. Still Karlo kept up his regular, steady pace, not looking back, his bow and quiver of arrows nicely dry just inches above the water, and he still puffed on a cigarette.

The river became even deeper, and we left it again and walked along the steep, muddy edge, stepping in precariously small and muddy niches. Karlo kept up his methodical pace, but the rest of us slowed down to one careful step at a time.

A mud trail took us away from the river and deeper into the forest again, and I began to think of Karlo as inexorable. He never paused. Even when vines or branches reached into the path before him, he did not hesitate in step but simply raised his left arm with the machete and let the arm drop, the vine or branch dropping in nearly the same motion. He never seemed to look down at the path, at his feet, even though spiny plants and graspingly prickly vines were all around us, but always seemed to be looking up, looking, I realized, for game. He began to seem not merely inexorable but mechanical, robotic even, and the small puff of smoke regularly rising over his head added to the robotic illusion.

We crossed the river several more times, and then at last Karlo went up the bank and started climbing a muddy, nearly vertical hill. There were trees and branches and roots to grab onto and climb with, yet the climb was slow and exhausting, partly because the hill was so steep, but largely because this was rain forest. It was hot and so perpetually humid that nearly everything except the mud — trees, fallen branches and logs, rocks, roots, even vines, lianas, and leaves — was covered with moss. Still, Karlo seemed to keep up the same pace, merely changing into uphill gear as he stepped up the hill ahead of me, still not looking back.

Finally, perhaps a third of the way up the hill, Karlo stopped at a small landing, a jutting knob in the earth, turned, looked down at me below and at Agustus and Antonius even farther below, smiled his broad, brown-toothed smile, and said, without apparent irony: "Tired?" He sat down while the three of us caught up.

So we rested there on the knob. We urinated down the side of the hill and wiped the sweat off our arms and foreheads. I noticed a stinging sensation on my legs and pulled up my soaking pantlegs to discover a few sticky, dark leeches, which I pulled off with my fingers and then had to pull off my fingers. We spoke in whispers. Agustus said, "I wonder if we see bilou. Because bilou is very clever, isn't it? Very trick." But Karlo reckoned bilous might be somewhere at the top of the hill. Since we were on the side of a hill, a nice breeze penetrated the forest there, cooling us off and vibrating some of the leaves and palm fronds around us.

Until then I had been concentrating on the heat of the forest and the difficulty of the walk, but now I had a moment to appreciate the grandeur and beauty of what we had entered. Noisy with frogs and insects and the occasional calling or twittering bird; always dripping wet; green everywhere, except for red mud and gray bark and the white, cream, and orange splotches of lichen; with, I thought, an unusual preponderance of feather-shaped palm leaves. Moss: here thin as a patina or thick as fur; there fine as billiard table velvet or rough as terry cloth; over there broadly and densely leafed, like small, wet feathers or outstretched frog fingers. Like other rain forests I had seen in South America and Africa, this also contained its great giants, so thick and tall their trunks disappeared into a sea of leaves dappled by light and small pieces of sky. Like the other rain forests, this too swirled up high with vines, while lianas thick as arms or legs dropped straight or twisted and wound in improbable, knotty crooks and spirals. Here, too, many of the trees were supported by high, thin root buttresses. And yet — I was no botanist and so knew only because I had read it — in spite of this rain forest's great physical similarity to those elsewhere in the tropics, I was sitting among a spectrum of tree and plant species unique to this part of Southeast Asia, many of them unique to the Mentawai Islands.

We eventually gathered ourselves together and climbed to the top of the hill, where we saw nothing other than a few birds. So we walked back down the hill, and up and down several others for another couple of hours before we sat down again. This time Karlo thought we were near the gibbon. "I smell bilou," he whispered. And he stood up, took off his shirt, dropped it to the ground, and motioned me to follow. Now he walked very slowly; actually, he stalked. Every step involved raising a foot and very slowly lowering it into place before shifting to the next foot. I followed. Finally he squatted, looking out into the forest a moment before turning to me and motioning me to come closer. I squatted next to him, and he pointed out into the forest.

"There. Coconuts. Look," he said. I looked, saw nothing but trees and then saw a coconut tree, perhaps 150 feet away, rising perhaps 100 feet from a base that was steeply downhill from us. I heard a coconut drop through the leaves with a tearing sound and then a thud on the ground. "You wait here," he said. "Watch." And so I waited there and watched while he pulled the bow away from its resting place on top of the quiver, unplugged his quiver and drew out an arrow, and slowly and quietly moved away. The next time I looked, he had disappeared into green.

Agustus had waited behind, but Antonius quietly moved up to where I was and looked also toward the coconut tree. Still I saw nothing, but Antonius said, "You see it?" "No." "It's monkey. Green." By which he meant, I imagined, that he was seeing a shape made green by a screen of leaves. But still I saw nothing, and heard only once in a while the dropping of coconuts. Some time passed, and then I saw in a slight opening of leaves a brown body with arms and legs appear, move up, and sit. Sun illuminated it from behind, producing a yellow tint around the edges, and I knew that this was no bilou, since bilous are black. We watched, and the monkey sat there.

Many minutes passed, a quarter of an hour, and then I heard a sound in the forest over and to my right: thwack! And I saw an arrow rising up and curving into a slight parabolic trajectory sixty feet up and over, precisely into the hole in the leaves where the monkey was sitting. I heard the monkey fall: tearing through the leaves down and finally striking the ground with a thud. But, no, it must have been a coconut that fell, since I could still see the monkey up there. Thwack! A second arrow rose up high and curved slightly, entering the hole in the leaves again with what seemed perfect accuracy, an astonishing shot, but again a miss, and nothing fell this time, and I saw the monkey still on its perch. Thwack! A third arrow, again perfectly aimed, and this time I saw it: the monkey falling, falling, arms and legs out, falling twenty, then thirty feet — but no, it's not a fall but a desperate leap, and I see now that the monkey has landed in a lower branch of the same tree. Thwack! A fourth arrow quickly rises into the lower branch, but I see the monkey quickly dive-leap away to yet another, lower branch. Quickly now: thwack! A fifth arrow flies, but then I see nothing but leaves and trees and shadows, and hear nothing except the background whining and buzzing of insects.

Antonius and I remained where we were, looking up to where the monkey had been. "Do you think he killed it?" I asked Antonius at last, and Antonius indicated with a few English words and a lot of pantomime that he didn't think so; if the monkey had been hit it would be screaming in pain now, as it died from the poison. Still Karlo didn't emerge from the forest, and after several minutes I began to hear chopping sounds. "Why is he chopping trees?" I asked Antonius, and he said, "To get the childrens." What children? "Didn't you see? The childrens of the monkey. The monkey has childrens." No, I hadn't seen. Finally, I asked, "What is he going to do with a baby monkey?" And Antonius indicated, again mostly with pantomime, that Karlo would tie the baby to a tree, and the baby's cries would lure the mother back so that he could kill it.

Half an hour later I heard a desperate squealing that sounded like an injured puppy, and then Karlo appeared with a baby monkey clinging to his chest, trembling, and clutching mightily with its tiny hands. This little monkey was wild with fear, and when Karlo approached the three of us — by then Agustus had arrived from the other direction — its eyes moved and focused in great anxiety from one to the other of us. Monkeys look most like humans when they are babies, and this baby monkey was about the size of a premature human infant, which it resembled, except for the thin brown fur that covered parts of its pale skin. Its face was bare, and the skin of its face was pale white, bluish around the nose; it had two little nose holes, a wrinkled forehead, yellow-brown eyes, and a little mouth that was open and constantly squealing. Karlo was smiling and stroking the monkey's head and body, trying to calm it down. Agustus had some biscuits with him, and he tried to place some small crumbs in the baby's mouth, but of course the monkey would have none of that, and the huge fingers and dry crumbs at its mouth merely served to distress it even more. It was noisy, and like the uncontrolled crying of a human baby, the sound became distressing and unnerving after a few minutes. Nonetheless, we remained there while Karlo and Antonius and Agustus smoked cigarettes and rested, squatting. Antonius took over the baby then, held it in his arms and stroked it, but it continued squalling violently. After a while I asked Karlo what he intended to do with it. He said he was going to keep it as a pet for his children.

It was, of course, impossible to look for the bilou now, with a crying baby monkey on our hands. Karlo looked at his watch and said it was going to rain. He retrieved his shirt then, wrapped the monkey in it, and we started back down toward the village. Soon it began to rain, and then to rain hard, and Karlo cut off a six-foot-long palm leaf and held it over his head as an umbrella, the tip of his machete poked into the leaf as a handle. Agustus and Antonius likewise held palm leaves over their heads as umbrellas, and I put on my poncho. We slid down and slogged through mud, waded again through the river, until at last we reached Rok-Dok, soaked to the bone.

There were no pet monkeys in the village, and our arrival with the squealing bundle provoked a good deal of attention. Karlo's children were indeed excited by the new pet and immediately began to play with it. Someone brought a wicker basket and placed the monkey inside that, and then hung it up beneath the roof of the porch. The monkey in the basket squealed and squawked much of the rest of the day. The children kept taking the basket down and pulling the monkey out, handling it by the arms and legs, then clumsily pushing it back into the basket. That evening a woman from one of the other houses came to visit; she took the baby monkey out of the cage and fed it a pasty gruel from her fingertips. It curled up peacefully across her breast, but after she put it back in the cage and left, it began crying again. It cried at least until I fell asleep and was crying when I woke up in the morning. I wondered if it would live very long, and Antonius wondered likewise: "I think maybe the monkey die soon. I feel sad when I see the baby monkey."

It rained hard the next morning, but after several hours it stopped, so the four of us went out again to look for bilous. We made the same kind of marathon journey through the forest and a slippery labyrinth of mud, swamp, stream, and river, and explored some other hills. At last we came to a place where Karlo looked over and pointed. "Bilou!" he whispered. There in the distance, in a gap among the leaves and branches and vines, crouched a black shape, clarified by a patch of light behind it, looking like a lonesome figure standing before a lighted window at night. As before, Karlo went on ahead, slowly, slowly, and disappeared into the forest; I was relieved to see he hadn't this time pulled out his bow or drawn forth an arrow. The rest of us continued to watch the small black figure in the trees, which didn't move. I watched it through my binoculars for several minutes, but the lenses kept steaming over. Suddenly it moved, appearing to reach forth and grab something, a branch perhaps, and then it was gone inside the leaves.

There must have been other bilous around us, but we never saw them. Karlo finally returned, we continued the search until it began raining again, and then we walked back down to the village.

That was the first and last bilou I saw in the vicinity of Rok-Dok, in the interior of Siberut. Continuing rains not only limited our further excursions into the forest, but, according to Karlo, limited the excursions of bilous away from their sleeping trees. They didn't like the rain any more than we did. Agustus later suggested to me that the bilous may have been overhunted in that region. "Many time in Rok-Dok they see bilou, they shoot" — he made a bow and arrow motion — "every day, so not so much bilou."

In spite of various legal prohibitions, gibbons are hunted as food over much of their southern and southeast Asian habitat. Bilous have been hunted by some of the indigenous Mentawai Islands people for centuries. Traditional Mentawai hunting equipment consists of what Karlo carried: a palmwood bow and poisoned, wood-tipped arrows. The arrow poison, incidentally, is mixed from four ingredients commonly grown in household gardens throughout the islands: toxic latex from the upas tree (*Antiaris toxicaria*), rotenone extracted from roots of the derris plant (*Derris elliptica*), chile peppers, and gingerroot juice. In some places in Mentawai, bows and arrows have been supplanted by more modern devices. Rifles are used on Siberut around trading outposts and the town of Muara Siberut. On the southern islands hunters who can afford them prefer .22 caliber handguns and air rifles; the air rifle pellets are coated with arrow poison, making them at least as deadly as lead bullets.

Human populations on Siberut have remained comparatively stable, but on the three southern Mentawai Islands — Sipora and North and South Pagai — the human population has expanded fivefold in the last century and a half. The population is still less dense than on Siberut and is generally concentrated in coastal villages, but during the last decade commercial loggers in North and South Pagai have cut roads into the islands' interior, and loggers now even provide free transportation for local hunters. Nonetheless the bilou is threatened by hunting primarily on Siberut, where natives are hungry for animal protein and intensively pursue all four primate types. The people of North and South Pagai hunt less than those of Siberut, partly because they obtain sufficient animal protein from the sea. When they do hunt primates, they usually seek the pig-tailed langur because they prefer its taste. Currently the bilou is protected by Indonesian law, but that law is not enforced in the Mentawai Islands.

Subsistence hunting threatens, yet the Deluge approaching the several gibbon species of southern and southeastern Asia today is primarily a Deluge of deforestation. Gibbons are almost wholly arboreal, absolutely dependent upon forest habitat, as they have been for millions of years. They are also exceedingly territorial and thus susceptible to even minor forest disturbance. The remaining forests with gibbon habitat in southern and southeastern Asia are declining and disappearing in our historical instant.

The forests of Bangladesh, once excellent habitat for hoolock gibbons, are being transformed into firewood and construction lumber. Hoolock gibbon habitat in both Bangladesh and Burma is already "greatly reduced." In the state of Assam in eastern India, hoolock habitat is threatened by large-scale commercial logging. Deforestation in China eliminated the lar gibbon from most of that country by the fourteenth century, and today in Thailand only a tenth of the original lar gibbon habitat remains. Forest clearing for timber, slash-and-burn agriculture, and road building threaten both the lar and the pileated gibbon in Thailand, and could reduce the lar gibbon to remnant populations within the next three decades. In peninsular Malaysia 10 percent of all forests were removed in a single decade, between 1960 and 1970; today forests harboring lar gibbons and siamangs in peninsular Malaysia may be disappearing at the rate of 1,000 square miles per year. In Sumatra commercial logging and clearing for agriculture is continually destroying gibbon and siamang habitat, while in Java the remaining forest habitat for gibbons consists of mere scattered patches. Based on current rates of deforestation, it is possible that nine tenths of the gibbon habitat that existed in the mid-1970s will have been destroyed or degraded by the 1990s.

As the Deluge rises throughout the larger region, so it rises in the Mentawai Islands. For the three millennia that they've inhabited the islands, local Mentawai people have selectively cut larger trees of one species to construct dugout canoes and houses. They clear small areas to plant household gardens and cash crops such as clove. In some regions local people also harvest two sorts of rattan vines, known as *manau* and *rotan,* raw material for wicker furniture. Rattan harvesting has blossomed recently, with 10 million yards of rotan (sold in ten-foot strips) and 860 tons of manau (thinner than rotan, it is sold in coils by weight) exported between 1977 and 1982. Since bilous and other primates eat fruit from rattan vines, this trade could conceivably become a significant threat.

Traditional cutting and forest degradation may occasionally remove crucial sleeping trees and food trees and reduce protective cover for gibbons, making them more vulnerable to hunters. Nonetheless, the overall impact of local Mentawaians is still minor. The Indonesian government claims possession of all the trees on the four Mentawai Islands, however, and in the early 1970s the Indonesian Forestry Department surveyed the islands to determine the locations and values of the best logging sites. By 1971 American, Indonesian, Malaysian, and Philippine logging companies had leased concessions for nine tenths of the land and had begun felling huge sections of forest for export. By the mid-1980s only two thirds of Siberut included undisturbed forest. In North and South Pagai the Minas Lumber Company (founded by North Americans, currently owned by Malaysians) leases a 360-square-mile concession. At present a Minas-owned sawmill on the islands yields some one hundred tons of cut wood per day, nearly all of which is shipped directly to wholesale markets in England and Holland.

Thus rises the Deluge. The only Ark for the bilou and the three other unique and endangered primates is on Siberut, in an area I never managed to visit. The Indonesian Ministry of Agriculture created a 25-square-mile game sanctuary on Siberut in 1976 and expanded it to include almost 50 square miles in 1977. But the sanctuary, known as Teitei Batti, included villages and mangrove swamps unsuitable for gibbons, and only a small fraction of the island's bilou population, about 400 to 500 animals altogether, lived within the reserve area. After the creation of Teitei Batti, Indonesia invited the World Wildlife Fund and Survival International to propose a more coherent conservation policy for Siberut. The resulting plan would have expanded Teitei Batti to about 190 square miles of strictly protected core area surrounded by a traditional zone of nearly 400 square miles, where commercial logging would be discontinued and Mentawai natives would theoretically continue their traditional way of life. Indonesia has not followed that proposal. The ministry of agriculture did add another 190 square miles to the original game sanctuary in late 1979, but the status of that particular Ark is still unclear; in Indonesia, a game sanctuary (*suaka margesatwa*) can either define strict protection of plants, animals, and general habitat, or it can allow traditional and commercial exploitation.

Eventually Agustus, Antonius, and I decided to go to another village to find bilous. We said good-bye to Karlo and his family, left some gifts, and took the dugout canoe back down the river all the way out into the ocean. I had not imagined the canoe would do well in open ocean, but it did, and we rose and fell and broke open wide white waves in swelling, purple waters for some time. Then we entered the green waters of a bay, passed over a bottom billowing with white coral (giant brains and cabbages beneath green glass), penetrated a channel through the black snaky roots of mangrove swamp, and finally beached the boat in an inlet lined with knee-deep mud.

From there we walked to a village called Limu, perhaps a dozen small cottages altogether. "All the people are family here," Agustus said. "Brother, uncle, sister. Not 'nother person." Few of the people we saw were tatooed, and none of the men wore loincloths. "The people here already get modern. You can find potatoes, because the missionaries get here."

We came to a small wood-and-bark, palm-thatched cottage with a view of the sea, decorated by seashells and two hanging wicker baskets that had exotic birds inside. It was occupied by a young man whose first name was Levi — I never learned his last name — and he wore a T-shirt that said: *Levi's. Levi Strauss of San Francisco.* Levi had a wife and three children, but his wife and two of the children were away visiting her sick father. The third child, a preadolescent boy, was at home, but he was terribly shy, as quiet and elusive as a shadow. Levi, too, was quiet, and our time there included many hours sitting on his small porch, enjoying or enduring long, long silences.

We went out to find bilous late one afternoon, out of the village into the high forest. The forest smell made me think of the residue of whiskey in old bottles, but small patches of air turned briefly sweet or musky or barnyardy. Everything not green and living was either rotting, rotten, or about to rot; and whatever was rotting or rotten (a great tree trunk, for instance, fallen shattered on its side) sprouted vertical new growth, flowers and mushrooms, green saplings and tender shoots curling toward the light above. We sweated profusely, which attracted many mosquitoes, and we slapped the mosquitoes carefully and talked in whispers and walked slowly, trying to maintain a perfect silence. When I stepped on a log that disintegrated with a loud crunch and crack, we could barely stifle our spirited laughter.

We came at last to the tree of the bilous. It was gigantic, incredible, surely one of the biggest trees I've ever seen in my life. I looked up at it and thought of a space rocket, the Apollo moon rocket. A great, gray trunk, nearly a dozen feet in diameter, rose perhaps a hundred feet before it was joined and obscured by a great crisscrossing cluster of many other gray trunks, which were all part of a single growth, all children of that great father (or mother) trunk. Levi spoke in whispered Mentawai, and Agustus and Antonius sometimes whispered translations to me. Agustus: "This tree special to sleep. Bilou." Antonius: "You see, they sleep in this big tree."

We stayed there, hidden behind ground-level vegetation, and watched while the light dissolved into dusk. In the distance I could hear the sound of running water and, nearer, the rattle, grind, and hum of insects; the whine of approaching mosquitoes; bird sounds: wheep, wheep, wheep, wheep who-tit, wheep who-tit; and, once, the cry of a monkey: ao-ao-ao-ao-ao-ao-ao-ao.

Then Antonius said, "You see. There are many, all over." And I saw moving leaves high up, and then a shadowy figure behind leaves that moved very quickly in a swinging motion from one branch to the next before it disappeared into the clustered leaves, branches, and trunks of the giant tree. Soon I saw another figure moving rapidly behind leaves that likewise quickly disappeared into the giant tree. "You see them? You see? There are many. Many," Antonius said. But that was all I saw, and then it became dark.

We went back to Levi's house and cooked rice, noodles, and fish over the bamboo fire in his kitchen. We talked about bilous. According to Levi, no one in the village of Limu hunted the bilou. People didn't like the taste of the meat, so hunters concentrated on the three monkeys of the forest but left the gibbons alone.

The next morning we got up before dawn and returned to the bilous' sleeping tree. It was still dark, so we made our way through the forest with the help of flashlights, following the whistling beacon of the bilous' predawn song. The song consisted of a series of whistlelike calls, three, sometimes four at a time, followed by a silent rest: the first long note quickly falling in pitch, a brief glissando; the second and third rising in pitch at the very end. We walked toward that whistling song very slowly and turned off the flashlights as the first crepuscular light appeared. As we moved closer to the sleeping tree, we began walking only during the whistles to mask our noise. Once a small branch cracked underfoot, and instantly the song broke in mid-note. A long silence followed. We waited. Then the song began again, and we began moving again.

At last we came to the sleeping tree, huge, looming above us, and walked slowly beneath it, beneath the whistles emerging from it, and slowly climbed a small hill and ridge to the far side of the tree, sat, and waited. "When they wake up," Antonius told me, "they will pass here." We waited as the leaves and branches above us began taking form in pale light. The whistling song ended. We waited. Suddenly I saw one. Or rather I saw a green wave with a dark shadow inside, as the leafy branches of tree after tree after tree dipped and swayed and rose, as a dark spirit of a gibbon moved swiftly behind them and then disappeared. I saw another, its form, arms and legs, land in a tree behind a screen of leaves, then disappear into thicker foliage. We continued to wait, but that was all I saw. They had all left the sleeping tree.

"We go down to the small river," Antonius said at last, so we climbed down a steep hillside down into a stream. This stream was beautiful, a watery gorge through a garden rain forest, mossy, with a series of clear pools and mossy waterfalls that we waded through and climbed up, moving upstream, sometimes hearing a rustling in the trees, pausing many times to look up, almost straight up at the high trees on either side of the gorge that rose above us.

The bilous saw us first, though, and started up a racket of alarm calls. First a whooo-whooo-whooo, then a rapid whohohohohoho-hoho. Then I looked up to see one appear from behind a curtain of leaves. It grasped a branch over its head and cast off in the manner of a circus trapeze artist, immediately to disappear behind another curtain of leaves.

We continued up the creek, hearing their alarm calls very distinctly now, but they were invisible still, spirits, so high up here, and between them and us a dense cloud of leaves. At last I saw clearly what I had only half observed before: a small black gibbon, illuminated from behind by a clear morning sky, hurling itself through the trees hand over hand, swift and straight, reaching as close to flight as a nonwinged creature will ever reach. The bilou. Then it was gone.

I was concerned about Antonius's six children, as he was — since he had left them to take care of themselves several days earlier, and since his wife was probably still away, caring for an emergency child-birth in Padang. Thus, after finding the bilous that morning, we left the village of Limu.

We loaded the canoe and headed back out to sea.

At sea the sky was stunningly dramatic, clouds everywhere washed by an expert watercolorist in basic colors. Seaward, where the sky reflected onto the sea, the water was molten silver, softly rippling. Shoreward, the island was still covered in places by long white fingers of fog, so I saw only a line of trees outlined by a white background, and standing behind that another line of trees and another white background.

Agustus wanted to know why you never find the body of a dead monkey in the forest. "Do they live forever?" He wanted to know why the bilou was found only on the Mentawai Islands. Couldn't it be yet undiscovered elsewhere? He wondered if I believed in the story of Adam and Eve.

We saw before us two schools of dolphins, perhaps sixty or eighty altogether, swimming in synchronized pairs and triplets, rolling up and under, up and under. They were spinner dolphins, I believe, my belief based on the following sight: one dolphin in the second school leapt straight out of the water in front of us, shedding a sunlit sheet of spray like shattering glass, rotated once, flipped upside down, and returned to the water without a splash. But the school kept right on swimming before us, cutting right across our bow, then disappeared beneath the surface just as we reached them.

When we came at last into the harbor of Muara Siberut, it was swarming with fishermen in canoes. We trolled a hook and lure on a long line, turned across the surf in circles, and hauled in one, two, three, four fat fish, mother-of-pearl on the outside, with white pointed tongues and maroon flesh showing inside their gills, that gasped and flapped at the bottom of the canoe. Antonius brought the fish home to his children.

By the end of the day Agustus and I were sitting on top of a sun-warmed heap of rough copra — split and dried coconuts — surrounded by long bundles of rattan, on a cargo boat bound for Sumatra. But the captain of this boat decided that the weather near Sumatra was not good enough to leave until the next morning, so we merely chugged along the edge of Siberut and then dropped anchor in a deserted bay.

Evening came, and the boat's first mate climbed on top of the cabin, put down a rug, faced Mecca, and began a long series of bows and murmured prayers. Night came, and I stayed awake on top of the sweet and oily-smelling copra, rolling beneath the sweep of the Milky Way, looking out a few thousand light years into the universe, listening to the occasional splash of jumping fish and the restless groans and snores of sleeping people, reflecting on my trip to Siberut and my vision of the bilou. I remembered the bilou, so quick and high and elusive, as a shadow behind leaves, a fleeting figure before light, a dream vision, a spirit, a child of the forest . . . I had other romantic notions, which were only temporarily dimmed when I noticed rats crawling around dark places on the boat.

Ten

WOOD

I WENT TO BORNEO, walked in a rain forest there, and listened to the pulsating whistle-ring of cicadas, the twitter of birds, and the occasional chuffing of hornbills taking flight.

I saw hornbills sitting in trees. They had scimitar beaks and white-tipped tails, and whenever they saw me watching them, they cried craawk craawk craawk, unfolded their big wings, chuffed into the air, and disappeared behind trees. Once I came across a troop of long-tailed macaques. They made a boiling in the trees, dived, and yapped with a hollow sound. I saw pig-tailed macaques also, and once was charged by a large, swaggering male, who approached on the ground as boldly as a nasty dog, hissed, showed his teeth, and rolled the skin and fur on his forehead in a most alarming fashion. Mostly, though, I watched butterflies flutter by. Amazing butterflies of all sizes, shapes, and colors: small yellow ones, moving fast and zigzaggy; large dusty brown ones with black spots, which flapped their wings and navigated around trees and brush, looping and swooping like slow birds; little bright blue ones, which showed their color only when their wings were spread, flashing blue on and off like lights; gold ones with white lace patterns; brownish gray ones with tattered wings that looked (when still) exactly like decaying leaves.

Bornean rain forests are famous for their butterflies and equally famous for their trees, which are among the world's biggest. In the Amazon and West Africa I had walked among trees with root buttresses a dozen feet high, but in this forest some of the buttresses rose twenty or thirty feet. Many of the trees towered like monuments, and, most distinctively, many of them were nearly perfect pieces of wood —

round columns that shot straight up to the sky before spreading a crown of branches and leaves. Many of those columns were as thick as I am tall, and they often reminded me of stacked oil barrels — which suggests not only how they looked but how much they were worth. The bark of those columnar trees was usually smooth, but sometimes striated, sometimes splotched with moss and lichen. Some were matted with vines, and many others were furry with mosses, bromeliads, ferns, and other epiphytes.

I had come to Borneo to find orangutans, and, on my second day there, I walked right under two, a mother and an infant, who were sitting on a branch thirty feet above the ground. Orangutans are big and reddish brown, but surprisingly hard to see in a forest. I didn't see this pair at first, but when I thought I heard a noise, I stopped and looked up into the trees around me.

There they were, still and stolid, once in a while making little pig grunts. The mother sat on a branch with both knees up. One foot grasped the branch while the other dangled loosely. Her long right arm was raised straight up so that her long hand could loop around a branch above her, while with her left hand she grasped the infant. I could see right away that the infant was a male, since the mother had him tossed over one knee, almost a spanking position, so that his little rear stuck out in my direction and his two frail legs were splayed out. I took out my binoculars and watched them, and the mother looked at me expressionlessly, then looked away and slowly blinked her eyes. The mother's fur was a dark brown reddened with a henna tint; the skin on her face was brown, but her eyelids were salmon-colored, so when she slowly blinked, salmon circles appeared in the midst of brown. The baby turned around then, lifted his head, and looked at me. They sat there, made pig grunts, and periodically chewed on something that looked like a banana. Eventually the baby wandered clumsily onto the mother's shoulders and then her back, and some time later the mother tucked him under an arm and began climbing up the tree, turning to look back at me once in a while, climbing just like a person — stepping on branches, pulling herself up with higher branches — until she was forty, then fifty feet up, and she and her baby were gone into green.

The name *orangutan* derives from a Malay expression meaning *man of the forest,* and most aboriginal mythology emphasizes this forest animal's strong resemblance to humans. One local story states that orangutans are descended from a man who, out of shame for some misdeed, left his own village to live alone in the forest. Another tale has it that orangutans were created after a pair of birdlike creatures learned to make humans and then forgot how they did it while they were asleep. When they awoke, they were able to produce only a strange and shaggy approximation to human form.

Early Europeans were likewise intrigued by the orangutan's interesting similarity to humans. A Dutch physician, Jacob de Bondt, wrote in his *Historiae Naturalis* of 1658 that the orangutan was a "wonderful monster with a human face," which he had seen walking upright. But with apparent exaggeration, de Bondt went on to describe a female of the species who was "hiding her secret parts with so great modesty from unknown men, and also her face with her hands (if one may speak thus), weeping copiously, uttering groans and expressing other human acts so that you would say nothing human was lacking in her but speech." According to de Bondt, the Javanese believed that these animals were able to speak but were smart enough not to, "lest they should be compelled to labor."

Daniel Beeckman, an English sea captain who briefly set foot on Borneo in 1712, considered "Oran-ootans" to be "the most remarkable" of the island's many primate types. According to Beeckman, they "grow up to be six Foot high; they walk upright, have longer Arms than Men, tolerable good Faces . . . large Teeth, no Tails nor Hair, but on those Parts where it grows on humane Bodies; they are very nimble footed and mighty strong; they throw great Stones, Sticks, and Billets at those Persons that offend them. The Natives do really believe that these were formerly Men, but Metamorphosed into Beasts for their Blasphemy."

Actually, Beeckman's brief description is more or less accurate. Female orangutans may stand three and a half feet tall, and males reach four or five feet in height (only exceptionally approaching the six feet Beeckman recorded). Although orangutans are by far the most arboreal of the great apes, they do sometimes walk on the ground, erect or on all fours. The orangutan's "tolerable good Face" is tolerably human-appearing in some ways. It is a bare face (light in infancy, becoming dark with age), with round, close-set eyes, small ears, and a high, smooth forehead uninterrupted by the brow ridge chimpanzees and gorillas possess. Although the orangutan's reddish hair is distributed on its body somewhat in the fashion of human hair, it grows much longer and becomes quite shaggy. Strands of shaggy shoulder hair can reach twenty inches in length, for instance, while finger hair grows up to four inches. In Sumatra adult male orangutans grow white or light yellow beards and mustaches over their protuberant jaws and upper lips. And in both Borneo and Sumatra they develop at least two distinctly weird and nonhumanoid features: protruding flanges of fibrous tissue on their cheeks and great inflatable pouches (*laryngeal sacs*) hanging from their throats.

Orangutans are mainly tree dwellers, and this preference or habit probably explains more about them than any other real or imagined quality. Anatomically they are well suited to the arboreal life. For reaching, hanging, and swinging, they possess immensely powerful arms that can stretch out well over seven feet from fingertip to fingertip. Their hands are built like great hooks, with long curled fingers and a small thumb placed low on the hand. Their prehensile feet are almost identical in form to their hands, making this primate quadrumanous rather than quadrupedal. The hip and shoulder joints are remarkably flexible, allowing an astonishing repertoire of postures in the trees — hanging casually suspended by an arm and a leg, for example; or supported by a branch beneath with two grasping feet, supported from above with one hand, the other hand reaching out for food; or stretched horizontally, grasping branches with one arm and one leg, the other arm and leg stretched out to reach food; or hanging entirely upside down, feet grasping a branch or two above, leaving the hands free to reach in several directions for food.

In spite of their several useful adaptations to life in the trees, however, orangutans are severely limited by one factor: weight. Only one primate species, the gorilla, is heavier. And of the primates that live, move, and feed mainly in the trees, orangutans are by far the heaviest. Females reach an adult weight of almost 90 pounds, while males can reach 220 pounds. As a consequence, orangutans tend to be careful, slow-moving, slightly ponderous — although in extreme circumstances they may dash, swing quickly, leap, or even crash head over heels to the ground, briefly grasping branches along the way in a semicontrolled fall. But normally they are careful to test the supporting capacity of a branch before burdening it with their full weight. They support themselves with more than one hand or foot at a time. And they usually subscribe to the insurance provided by gripping two or more branches rather than one. Larger males eventually succumb altogether to gravitational reality, in fact, and spend much of their time at lower levels or even moving and foraging on the ground.

The forest I walked through, the Sepilok Forest Reserve in the Malaysian state of Sabah in northeastern Borneo, supposedly contains around three-hundred orangutans. Many of them are entirely wild animals, but many are only semiwild; they are rehabs, orangutans that have been orphaned by logging operations or captured to serve as pets, then confiscated by the government and returned to the forest at Sepilok's Orangutan Rehabilitation Center.

At the center baby orangutans are kept in cages and gradually habituated to the forest until they are ready to cast out on their own. The rehabs never become entirely wild, however. Some disappear for weeks or months and even mate with the wild orangutans, but many of them just hang around and return twice a day to a feeding tower in the forest, where bland food — bananas and porridge — is provided.

Thus the easiest way to find orangutans in Sepilok is to go to the feeding tower when the food is brought out, and wait. I did that, and shortly after two workers from the center placed a bucket of porridge and several bunches of bananas on the tower platform, orangutans started appearing.

Having seem them previously only in zoos, I had always thought of orangutans as dull and ungainly. In zoos they will sit for a long time, hunched, looking like shaggy hummocks or used cat furniture. Orangutan babies and toddlers can be very active and playful, excitable and full of expression even in zoos; but the grown-ups in zoos seem jaded with life, and they take on the zonked expressions of factory workers on an assembly line or patients recovering from electroshock treatment. Dull and ungainly I thought they would be, so I was wonderfully surprised, illuminated actually, to watch young, adolescent, and adult orangutans at Sepilok move through the forest and gradually roll toward the feeding station. They moved not with the leaping drama of black-and-white colobus, not with the four-footed purposefulness of lion-tailed macaques, and certainly not with the brilliant speed and agility of bilous. But in their own unique way, they were perfectly dramatic, purposeful, and agile.

They could climb anywhere, it seemed, and they moved through even the smallest trees and across even the thinnest network of vines. I saw one climb a vine hand over hand, foot over foot, and then stop, loosen one hand, and put his foot right up where the hand had been. He hung down, suspended by one hand and one foot, and used the other foot to grasp another vine for balance, the other hand to scratch. He puffed up his lips and blinked slowly. I saw another one swing on a vine across a gap in the trees, Tarzan-style. She swung until the branches of a small tree were within grabbing distance, then grabbed the branches with a hand and foot, still held the vine with the other hand and foot, and slowly transferred herself into the tree. Once in the tree, she stood right up on a branch, clutching the trunk, and began heavily shifting her weight, back and forth, back and forth, until the whole tree was oscillating back and forth in an increasingly wider swing — until she could at last reach over to the branches of another tree, transfer herself into it, climb higher, and begin shifting her weight again, repeating the process. I saw another who had been nicely climbing across a matrix of vines and branches, perhaps sixty feet above the ground, when she suddenly flipped upside down into some lower vines and branches, and then flipped right side up again to reach some other vines and branches.

Slowly, slowly, a half dozen or so reddish brown, furry creatures rolled in from the deeper forest from all directions and slowly climbed, swung, clambered, shifted, descended, and ascended until they arrived at the feeding tower platform, where they were met by the two Sepilok workers offering a bucket of slop and many bunches of bananas. An adolescent male arrived at the platform first, looking grumpy, grabbed a cluster of six bananas, held it in his mouth — the bananas looked like big buck teeth — and took off again, climbed into the top of a half-dead tree, and sat by himself eating the bananas. A pair of young females, traveling together, arrived at the tower next and ate side by side, arms thrown across each other's shoulders. A cup of slop was held out for these two, and together they pursed their lips into tubes and slurped the slop. One reached up to scratch her back, and her arm was so very long it seemed to have several extra inches of elbow, like bad plumbing with fur.

Others arrived at the feeding tower, and there was some fond hugging and, later, some real squabbling, including squeals, pushes, punches, and bites. One of the animals, having eaten, lay down on the tower platform and was groomed by another. The groomer worked over the lying one's fur very intently, as if she were looking for a lost needle in a pile of yarn.

Later, after it seemed that feeding was over, a few other orangutans began appearing in distant trees and slowly rolling toward the tower. A larger, subadult male swung his way over to the tower and tried to steal all the remaining bananas, two or three dozen. The workers grabbed some away, and this orangutan took off.

I noticed some movement in a tree, and saw the mother and baby I had been watching earlier in the day; the mother, with the baby clutching her side, was slowly negotiating tree branches. Then she sat for a long time, considered the situation, and began moving again, the baby at her belly now, and eventually worked her way across vines onto a small, polelike tree. She swung it back and forth several times, each time a little wider, until she could grab the railing of the feeding platform and climb aboard. The baby had a little flag of hair on its head. The mother grabbed a bunch of bananas and cast off into a tree, ran up the tree, and sat, holding one vine with her left hand

and another vine with her left foot, clutching a slim branch of the tree with her right foot, and cradling the infant and bananas across her right knee with the help of her right hand.

I confess I was mildly disappointed to see the mother and baby at the platform, since I had allowed myself the illusion, when I first saw them, that they were wild orangutans, not part of the semiwild rehab population that will approach humans and eat the food they offer. But later one of the workers pointed out to me a genuinely wild male orangutan, who had come near the feeding area to watch. He lurked way in the distance, appearing and disappearing behind a high, dense clump of branches and vines in a tree. He seemed very big and very dark, and he was leaning far forward and peering out at the feeding animals from his shadowy vantage point: a mournful voyeur, an aging playboy spying on a high school prom.

Orangutans reach sexual maturity at seven to ten years, although males then enter a subadult phase that may last until they are about fifteen. At that age, or sometimes later if a dominant adult male resides nearby, the young males reach their full size and acquire the physical characteristics of adulthood, including the thick, fleshy pads on their cheeks and the inflatable laryngeal sac. This pouch functions like the bag of a bagpipe, and the adult male will sometimes use it to announce his existence with a long call.

According to John MacKinnon, who carried out major field studies in both Borneo and Sumatra, the adult male's long call lasts from one to three minutes and consists of "a series of groans, to which the partially inflated laryngeal pouch gives a deep and resonant tone. The call rises to an early climax, at which point the animal can aptly be described as roaring, then tails off gradually into soft sighs, and ends in a glorious string of deep bubbling noises as the pouch deflates." Actually, orangutans are relatively quiet animals, and the long call is their only vocalization that carries over any distance. It can be heard for well over half a mile in the forest.

What is the function of such noisy behavior? Females usually ignore the call, and we can presume that it is not primarily a sexual invitation, although according to orangutan expert Herman Rijksen: "For a female in heat the long call of her particular male(s) serves as a beacon of love." But mostly the call is a male-to-male communication: the resident adult male announcing his possession of territory. Other males — nomadic, nonresident adults that happen to be in the area — will respond to the call by making themselves scarce, or, alternatively, by challenging the caller in extended, fierce, and sometimes fatal battles. Older males carry with them the evidence of past battles in the form of deep scars, torn nasal septa, broken or missing (bitten-off) fingers, and so on.

Orangutans are the most solitary of the apes, and adult males are the most solitary orangutans. They spend most of their time quite alone, either wandering through the forest or residing in a more defined home range. Unlike the adult males of most other primate species, they are entirely intolerant of juveniles and other adult males, and they seldom or never maintain an association with adult females beyond brief periods of courtship and mating.

Orangutans mate under two very different circumstances. One is the forced mating, or rape, usually perpetrated by a subadult male who is sexually mature and interested in, but not yet interesting to, females. John MacKinnon observed matings eight times during his early field study in Borneo. Seven of the eight were forced; the female screamed and otherwise expressed fear and continually attempted to escape, while the young male violently pursued her, sometimes striking and biting, and grasping her thighs or waist with his prehensile feet as he attempted penetration. Those were, MacKinnon insists, genuine rapes.

The second sort of mating is mutual and occurs after an extended period of peaceful courtship. When, after the weaning of the youngest offspring, the female becomes fertile and develops sexual inclinations, she will seek out or make herself available to a fully adult male, usually the resident male, with whom she has had a long, though ordinarily distant, relationship. Typically the two will conduct a courtship lasting days or even weeks, during which they may forage together and engage in considerable sexual foreplay. The foreplay may include a series of embraces or much more directed acts, and it may be initiated by either male or female, although females are often the more active. The female may attempt to arouse her partner by stroking his penis with her hand, for example, or by lounging against branches with her legs spread apart, in provocative proximity to his face. When at last the courtship is consummated, the couple engages in a leisurely and apparently affectionate intercourse high in the trees. Intercourse may occur with the female facing away from the male, or with the pair facing each other — a position sometimes favored by only one other primate species. According to one expert, the position most likely to result in a pregnancy is with the male lying on his back or suspended across branches with his back toward the ground and the female above, straddling him.

Fossil remains indicate that orangutans were once common, widely distributed in Borneo and Sumatra, and ranging on the Asian mainland into India and as far north as Peking, as far south as the islands of Java and Celebes. Today this beleaguered great ape survives only in the declining lowland forests of Borneo and northern Sumatra. In the early 1960s it was estimated that only around 4,000 wild orangutans were left, an alarmingly low figure. More recent and probably more accurate estimates suggest that several times that number still remain. But even in the best of circumstances, orangutans reproduce very slowly: a female may bear only four or five offspring in her lifetime, of which only two to three are likely to survive to adulthood. These reproduction and mortality rates are perfectly adequate for a stable existence in a stable environment, but under present circumstances they could prove disastrous.

Orangutans have been and are still occasionally hunted. But equally destructive in this century has been the live animal trade. Until recently, orangutans of Borneo and Sumatra were captured and sold in comparatively large numbers. Some captured animals were sold locally as pets; wealthy Indonesians or expatriates would pay well for an infant or juvenile orangutan.*

Most of the captured orangutans were exported. The first European to possess a live orangutan was probably British sea captain Daniel Beeckman, who purchased one for a pet in 1712; it lived for seven months. The first captive orangutan to survive the voyage to Europe arrived there in 1776 and was kept in the private menagerie of the Prince of Orange briefly before it died. The second captive orangutan in Europe survived for five months in 1808 in the private collection of the Empress Josephine, at Malmaison, France. Two arrived in

*Herman Rijksen and Ans Rijksen-Graatsma, who searched northern Sumatra for illegally kept orangutans in the 1970s, found that many of the captive animals were fed only cooked rice, even though infants must have milk to survive. For many owners, especially those hoping to make a profit, the price of canned milk was too high, and they preferred to risk the malnutrition and death of their animals. The researchers also found that most of the captive animals were kept in cruelly small quarters. One young female they discovered had been kept for six years in a cage slightly larger than three by three by five feet. Another young animal, after spending six years in a small cage, which became too weak, was placed in the front engine compartment of an old car and fed through a hole. Several of the illegal pets served as substitute children for expatriate couples. Some owners dressed their orangutans in baby clothes. One American bachelor regularly took his pet out drinking at night; she became an alcoholic. Rijksen and Rijksen-Graatsma never found an adult orangutan, though. "Before reaching this age, the animal becomes unmanageable and will be shot at the owner's request by a policeman or a soldier."

London, in 1816 and 1818. Another was taken to Boston in 1825. Altogether, from 1776 to 1925 more than three hundred living orangutans were exported to major collections in America, Australia, Europe, and India. But the animals — tree-living tropical apes imprisoned in steel cages and subjected to cold climates — endured captivity for less than a year on average.

In 1926, however, a Dutch collector named Mijheen Van Goens captured a mature adult male orangutan in northern Sumatra, which he sold to the Dresden Zoo for a handsome sum. This dramatically large animal, named Goliath, proved to be a valuable attraction for the zoo and lived for two years in captivity before dying from what his keepers euphemistically called "old age" — and the public wanted more. Thus, in 1927 and 1928, Van Goens delivered three boatloads of orangutans, 102 animals in all, to Europe, all of which were promptly sold. (The entire second shipment of 33 animals was purchased by American circus owner John Ringling, who shipped them across the Atlantic. Two died during the voyage. The fate of the rest is not entirely clear.) Overall, the world trade in live orangutans expanded during the 1930s and began declining in the early 1960s, when legal controls improved in Borneo and Sumatra. Nonetheless, around 340 wild-caught orangutans were purchased by zoos between 1964 and 1974. Of course, the official records do not include statistics on the substantial illegal trade, and probably the number of animals captured and traded during this time was much higher.

Indonesia declared the killing and capture of orangutans illegal as early as 1931, but such protection was not seriously enforced until the early 1960s. Malaysia legally protected its orangutans in northern Borneo beginning in 1958, but the law is still difficult to enforce in remote places. Both Indonesia and Malaysia have also diminished the local pet trade with a recent policy of confiscation; the confiscated animals are placed in zoos or taken to rehabilitation centers inside protected forests. And the international trade in live orangutans has been nearly ended by several agreements and regulations. For instance, in 1969 all major zoos in the world agreed not to import orangutans without official permits issued by the International Union for the Conservation of Nature. Britain further restricted its import practices with a 1972 law; and an international animal trade treaty, signed by several major importers during the mid-1970s, also reduced the traffic. In spite of such actions, a minor illegal trade in live orangutans continued at least through that decade. As late as 1977 one expert reported that pet orangutans were still great status symbols in Indonesia, while illegally caught orangutans were yet being smuggled out of the islands on Japanese lumber ships and oil tankers.

Nevertheless, continued hunting and a discontinued live-animal trade represent only a trivial portion of our century's Deluge for orangutans. By far the most serious threat appears in the form of human primates riding giant bulldozers and wielding powerful chain-saws: people seeking wood.

WOOD. Worldwide, people now consume around 4 billion cubic yards of it annually, from both temperate and tropical forests. Enough to build *each year* a solid wood highway, two feet thick and four to six traffic lanes wide, to the moon! This is roughly four times the amount used yearly at the start of this century, and certainly many dozen times greater than the yearly consumption during any other century of human history. By the end of this century the amount of wood consumed per year will probably rise by about a third of today's volume. All that wood: what happens to it? About half is burned as fuel, and about half is used commercially as paper pulp and industrial wood.

Fuelwood. Of the approximately 2 billion cubic yards of wood burned as fuel each year, some four fifths is burned in the tropics, often for the basic purpose of cooking food. The demand is huge. Half of all the trees cut in the world are cut for fuel. The demand also appears to exceed a sustainable supply. In the developing nations a minimum of 1.5 billion people meet their fuelwood needs only with difficulty, and another 110 million are perpetually short of fuelwood. And the demand is growing very rapidly. By the year 2000 some 2.3 billion people will have difficulty finding enough fuelwood, and another 350 million will suffer severe shortages; by the year 2025, about 3.5 billion will be short of fuelwood, and perhaps another 1 billion will have to deal with severe shortages.

Fortunately for the tropical forest primates, gathering wood for fuel most intensively affects woodland and savanna regions. Altogether, perhaps only a seventh of the developing world's fuelwood is taken from tropical forests, most obviously in parts of southern and southeastern Asia, West Africa, eastern Madagascar, and Central America. For the most part we don't know whether it is possible to sustain the current rates of fuelwood harvesting in tropical forests. We do know, however, that the rapidly increasing demand for fuelwood in the developing world represents a significant pressure on tropical forests.

Commercial wood. The 2 billion cubic yards of trees cut for commercial purposes each year are processed into paper pulp (20 percent) and sliced into the boards and sheets of industrial wood (80 percent).

Both temperate-zone softwoods and tropical hardwoods are used in this enormous worldwide trade.*

Paper pulp begins when logs are passed through a mechanical chipper, a sort of meat grinder for trees. The resulting chips are later pulverized into pulp, and the pulp eventually ends up in the form of reading material (such as newspapers and books), packaging (such as boxes), and convenience products (such as tissue paper and diapers). Most pulpwood is consumed in industrialized countries. The average U.S. citizen, for instance, uses 660 pounds of pulpwood each year, and the average British citizen requires half that, whereas the average third world resident uses only about 11 pounds — the equivalent of two or three issues of the Sunday *New York Times.* Nonetheless, the demand is expanding everywhere, at a rate about double that of world population growth. Over the last three decades world consumption of pulpwood has risen from 40 million tons per year to 180 million tons, and is projected to reach 400 million tons by century's end and perhaps 800 million tons by the year 2020.

Historically, pulpwood has been a product of temperate softwoods. Until the mid-1970s it seemed impractical to pulp tropical hardwoods because no mechanical chipper could chew its way through the tremendous variety of tropical species. The very big chippers of today, however, with eight-foot bladed disks revolving five times a second, can grind up hundreds of species with equanimity and can transform an average-sized tropical hardwood tree — trunk, limbs, and branches — into chips within a minute or so.

The world's tropical forests still feed only a small portion of the market for pulpwood chips; the great majority still comes from temperate-zone forests. The second largest consumer of chips, the United States, still is self-sufficient, as are several West European countries. But Japan, the world's biggest pulpwood chip consumer, imports about half its yearly dose and increasingly relies on tropical suppliers. Japan's Jant Paper Company, for instance, is right now hard at work turning a large bit of superb Papua New Guinea rain forest into five hundred kinds of cardboard boxes. Strangely, perhaps, the tropical nations still satisfy a lot of their own pulpwood needs with imports from temperate forests. But, given rising demand in the third world, along with the size of the bill for imports from the temperate zone

*Hardwood, the product of broad-leafed trees, is generally harder, denser, and more resistant to insects and decay than softwood, which comes from needle-leafed or evergreen trees. The great majority of tropical forest trees are hardwood. Temperate-zone forests include both softwood and hardwood, but temperate hardwood trees are increasingly protected from commercial exploitation and thus expensive.

(about $4 billion a year), we can be sure that the tropical nations will increasingly turn to their own forests for chips. And chipping can be one of the most devastating forms of forest exploitation, for it is very efficient. Commercial logging in the tropics typically extracts only a few tree species and leaves many others standing, but a chipping operation will take almost everything.

Big as it is, the world trade in pulpwood represents only a fifth of the entire commercial trade in trees. The remaining four fifths involves industrial wood — for house construction and paneling, furniture manufacture and veneer, and other solid wood incidentals. Again, the bulk of the demand arises from the tastes and needs of people living in the wealthy, developed nations, which currently account for about three quarters of the world's consumption. As with pulpwood, the world demand for industrial wood is growing very rapidly: between 1960 and 1980 demand for industrial timber increased threefold, and it will likely increase another threefold by century's end.

Most of the world's industrial timber still comes from the softwood trees of temperate-zone forests. But consumers in the developed nations have come to value the close-grained, beautifully textured, exotic, and durable tropical hardwoods for furniture and paneling and trim, for salad bowls and pleasure boats, cribs and coffins. And as the stands of temperate-zone hardwoods have declined and increasingly been protected for environmental reasons, the tropical hardwood forests have increasingly been drawn into international trading.

Consider this: if you buy a house in the United States that was built before World War II, chances are it is finished with oak or some other temperate-zone hardwood. But if your house was built after World War II, it probably includes some tropical hardwood — mahogany paneled doors, for example. Only after World War II did the United States and other temperate-zone nations begin importing tropical hardwoods on any significant scale. The 5 million cubic yards of tropical hardwood sold internationally in 1950, though, had grown to around 92 million cubic yards by 1980, and is expected to have increased by half again by the end of this century. The biggest consumer, as with pulpwood, is Japan, which today accounts for half the import trade. The second biggest consumer once again is the United States.

For several tropical forest nations the nearly sixteenfold increase in hardwood exports since World War II (along with a threefold increase in internal consumption) has meant money. The export trade is now a $7 billion a year business, which places hardwoods among the top income producers for third world nations — triple the value of cocoa, double that of rubber, about equal to the income from cotton.

The pattern of exchange is simple. Developed nations have money and want wood. Developing nations have trees and need money. For many impoverished tropical nations, the forests seem to represent raw wealth. An average-sized (five-ton) tree contains perhaps $750 worth of timber; a single acre of forest might contain ten marketable trees of that size. So why not cut them down?

It is possible, of course, to cut down trees sensibly, to harvest on a sustained-yield basis, extracting timber moderately and carefully enough that a forest can be counted on as a perpetual source of wealth, a renewable resource. Unfortunately, many tropical nations have harvested their forests not as if they were gardens, capable of renewing themselves, but as if they were mines, to be exploited as a one-time resource. The pressing needs of the present are real. The facts of the future are hypothetical. Very much like deficit spending in the industrialized world, deficit forest harvesting in developing nations has its own meretricious beauty, especially when present circumstances are so dire. "It is not so much that we want to live today rather than invest in tomorrows," said one Southeast Asian forestry official to British conservationist Norman Myers. "We want to survive today, then live a little tomorrow, and think about the future next week." Thus Nigeria, the Philippines, and peninsular Malaysia are already forced to import more timber than they produce, merely to meet domestic needs. Thus nearly all of West Africa could become a net importer of timber by 1990. Thus Indonesia, which is today one of the biggest exporters, may have another two decades or so before it too becomes an importer.

Sustained or not, commercial logging in the tropics is very different from temperate-forest logging. For one thing, the very high diversity of tree species in a tropical forest means that industrial wood harvesting selectively extracts only perhaps 5 percent of the available trees. Unhappily, the high diversity of species in tropical forests means that there are few trees of each species in an area, so concentrated extraction of certain species can eliminate them once and for all from an area (along with great webs of codependent, highly specialized plant and animal species). Even very careful logging of primary tropical forest will diminish the original species diversity for many years, if not permanently.

Moreover, since the large trees in a tropical forest form a continuous upper-story canopy, matted and intertwined with climbing vegetation, when one big timber tree crashes down, it damages and destroys several neighboring trees. Loggers in a tropical forest usually destroy many more trees than they remove. Furthermore, opening up the canopy exposes the forest floor to intense sun and rain; and where heavy logging machinery has compressed the earth, the capacity of the soil to absorb rainwater, and thus control erosion, declines drastically. A standard logging operation in Southeast Asia will damage beyond recovery one third to two thirds of the unharvested trees, and may leave a third of the ground bare and compressed by machinery. The particular and immediate effects of logging vary from place to place, to be sure, minimally disturbing some areas while severely degrading others. But the most damaging effect of commercial logging is usually indirect; the carving out of logging roads and skid trails often opens up previously impenetrable forests to an influx of squatters, landless peasants who proceed to slash and burn the loggers' leftovers, making forest regeneration impossible.

A primary forest can be harvested only once, by definition — and by common sense: tens of thousands of plant and animal species intermeshed in unparalleled intricacy will not return to their original state within any time span relevant to humans. Yet logging within the world's approximately one million square miles of secondary (already harvested or otherwise damaged) tropical forests might, given reasonable care and management, meet the demands of the international hardwood market indefinitely. Tree species that populate secondary forests are generally not as hard as hardwoods that have matured in primary forests, but they are sufficient for paper pulp, plywood, veneer, particleboard, and so on. In addition, the trees that first colonize secondary forests tend to be opportunistic, fast-growing species, which yield more wood faster than primary forest species. Secondary forests have the further advantage of accessibility; they are usually located near existing roads and processing centers.

It can be argued that no secondary tropical forest has ever been harvested regularly. However, secondary forests in temperate nations are now regularly harvested. Theoretically, at least, only two conditions need to be met before secondary tropical forests would likewise become perpetual timber resources. First, the time between harvests must be long enough to allow complete regrowth. Second, the damage caused by each harvest must be minor enough to allow complete regrowth.

The Republic of Indonesia, according to a 1978 report, officially possessed almost half a million square miles of forest, most of which would be extremely prolific and diverse, biologically rich and commercially valuable rain forest — if it were really there. Indonesia's official figures, unhappily, are based on information thirty years old. It is more likely that this archipelago nation's forests today cover about a third of a million square miles, in some combination of primary and depleted secondary forest. In any event, Indonesia, like Zaire in Africa and Brazil in South America, is a major player in the tropical forest game, responsible for roughly half of Southeast Asia's forests and perhaps a tenth of the world's total. How does Indonesia handle this responsibility?

Most of Indonesia was for a long time controlled by the Netherlands, which was persuaded to grant its possession independence directly after World War II. Following an attempted coup in 1965, the Indonesian military, led by General Suharto, took over, and in 1968 Suharto was elected president. Under his leadership the archipelago lurched toward Western-style development — which meant, among other things, sudden and massive exploitation of the forests. Indonesia's 1966 Investors Incentive Act gave foreign investors tax breaks, financial guarantees against nationalization, and several other incentives, which resulted in an invasion of foreign capital, expertise, and technology. Timber corporations from Japan, the Philippines, Singapore, South Korea, Taiwan, and the United States had by 1978 acquired almost 700 timber concessions throughout the islands, covering more than a quarter of a million square miles of forest, including just about all of the accessible lowland rain forests. In 1965, the year before this influx of foreign loggers, Indonesia was exporting a mere 180,000 cubic yards of timber, but during most of the 1970s, the nation exported between 19 and 24.5 million cubic yards yearly and consumed half again that amount for domestic purposes. Indonesia expects to harvest 65 million cubic yards of hardwood by 1990, of which 40 million cubic yards might be exported.

As early as the 1930s, major portions of orangutan habitat on Indonesia's large island of Sumatra had been cleared for the creation of rubber plantations and other forms of agriculture and development. Commercial logging and agriculture combined to destroy about half of the remaining orangutan habitat in Sumatra by 1974. "Considering the enormous human population in North Sumatra," wrote John MacKinnon in 1973, "the amount of natural forest remaining is surprising and gratifying. This is largely due to two factors: the steepness and difficulty of terrain and the relatively disorganized fashion in which timber felling has been conducted." But even as he wrote, such "relatively disorganized" logging was past history. In the province of Atjeh alone, three major timber companies were given licenses to log 780 square miles of previously untouched lowland forest, which happened to include most of the major remaining orangutan habitat outside of Sumatra's reserves. The lowland forests of Sumatra are expected to be logged out by 1990.

And Borneo? When William T. Hornaday arrived in Borneo a century ago to kill orangutans for American museums, he saw "a vast island clad from centre to cirumference with a wonderful and luxuriant growth of unbroken forest" that was "teeming with animal life." Today the great forests of Borneo are no longer "unbroken." In southern (Indonesian) Borneo, extensive forests that include orangutan habitat remain, but mechanized logging and road building are fast depleting many such areas. Commercial loggers arrived en masse in the 1970s — in the large eastern portion, more than a hundred logging concessionaires were granted rights to harvest 32 million acres, or about two thirds of the forests.

The lowland forests of Indonesian Borneo and many other Indonesian islands will probably be finished for commercial purposes before 1995. In the meantime the ecological effects of such massive enterprise may already have begun manifesting themselves. A section of prime orangutan habitat in southeastern Borneo the size of Connecticut and Massachusetts combined — around 13,000 square miles — was entirely burned over by a slow, gradual fire in 1983. News of the disaster was very slow to emerge (though the smoke could be seen in satellite photographs), and it has been suggested that Indonesia was reluctant to reveal its existence and extent. But an article in the *New York Times* described the event in extreme terms, as "perhaps the most severe ecological disaster the earth has suffered in centuries": uncounted "hundreds of thousands of mahogany trees, evergreens, plants and vines were destroyed. Countless numbers of birds, insects and animals such as leopards, bears, deer, pigs, civets, forest cattle and rodents were killed, leading to the extinction of many species." Scientists who were finally allowed to enter the area after the fire found great skeletal remains of burned hardwood trees. Most of the larger trees were gone. Many fruit and nut-bearing tree species were gone, and many of the birds and animals dependent on them had disappeared as well. Leeches, once very common in the forest, were entirely absent, while mosquitoes swarmed in great numbers. Spiders had constructed vast silken networks of webs in pursuit of the mosquitoes.

No one had ever reported the burning of a rain forest before. The event was unprecedented. Rain forests don't burn. Normally southeastern Borneo receives three or four yards of rain a year and is continuously wet and humid. However, a rare but severe drought had earlier dessicated many of the trees, littering the forest floor with dry leaves. Mechanized logging had added waste wood to the flammable detritus and had broken major portions of the forest canopy, opening up the floor of the forest to unusual drying from direct sunlight and increased heat. The result: a rain forest ready to burn.

Malaysia's portion of Borneo, roughly a third of the island in the north, is divided into two states — Sarawak to the west and Sabah to the east. Habitat destruction is not yet a severe problem in Sarawak; about three quarters of the state is still forested in some fashion, and only one partly paved road slices across the state. But a good deal of Sarawak's forest land is legally open to various forms of small-scale exploitation by tribal groups, and at least one expert claims that shifting cultivation has already affected some 10,000 square miles of once primary forest. Commercial logging is also important; in fact, Sarawak's timber exports provide a regular income second only to that of petroleum. By 1979, nearly all of Sarawak's lowland forests had been leased to commercial loggers.

Officially the state of Sabah is still covered by forests over 86 percent of its territory; and if one considers the government figures uncritically, the situation appears good. However, logging is crucial to Sabah's economy, and the once conservative state policy of allowing for sustained timber harvesting over an eighty- to one hundred-year rotation has recently been altered to permit much speedier and more complete exploitation. About half of Sabah's remaining forests, including exceptionally rich rain forest, have been altered by logging, shifting cultivation, and permanent agricultural development. Some of the best remaining orangutan habitat, in Sabah's eastern lowland forests, is now targeted for agricultural development.

The largest and most significant Ark for orangutans is defined in Sumatra: a U-shaped piece of land spread over more than 3,000 square miles, known as the Gunung Leuser National Park. Covering a wide variety of habitats, Gunung Leuser contains samples of more than half of Sumatra's land vertebrate types, including 105 mammal species, 313 bird species, and 94 reptile and amphibian species. The park protects specimens of the world's largest and smelliest flower, a root parasite with three-foot-wide flowers, informally called the stinking corpse lily, as well as several endangered mammals, including the clouded leopard, the Sumatran tiger, Sumatran rhino, siamang, and orangutan. Gunung Leuser probably covers about a third of the total remaining habitat of orangutans in Sumatra, and all told may protect some 3,000 of the red apes. However, most of the park is too high (over 6,500 feet) for orangutans, and some land inside the park (the Sikundur area) was recently logged; extensive habitat just outside the area's boundaries is rapidly disappearing through commercial logging and intensive cultivation.

In southern Borneo, Indonesia protects two significant areas of orangutan habitat, the 1,000-square-mile Tanjung Puting Reserve to the far south and the 800-square-mile Kutai Reserve to the east. But only about half of the Tanjung Puting Reserve includes suitable orangutan habitat, while about a third of the Kutai Reserve was cleared by commercial loggers in the 1970s. One scientist who spent several months in Kutai during that time described the situation to me: "The Kutai Reserve was a huge area that was beautiful, a pristine habitat when I got there. I was fortunate to see it, because today large parts have just been destroyed. The Kutai Reserve was part of a much larger area of forest that, until about 1970, had only been touched by selective logging done with hand tools. Within the time I was there, literally hundreds of Caterpillar tractors were shipped in. Major logging operations started up all around, to the west of the nature reserve, to the east, to the north, and in fact *in* the south of it — because eventually a logging company was given a concession in the southern third of what used to be the reserve, and they started cutting there in 1973. To the north of my study area, just up the river, you see trees along the river, but if you go through the trees, on the other side the place is a wreck."

On the positive side, it appears that the Indonesian government is increasingly motivated to conserve some of its great natural heritage. For one thing, it has been steadily expanding its national park system. For another, the director general of forestry publicly promised in 1979 to ban any further logging in reserves or proposed reserves. Most important, the Indonesian government is now considering the establishment of two more reserves in southern Borneo that would include superb orangutan habitat: the proposed 560-square-mile Bukit Raya Reserve, which might be expanded to include a total of 4,800 square miles, and the proposed 6,400-square-mile Kayan River Reserve. If legally established and properly protected, the ecologically rich Kayan River Reserve could become the most significant forest reserve in that portion of the world.

Until little more than a decade ago, only one tiny piece of the Malaysian state of Sarawak had been set aside as an Ark: the 10-square-mile Bako National Park. Since 1974 Sarawak has added five national parks and two wildlife sanctuaries to its conservation system.

At the state's eastern edge the 23-square-mile Samunsam Wildlife Sanctuary specifically protects the highly endangered proboscis monkey. At the southern edge the Lanjak-Entimau Orangutan Sanctuary was announced in 1983 to protect the endangered hornbill as well as the orangutan. Since they are not open for human recreation, those two sanctuaries will never bring economic returns in the form of tourism, yet they provide other important benefits. The forests of Lanjak-Entimau, for example, protect the watersheds of eight important rivers and will soon temper the flow and quality of much of the water draining into a hydroelectric project being constructed on the Batang Ai river. Most fortunately, the large Lanjak-Entimau Orangutan Sanctuary (at 652 square miles, it covers more land than all the other conservation areas combined) is said to contain "good numbers" of orangutans over a "wide area." It is not clear that any of Sarawak's other conservation areas include orangutan habitat, however. And in any case, all the conservation lands in Sarawak still account for only 2 percent of the state's territory, and most of those lands are so small (with two exceptions, all are under 26 square miles) that if they became isolated from the surrounding forests, their biological contents would too quickly diminish.

Mount Kinabalu National Park in Sabah, northeastern Borneo, may have harbored fewer than twenty orangutans two decades ago. And Sabah's Sepilok Reserve is said to contain, as I mentioned earlier, three hundred orangutans. A 1965 conference of the International Union for the Conservation of Nature recommended that Sabah create a permanent sanctuary for orangutans in its Ulu Segama area. In the late 1960s John MacKinnon surveyed part of the Ulu Segama forests and discovered an unexpectedly high concentration of orangutans; he believed it to be one of the largest remaining wild populations anywhere. But since that time the government of Sabah has allowed the area to be cleared for its timber. Some recent surveys suggest that a dense population of orangutans may yet survive in Sabah's Danum Valley.

Three Arks in Borneo and Sumatra now include rehabilitation centers for orangutans confiscated as illegal pets. Although the policy of confiscation has helped eliminate the market for baby orangutans, orphaned wild animals will certainly continue to appear as a consequence of the continuing massive destruction of habitat. In any case, the wisdom of introducing ex-captive animals into areas inhabited by wild animals has been questioned by many. Why? First of all, past experience in rehabilitation suggests that only adolescent females will integrate into the social systems of wild populations, which are otherwise hostile to newcomers. In many instances, then, reintroduction seriously stresses both the introduced animals and the resident wild groups. Second, their similarity to humans in physiology and biochemistry means that orangutans are vulnerable to a host of human diseases. Introducing formerly captive animals, often former pets, into wild groups could also introduce such diseases as tuberculosis and polio to the wild populations. The fact that two reintroduction centers, at Sepilok in Borneo and at Bukit Lawan in Sumatra, have become tourist centers only compounds the threat of disease.

I visited both Sepilok and Bukit Lawan. Bukit Lawan was indeed touristified — swarming with Europeans and Americans — but still beautiful. And the introduction project was still, I thought, well managed and tightly controlled.

Sepilok likewise was crowded with tourists — a whole flock of British bird watchers landed there during my visit — although the forest itself is big enough that you could wander off by yourself for several hours; if you were brave or foolish enough to leave a trail, you could even get lost for a day or two.

Sepilok had a small, partly forested area where baby orangutans, too small to be released overnight in the forest, could play during the day, and I went there. To my delight, the babies began untying my shoelaces, climbing over my shoes, chewing at my pants, tugging at my pack, trying to steal my pens, hugging me around the legs as if I were their mother, climbing up my body as if I were a tree. They were, of course, very cute — inquisitive and potbellied. But to my dismay, I later saw how many, many visitors to the center entered that area, with almost no supervision from the Sepilok staff. Once I arrived at the baby orang area to find a group of visitors playing with, laughing at, and indeed teasing the animals — animals supposedly destined for return to a wild existence. To my disgust, I observed one baby climb a tree carrying some visitor's Coca-Cola can; I quickly left the area.

The Sepilok Rehabilitation Center's official information building included a small museum and auditorium, where an informational videotape on the rehabilitation project was shown. I watched it. "When the far-sighted government outlawed private ownership," a mellifluous male voice informed me, "it suddenly had a lot of orangutans on its hands. So it decided to try a bold experiment. Return them to the wild! Free!" There were scenes of the legal confiscation of a pet orangutan, as well as scenes of logging, in which a tree fell and a frightened baby orang was discovered by hard-hatted but soft-hearted loggers. The orphaned animals were taken to a vet, given an examination, and cleared for entry into "Camp Sepilok." The videotape showed comic scenes of baby orangutans eating slop. "Hmmm! Looks like Emily Post hasn't been through this part of the jungle!" But ultimately, I was informed, all is well. The voice continued: "Once in the jungle, the orangutans always have a wonderful time in the trees! This is their home." I was shown orangutans gaily tumbling in the trees and was finally assured that "the program of putting the animals back in the jungle is a noble experiment." The animals are, of course, endangered, yet "they may still have a future, swinging freely and happily through the trees of Southeast Asia."

Orangutans, especially young ones, make superb film subjects. They can be very appealing and dramatic, and it is nearly impossible to put together a film about young orangutans that doesn't stroke the funnybone and pluck the heartstrings. Nonetheless, I was put off by that video, by its language and tone, and I began to see it as a piece of self-serving propaganda. Indeed, I began to see Sepilok itself — in spite of the friendliness and kindness of the staff — as essentially a public relations operation for the government of Sabah. After all, almost all the orphaned orangutans are orphaned precisely because of government-supported logging operations, so why should the government congratulate itself for finding a home for the orphans it has just created? Indeed, as far as I could figure it, the information center had been built by money from a logging concession. Moreover, Sepilok itself was now nothing but a minor island of forest, about fifteen square miles of what the government calls "virgin jungle," surrounded by entirely cleared land, some of it so recently cleared by burning that tree stumps were still smoldering when I was there. See the orangutans at Sepilok! But Sepilok was becoming no more than a large public zoo. One of the staff members told me that several thousand visitors had come there the year previously, and I believed him, after discovering in some of the comparatively accessible parts of the reserve more litter — plastic bags, bottles, cans, plastic and paper wrappers describing ingredients and brand names in Chinese, Malay, and English — than I was used to seeing in the streets of Boston. And it became increasingly difficult for me to believe that anyone took the reintroduction project seriously, since the baby orangutans were so commonly and freely handled by so many visitors.

AT ONE TIME, when I lived in California, I made a modest living as a carpenter, working first for a company called A.R.S. Construction, which specialized in turning one-story houses into two-story houses, and then for Rainbow Designs, which specialized in remodeling rich people's houses. Thus I came to know wood as a carpenter does, intimately. I came to know the swirling poetry of it, the feel and smell and even taste of it. I climbed triumphantly over it and sweated painfully under it; carried it on my head, in my hands, under my arms; had its fine dust settle in my throat, its thin splinters pierce my palms, its solid pieces knock me over the head. I felt its texture and flex, its give and resistance. I discovered how it can be just right, perfect enough for a surgeon — and how it can be long when you want it short, and, worse, short when you want it long. I have both praised and cursed wood.

Once, after retiring from carpentry and moving to the Boston area, I built some book and toy shelves for my children. Pine is the wood I prefer to work with. It's beautiful, light, easy to manipulate. Best of all, it's comparatively cheap, coming as it does from large, managed reservoirs of coniferous forests in North America. So I cut, squared, trimmed, glued, and nailed the pine shelves into functional form. To stabilize the shelves, I wanted to use a quarter-inch plywood backing. My lumber store didn't have pine plywood in that thickness, so I bought a dark, fine-grained plywood described to me as Philippine luan. As I glued and nailed in place the quarter-inch Philippine luan plywood, I fantasized about the history of that wood. If I could crawl inside it, or if I could, through some convenient miracle of molecular manipulation, become the wood and also ride backward in time, where would I go?

To assist in my fantasy, I contacted the supplier of my lumber company and was told that they don't import wood from the Philippines anymore. "All the trees aren't there anymore," I was told. "They hadn't planted any trees, and now the Philippines is out of trees." Thus the dark plywood sold to me as Philippine luan actually comes from Indonesia. Directly? Five or six years ago raw Indonesian timber was sent to factories in Taiwan (or Japan or South Korea), where it was sliced thin, pressed and glued together, and shipped out as strong, light plywood. Now, however, Indonesia has built its own plywood factories, and the Philippine luan plywood I purchased came directly from Indonesia.

So, my backward fantasy went like this. I am a sheet of luan plywood, sliding out of the shelves in my house, floating downhill to the stacks of plywood in my lumber supplier's barn. From those stacks I am jounced by truck back to the importer, still in Massachusetts. From the importer's warehouse I am transported by truck to a dock, and then hauled by crane in a banded unit with 114 other pieces of quarter-inch plywood and stacked on a ship. The ship, going backward, moves down the Atlantic coast, passes through the Panama Canal, into the Pacific, and takes me after several days on that rolling body of water to a factory somewhere in Indonesia — the vision grows dimmer here — perhaps in Jakarta or some other major industrial city. What happens then? I am tortured in the bowels of the plywood factory and magically returned to cylindrical form, a chunk of timber. And from there, backward once again by ship and truck, my cylindrical self is hauled to — where? Indonesian Borneo, as likely as not, and I eventually return to my original, upright, and living existence as a tree in the steaming midst of a rain forest.

The barriers of time and thought, distance and transformation, firmly limit our vision, stop us from seeing the immensely intricate structures of everyday life at the end of this century. But the fact is that the Deluge in Borneo now begins in Boston as much as anywhere else.

~ *Eleven* ~

CAGES

I DROVE WEST from Boston on a six-lane highway through the rain, behind trucks that trailed clouds of mist and flung blankets of water across my windshield. I listened to music on the car radio and once in a while consulted a sheet of directions on the seat beside me, as I swoopingly discarded six lanes in favor of four, then four in favor of two. After several miles and turns along the two-lane road, I entered gray, wet countryside, passing scattered houses and farmers' fields. At last I turned onto a narrow road and entered a sweet-smelling pine forest. I parked the car within the forest not far from a modern, two-story edifice — rectangles and cubes in brick and concrete, glass and aluminum — walked across the lot and passed through glass doors. I tossed my dripping umbrella next to two others, then passed through more glass doors into a capacious central lobby. I approached the reception desk, and the receptionist wrote my name on a temporary visitor's badge to the New England Regional Primate Center.

Soon my guide arrived, a nutritionist named Lynne Ausman, who wore a white lab coat and smiled pleasantly. Dr. Ausman and I had already spoken briefly on the telephone about my visit, so she had a general idea why I was there. After some extended preliminaries, including a discussion with the director of the center, she led me into another wing of the building, where we paused at a window of the nursery. Beyond the window three infant monkeys lay fast asleep in glass and metal cribs. We met there Andrew Petto, an anthropologist who was managing the group of monkeys I sought. The three of us proceeded along a bright corridor, pausing in a small vestibule between two sets of doors to dress ourselves in surgical gowns, face masks, and slipperlike shoe covers. As we dressed, a set of doors opened to admit simian sounds and smells and a woman dressed in gown, mask, and slippers. She pushed a metal cage on wheels, which looked rather like a shopping cart.

Inside the cage sat a small monkey, tan-furred on its back and tail, white on its legs and belly, with a bare brown face and an amazing pompadour of snow white hair topping its head. It looked at me. I looked at it and recognized a member of the species I had come to see: the cotton-top tamarin of northwestern Colombia. I was delighted. The woman paused to talk to Dr. Petto, then pushed her cage-cart through another set of doors.

After the three of us had gowned, masked, and slippered ourselves, we proceeded, Petto leading the way, until we entered a blue-painted cinderblock room lined on either side with several wire cages. Each cage contained a branch and a wood nesting box, a couple of metal perches, and a half dozen or so cotton-top tamarins.

They were the size of small squirrels. Indeed, their bodies and tails were squirrelly, but their faces not at all. A squirrel's eyes are widely spaced, on either side of its head, and it looks at you sideways with a nervous twitchiness. These monkeys also looked at me twitchily, but their primate eyes — set close together and facing straight ahead without the interference of a snout — stared at me straight on. Their faces seemed remarkably expressive, verging on the humanoid, particularly given that plume of pure white hair: a pompadour dramatically rising at the forehead, curled back and out, reminiscent of the dramatic hairdo of composer Franz Liszt.

"They're family groups," Andrew Petto said of the groups in cages. "Mother and father and litters, usually twins, often triplets. It has turned out that it's very, very important for the males to have experience with siblings. When that's the case, we get about 75 percent of males fathering babies, which is just about the same fecundity rate as we get from our males that were born in the wild. If the males aren't raised with any brothers and sisters, then we have only about a 25 percent fecundity rate, which is really awful." Why? "What happens is that the males are very, very important. That's a male carrying that baby." What baby? I looked at the cotton-top he was indicating and only gradually noticed a small extra piece of fur, a tail perhaps, protruding from the thick, ruffled fur of his back.

Petto continued, "Males raised with siblings get better at child care because as older siblings they are required to do that." But why, I persisted, are males raised without siblings less fecund? "In terms of why they don't get anybody pregnant, there is no good evidence except that in other monkey species, rhesus monkeys for example, that are well studied, males who are not raised with older males and females never learn the proper position for intercourse, so they just never get into the right place. They don't pick up the signals. They don't figure out when the female is soliciting sex and when she is not."

The monkeys had an explosive way of leaping, an explosion of the back legs that would suddenly propel them to the top of the cage, or over to the side, or over to the climbing branch. One of them, standing four-legged on a platform, exploded up to the cage top, flipped upside down, and flipped again back down to where he (or she — it was hard to tell without examining a crotch) had begun, and then repeated the process again and again. The regular explosions of these three or four dozen monkeys, a burst of feet away from the metal cage, a sudden landing on metal, created, I thought, the effect of popcorn perpetually popping. And their regular chirping created the sound of a whole treeful of very noisy sparrows.

Until comparatively recently, North American and European pet dealers and biomedical researchers were well acquainted with the cotton-top tamarin, largely because its range is near what used to be one of three major live animal export centers in South America — Barranquilla, Colombia. The monkey was often captured by locals (rather than professional trappers) who, upon sighting one would simply chase it or throw something at it, hoping to corner it or knock it off a branch so that the monkey could then be captured by hand. Sold in Barranquilla, a single cotton-top tamarin would bring a Colombian farm worker the equivalent of two or three days' wages.

From 30,000 to 40,000 cotton-tops were exported from Barranquilla between 1960 and 1975, mostly to the United States. During this same period, however, the species' wild habitat was rapidly disappearing. By 1966 around 70 percent of its original habitat had been converted into farms and ranches to support Colombia's burgeoning human population. By the mid-1970s the northern three quarters of the cotton-top's range was almost entirely destroyed, with only about 5 percent of the original forests remaining in scattered and isolated patches of privately owned land. Eventually it became apparent that cotton-tops were threatened with extinction. Colombia banned their exportation from 1969 to 1972, at which time it declared the animal endangered. In 1973 Colombia prohibited the exportation of all primates except for use in biomedical research, and a year later it banned the exportation of primates altogether. Although it was illegal in the United States to import animals from a country prohibiting export, about 2,500 cotton-tops were still sold each year in the American market after Colombia's prohibition. American animal importers were able to circumvent the legal ban by trading through intermediate nations, mainly Bolivia and Panama, that did not prohibit the export of primates.

The New England Regional Primate Center is one of seven regional centers in the United States, created and funded by the National Institutes of Health primarily to support biomedical research using primates. After it opened twenty-five years ago, the New England Primate Center focused its interests on New World monkey species, while most other biomedical research in the United States involved Old World primates. At that time several monkey species were coming into the United States from South America, including the cotton-top tamarin. And since cotton-tops were inexpensive and readily available, the New England Primate Center acquired many.

While they were caring for and learning about this monkey species, workers at the New England Center and elsewhere discovered that it was susceptible to several herpes viruses, as well as to a more mysterious disease called, vaguely (since no one really knew what was going on), marmoset wasting syndrome. Eventually it became clear that these monkeys were uniquely susceptible to two important and related human diseases, ulcerative colitis and cancer of the colon.

Lynne Ausman, my guide, described her interest in the cotton-tops as models for studying human disease. "Japan," she began, "eats about 9 percent fat. Our country eats 38 to 40 percent. Their rate of colon cancer is very low. So it looks on these cross-country comparisons like fat is important, but guess what? There is also a big difference in fiber and amount of protein eaten — forget about the fat that's included in the protein — et cetera. So that you have these multiple differences."

She continued, "One of the things I would like to do with humans is to find out if indeed you feed them a low-fat diet, can you cause a lower cancer rate?" However, since only a small percentage of humans will ultimately develop colon cancer, no matter what the diet, and since the subjects would probably need to be on carefully controlled diets for a significant part of a lifetime, such an experiment is logistically impossible. "It's the kind of trial that you really can't do. You can't take 20 million Americans and say, 'OK, you're going to be on a 20 percent fat diet for the next twenty years.' Because you can't. It would cost billions of dollars, and you can't control it. Yet if you take too small a group of people, what finally happens is you get negative results and they don't mean anything. They're not conclusive, and that's a waste of scientific effort and tax dollars."

Enter the cotton-top tamarin. Around half of all captive cotton-tops will spontaneously develop ulcerative colitis during their ten-to-twenty-year life spans, and of those a large proportion will eventually die of cancer of the colon. Why? We don't yet know. "Noboby knows if they get it in the jungle. So we may have an absolute lab artifact. All we know is we haven't gone and given them dimethylhydrazine or any of the other potent carcinogens that you give to rats to give them cancer."

It could be that the cotton-tops have a genetic tendency to develop cancer of the colon. It is already clear that some humans do. But beyond such an inherited tendency, what is the effect of diet? The central portion of Ausman's experiment will test the effect of particular diets on the development of colitis and cancer in laboratory cotton-tops. Roughly 120 animals will be involved altogether, over a five-year period. About 40 of the animals, a control group, will receive the standard laboratory diet for these animals, which includes crickets, worms, quail eggs, yogurt, cottage cheese, many kinds of fruit, and a commercial preparation called ZuPreem. "This is the standard diet. The monkeys grow well on it, and they thrive. They also get colitis and cancer on it." The remaining 80 animals will be given two sorts of experimental diets. Half will eat a high-fat, low-fiber diet; the other half will eat low fat and high fiber. If this experiment confirms its working hypothesis, at the end of five years the cotton-tops on the high-fat, low-fiber diet will have a significantly higher rate of colitis and cancer than those on the other two diets; the ones on the low-fat, high-fiber diet may have the lowest colitis and cancer rates of all.

I don't wish to deny the importance of cotton-top tamarins to biomedical research, for the study of ulcerative colitis and colon cancer and of particular viral diseases. However, most biomedical research with this primate species before the mid-1970s had nothing whatever to do with its unique anatomy or biochemistry. Rather, the animal was widely exploited for a number of research projects simply because it appeared to be a common and cheap living resource. Unquestionably the 30,000 to 40,000 cotton-tops exported from Colombia during a decade and a half of international trade helped deplete this now endangered species. Perhaps paradoxically, though, the biomedical research industry, once part of the Deluge for cotton-top tamarins, has lately become part of the Ark. Biomedical research institutions do not consider themselves in the business of conservation, or even, particularly, of breeding primates in captivity. Breeding primates is expensive. But the cotton-top tamarins are intriguing enough and valuable enough that some research institutions, including the New England Regional Primate Center, have initiated expensive breeding programs for the species.

New England had about seventy cotton-tops in its colony in 1976, after the last shipment from Colombia arrived, but they weren't reproducing successfully. Cotton-tops mature rapidly and are capable of having two litters a year. They are fertile during lactation, and, since they normally produce twins or triplets, the center could conceivably have had many animals. But the captive mothers were, unfortunately, neglecting their young in the first days after birth. Ultimately it was decided to place cotton-top newborns in the center's nursery, where they were bottle-fed and reared by humans. This caged colony, now in its third captive generation, includes around 280 individuals, making it probably the largest captive group of cotton-top tamarins in the world. (Oak Ridge Associated Universities in Tennessee maintains a comparable-sized colony.)

Because they are members of an endangered species, cotton-tops are today used conservatively in biomedical research. That is to say, they are neither sacrificed in "terminal research" (as the standard expression has it), nor more generally threatened by dangerous or intrusive experiments. Because of its value to the biomedical research community — indeed, its value to humanity at large, as a unique disease model — the cotton-top tamarin today is probably as soundly preserved in captivity as any other endangered primate. More than 1,500 individuals of the species exist today in Europe and North America within a genetically healthy, numerically expanding Ark of cages.

THE NINETEENTH CENTURY marked, by and large, the end of human slavery around the globe, and any true believer in Progress will point to this remarkable event in the history of human ethics. Interestingly enough, though, one nineteenth-century naturalist suggested that nonhuman primates ought to be domesticated on a large scale, to serve humankind as a new class of slave. In his book *Les Singes Domestiques* (1886), Victor Meunier argued that domesticated monkeys and apes could be trained to perform the most monotonous, difficult, degrading, or dangerous tasks that burdened humans. With plentiful supplies of primate slaves, humans would forever be freed from unpleasant work and would enter a new era of leisure and creativity: "With the ape we will found the happy society." Because of their exceptional physical strength, gorillas might perform heavy labor in factory and farm. (Meunier's book includes the impressive sketch of a gorilla farmhand, dressed in overalls but otherwise naked and hairy, wrestling a mad bull to the ground as a few human farm workers look on from a safe distance.) Given other primate species' climbing abilities, Meunier proposed training some to work with fire departments; monkeys could carry ladders to great heights under dangerous circumstances and rescue stranded humans. Still others might serve as cobblers, sailors, construction laborers, painters, and guards. Nonhuman primates could also solve the nineteenth-century domestic servant problem, tending to their masters' personal needs with the honest humility and dog-eyed devotion that had become distressingly uncommon among human servants. Although cooking was too high an art to leave in the hands of a monkey or ape, the animals could be cooks' assistants — washing dishes, scrubbing vegetables, clearing and setting tables, serving food, perhaps even announcing the start of a meal.

Of course, this new slave class would have to be produced on a large scale. So Meunier proposed that breeding stations be established in the tropical nations to produce large numbers of domesticated primates selectively bred for high intelligence, good morals, and satisfactory overall appearance. Breeding primates to improve their appearance seemed important, since many of them, "including some of our most useful types," were "constructed in a manner to be a formidable trial to our aesthetic sense." Although this newly domesticated animal was also to be bred for "morals," Meunier believed that ultimately only females and castrated males might serve the general public. Uncastrated males, because of their "potential danger to women," were to be used for stud service only.

Strange or outrageous as Meunier's ideas may sound, in actuality nonhuman primates have served the human primate for at least five millennia.*

Ancient Egyptians trained baboons to gather figs from trees too delicate to support people. Two thousand years ago apes harvested pepper trees in the Mideast. And for centuries in Southeast Asia, trained pig-tailed macaques have climbed palms to gather coconuts for their human masters. In South Africa baboons, fed salty food to make them thirsty, have proved to be excellent finders of water. In

*See Morris and Morris, *Men and Apes* (1966), for a more thorough consideration of the historical relationship between the two-legged and four-legged primates. Much of the following derives from that source, as does the discussion of and quotations from Meunier.

West Africa, according to R. L. Garner's nineteenth-century account, one African village possessed a tame chimpanzee that sometimes gathered firewood. (Garner considered buying the ape but was quoted a price nearly twice that of a human slave, which seemed excessive.) Elsewhere in Africa, also in the nineteenth century, Sir Gardner Wilkinson described groups of monkeys serving as torchbearers during a dinner party. In other times and places, trained primates have served as golf caddies, railroad and farm laborers, goat and cattle herders. More recently South American capuchins have been trained specifically to assist handicapped people in the United States.

A few nonhuman primates served us heroically as pioneer astronauts. America's Spacechimp program began in 1957, when NASA acquired forty chimpanzees and began four years of intensive training. The animals were trained to carry out appropriate tasks through a conditioning program that included electric shock punishment. They were whirled about in a large centrifuge at Wright-Patterson Air Force Base to prepare for high-gravity forces. They went for parabolic rides in jet planes to gain a foretaste of weightlessness. Finally a chimpanzee named Ham was selected to become the first primate of any sort ever to leave the earth's atmosphere. Dressed in a space suit and diapers, strapped into a seat atop a rocket the size of a small skyscraper at Cape Canaveral, Ham was fired 155 miles into the sky on January 31, 1961. He survived the gravitational stresses of ascending and descending to splash down in the ocean and bob in a leaky capsule for two hours before the boats arrived. Pulled out of his one-ton, water-logged space capsule, an excited and no doubt relieved Ham was welcomed back to earth with handshakes, lettuce, an apple, and half an orange. He retired from the Air Force in 1963 to become a celebrity display — enduring almost two decades of solitary confinement in a cage at the National Zoo in Washington, D.C. In 1981 he was placed in better conditions at a zoo in North Carolina and introduced to two female chimpanzees. He died in 1983. NASA's original Spacechimp project has long been abandoned, but a 1985 flight of the space shuttle *Challenger* included two squirrel monkeys, as well as two dozen rats, to test the potential for future animal experimentation in space. The monkeys were named 384-90 and 3165 to discourage any tendency to sentimentalize or humanize them. According to Mission Commander Robert Overmyer, "They are vicious and there is nothing lovable about them."

Monkeys and apes have also served humankind more passively, as objects of affection and substitute children — in other words, as household pets. Five-thousand-year-old paintings indicate that ancient Egyptians kept monkeys, restrained by collar and leash, as pets. Baboons were worshiped in ancient Egypt, and the most respected of them lived pampered lives inside temples, eating meat and drinking wine and undergoing elaborate mummification after death; some may have lived less sacred lives as household pets. Wealthy Greeks of the seventh century B.C. imported monkeys as pets, and wealthy Romans likewise frequently kept primate pets.

Trade with North Africa in the eleventh and twelfth centuries A.D. introduced Barbary macaques and other species into Western Europe. Imported monkeys were plentiful enough in medieval France that Paris levied a small tax, four deniers, on monkeys brought into the city. Comtesse de Guiches, consort of France's Henry IV, habitually took her monkey to mass; and in the early sixteenth century, Pope Julius II kept his own pet monkey. In seventeenth-century England monkeys dressed up as pages in the court of Charles II were distracting enough that the king's advisers considered taxing the animals' owners. Primates were often kept as pets by the well-off middle class in Victorian England, while in our century owning a pet monkey or ape (of the apes, mainly chimpanzees) has been just about anyone's option, with prices ranging from several dollars to several thousand.

Infant monkeys and apes often resemble human infants, and perhaps their expressive faces and cuddly dependence trigger a strong maternal or paternal response in human adults. Be that as it may, monkeys and apes generally do not make good pets. Often they cannot be house-trained; they require extensive nurturing; and they are expensive to feed and keep. Monkeys can also transmit diseases to humans. Macaques, for example, can pass on a fatal herpes B virus. Chimpanzees can carry hepatitis. But above all, baby primates grow up: they tend to outgrow their appealing cuteness and dependence, maturing into powerful, agile, sharp-toothed animals, whose sexual and social needs are typically frustrated and twisted by their isolation from their own kind. Upon reaching sexual maturity they can become aggressive, even to the point of attacking their owners.

Some pet monkeys are captive-bred, but most have probably been captured in the wild, either caught in a trap or pulled away from a dead mother. In 1975, however, the U.S. Public Health Service amended the Code of Federal Regulations 42 (Public Health), declaring it illegal to import primates into the United States for sale as pets. Since there are no wild nonhuman primates in the United States, the new regulations substantially reduced the overall pet trade in these animals. (Similar disease-control regulations in the United Kingdom have likewise reduced the pet market for primates there.) Nonetheless it is still possible to buy pet monkeys and apes in the United States through pet shops and mail-order houses that offer animals supposedly imported before October 1975 or bred in captivity.

For perhaps as long as they have served humans as household pets, monkeys and apes have also entertained. Trained monkeys entertained citizens of ancient Greece and Rome by dancing, playing musical instruments, shooting with bow and arrow. The Roman poet Juvenal tells of a goat-riding monkey who, wearing a helmet and carrying a shield, would hurl a javelin for the general amusement of the public. In medieval Europe traveling minstrels and jugglers commonly kept an entourage of trained animals, including monkeys (usually Barbary macaques), to entertain audiences. The macaques would dance, juggle, tumble, walk on stilts, and play such instruments as the lute, harp, bagpipe, trumpet, and drum. Medieval and Renaissance monkey shows also came to include displays of the animals riding on the backs of dogs and horses.

The performer of preference in our century is the chimpanzee, with its tremendous athletic abilities and astonishing strength, its high intelligence, unusually expressive face, overall resemblance to humans, and, perhaps most important, its extroverted temperament. A six-year-old chimpanzee named Peter, who amused audiences at the New York Theatre in 1909, may have been among the most talented of his kind. Peter (according to an account in *Men and Apes*) would come on stage, bow, take off his hat, sit down at a dinner table, and expressively eat a multicourse dinner. Finished with the meal, Peter proceeded to smoke his after-dinner cigar, spit in a spittoon, brush his teeth and hair, powder his face, and — at last ready to leave the dinner table — give a small tip to his trainer. Continuing the act, Peter prepared for bed, lighting a candle, taking off his pants, putting out the candle, and finally climbing under the covers — but only to get up again, put his pants back on, put roller skates on his feet, and chase a woman around the stage. After the roller skating performance, Peter climbed onto a bicycle and did fifteen frenetic minutes' worth of bicycling tricks before leaping off and jumping about the stage, waving a flag.

Chimpanzees apparently love to perform, and it might be argued that these and other performing primates are better off, at least in a strictly physical sense — better fed and healthier — than their wild counterparts. But not necessarily. Until comparatively recently, few primates survived captivity for very long. And even with today's advances in animal husbandry and veterinary science, the careers of some performing primates are brief. The more than two hundred infant and juvenile chimpanzees currently employed by beach photographers in Spain and the Canary Islands, for instance, commonly work for a very short time — their feet deformed by human shoes, dependent on tranquilizers, psychologically disoriented by isolation from other chimps and the constant handling of tourists — before outgrowing their useful youthful cuteness and docility. Meanwhile, the photographers' profits easily support a continuing trade in baby chimpanzees smuggled from West Africa.

Some people believe that transforming any naturally wild animal into a trained clown for the peculiar whimsy of humans degrades both the animal and its spectators. In any case, it is certain that performing primates are wild animals cut off from their natural lives and forced to live in a state of often pathetic and sometimes precarious dependence. The precariousness of a caged animal's life becomes especially obvious when a trusted trainer dies. For instance, four circus chimpanzees trained by Mickey Antalek of Ringling Brothers–Barnum and Bailey Circus were packed off to the White Sands Research Center's toxicology laboratory in New Mexico, immediately after Antalek died in 1984. It seemed that no one else in the circus could manage them, and the research center promised to feed and house them; the chimpanzees would pay for their keep like the approximately six hundred other laboratory primates at White Sands, as experimental subjects testing cosmetics, pharmaceutical drugs, and insecticides. Only massive unfavorable publicity persuaded the circus to reverse its decision, retrieve the animals, and send them to the Wild Animal Retirement Village in Waldo, Florida.

Although some zoos have trained their primates to be popular entertainers, generally primates in zoos have served more sedately to stimulate or satisfy public curiosity, and, ideally, to educate. Zoos in one form or another have existed for millennia, as far back as the ancient civilizations of China, India, Assyria, and Egypt. Egypt's earliest zoo was created in Thebes in 1680 B.C. and included several African monkeys owned by Queen Hatshepsut. King Solomon maintained collections of wild animals, including monkeys, as did various Babylonian nabobs. In the thirteenth century A.D., according to Marco Polo, Genghis Khan kept a collection of monkeys.

Modern zoos appear to be descended from the menageries more recently maintained by European royalty and nobility — the early eighteenth-century collection of Frederick II of Germany, for example, and the menagerie established by Louis XIV at Versailles. But the general democratization of Europe, beginning most obviously after the French Revolution in the late eighteenth century, was accompanied by a decline in private menageries and, during the early nineteenth century, the emergence of public zoos. Like the private menageries, these public zoos included in their collections any primates available — first monkeys, soon apes. In 1835 the recently founded London Zoo exhibited its first chimpanzee, Tommy, who survived captivity for about six months; he was not replaced until a half century later. The first orangutan of the London Zoo, a young female named Lady Jane the First, arrived in 1837. According to an article in the *Daily Mail,* she "spooned and sipped her tea in a most lady-like way for the edification of Queen Victoria and Prince Albert when they visited the London Zoo at Christmas in 1842."

Unfortunately, zoos have often cared poorly for their animals. The pathetic situation of Tanga — a 450-pound male gorilla who was caged indoors for twenty years at the Milwaukee Zoo, until the completion of an outdoor facility in 1984 — is not exceptional. And many zoos, including at least a dozen in North America, have at times offered their "surplus" primates to laboratories. The Detroit Zoo once sent more than thirty crab-eating macaques to a terminal research project at Washington University in St. Louis. The San Diego Zoo sold one of its gibbons to a cancer research project at the University of California at Davis. During the late 1970s at least four "safari park" zoos in England were selling their surplus primates to animal dealers that supplied laboratories. Although the sale provoked some public clamor, one of the dealers, Richard Hackett of Shamrock Farms, defended the practice by noting that "animals are killed to provide bacon and so on, what's the difference?" Hackett later assured the public that not all the animals would be placed in painful experiments, and that some were fed "chocolate cake and ham sandwiches."

Their resemblance to humans undoubtedly contributes to the attractiveness of monkeys and apes as pets, entertainers, and zoo exhibits. But for medical and scientific research, that resemblance — in many cases extending beyond mere superficial resemblance to similarities anatomical, behavioral, and biochemical — has made them seemingly indispensable. Today the international trade in live monkeys and apes serves *primarily* to satisfy the demands of research laboratories in developed countries.

When dissection of human cadavers was forbidden by tradition or law, examining the bodies of nonhuman primates was an important part of the anatomical education of medical students. At least until the thirteenth century, when European authorities finally tolerated the cutting of cadavers, the medical understanding of human anatomy was largely based upon animal dissections carried out by the Roman court physician Galen (130–200 A.D.). The modern use of primates in medical and scientific research traces back to the work of a British physician, David Ferrier, who in 1876 published a study of localized functions of the brain that included comparisons of human and monkey brains. Following Ferrier's work, the neurophysiologists V. Horsley and C. E. Beevor studied the brain structure of a bonnet macaque and then proceeded to work on a more humanlike primate, the orangutan. By 1917 Sir Charles Sherrington had published a major study of ape brains, based on his experimental work with twenty-two chimpanzees, three orangutans, and three gorillas. About the same time as these early brain studies were carried out, scientists began using nonhuman primates to study the transmission of such infectious diseases as tuberculosis, scarlet fever, enteric fever, and, by the early twentieth century, syphilis and poliomyelitis. Research on the latter disease eventually required immense numbers of live primates.

Like other viruses, the poliomyelitis virus cannot live very long on its own. A host organism — a human or some other mammal — provides the virus with a warm and temporarily nurturing home. The polio virus enters its host through the mouth and descends into the digestive system, where it reproduces and soon announces its presence with such comparatively mild symptoms as fever, headache, and stiffness of the neck and back. The host responds to this early infection by producing antibodies, which attack the virus; they may succeed in completely eliminating the virus at this point. If the infection continues, however, it eventually spreads into the bloodstream and migrates to a portion of the host's spinal cord that controls muscle function. This secondary infection will destroy parts of the spinal cord and cause paralysis of the limbs or of the muscles that control breathing and swallowing. The human or animal host may be permanently crippled or, unable to breathe, may die.

Perhaps the most important task in the early stages of research was simply to isolate the polio virus, so that it could be examined and tested inside a laboratory. In 1908 Karl Landsteiner and Erwin Popper of Vienna, Austria, removed some infected matter from the spinal cord of a boy who had died of polio, ground it up, passed it through a very fine filter, and then injected the filtered substance into two monkeys. One became paralyzed in both legs, and both died; their spinal cords showed changes similar to those of humans infected by the disease. Landsteiner and Popper had thereby demonstrated that the disease was caused by a filterable virus — and also that at least some nonhuman primates were susceptible to the disease. Soon monkeys and apes became preferred laboratory animals for polio research, and monkey tissue became a standard medium in which to cultivate the virus.

In the United States, defeat of the disease became something of a national crusade, obviously stimulated by the tragic image of a polio-crippled president, Franklin Delano Roosevelt, and financed after 1937 by the National Foundation for Infantile Paralysis, an organization that for over two decades raised through charitable donations an average of $25 million a year. By 1942, largely because of the campaign against polio, large numbers of primates were being imported into the United States; toward the end of that decade, one research project alone required 30,000 monkeys.

Conceptually the task of creating a vaccine against polio was simple: to develop a poliolike virus that would be virulent enough to cause the recipient's body to produce antibodies, a natural immunity, but not virulent enough to infect the recipient's spinal cord. In actuality the task was tremendously difficult. Vaccines were tried as early as 1935, but not until 1953 did Jonas Salk create an effective and safe vaccine based on a killed polio virus that, emulsified in mineral oil, could be injected with a hypodermic. (The dangers of a vaccine became apparent when one batch of Salk's killed-virus vaccine, manufactured by Cutter Laboratories, began causing rather than preventing polio in about 10 percent of the recipients. Obviously the Cutter Laboratories' "killed" virus had not been killed well enough, and the batch was quickly withdrawn.) Toward the end of the decade, Albert B. Sabin developed a second sort of vaccine, based on a live but weakened virus, that could be administered orally; after tests on many monkeys and millions of children, Sabin's vaccine was licensed for open use in 1961 and then distributed very widely. By 1974 the two vaccines had virtually ended polio in the United States; only seven cases were reported in the entire country during that year. Both sorts of vaccines were produced in cultures made from monkeys' kidneys, and both required continual testing to assure a safe level of virulence. Production and testing of the vaccines were fatal for the primates involved, and during the peak years of vaccine production, lots of monkeys were used. For about six years during the 1950s the United States alone was importing from India perhaps 200,000 rhesus macaques annually to do battle with polio.

We made, in short, a trade: one, two, possibly three million monkeys were sacrificed in exchange for control over a malignant human disease. Few reasonable people would wish to reverse the exchange. As L. Harrison Matthews, scientific director of the Zoological Society of

London, expressed the matter in a 1963 speech: "Mothers could not care less that every three doses of vaccine cost the life of a monkey when they queue up to have their children immunized during a polio scare."

THE EXTENT of this century's international trade in live primates has been reasonably well documented. The trade peaked in the 1950s, when well over 200,000 (possibly as many as 400,000) monkeys and apes were shipped in cages each year from the tropics to the temperate zones. Since then the volume of the trade has steadily declined, and by 1979 about 64,000 live primates were exported to the consumer nations. The trade appears to have continued declining since that year, and a rough estimate for 1984 suggests that around 43,000 live primates were shipped, mostly to serve the biomedical research industry.

Primates used in laboratories are often killed or injured in the process. Most of the approximately 50,000 primates used annually by laboratories outside the United States in the late 1970s and early 1980s were, as the scientific euphemism has it, "utilized terminally" — that is, killed in research or the production of pharmaceuticals. A smaller but still significant proportion of the 50,000 to 60,000 primates used annually in American laboratories during that same period was also sacrificed.*

Of those laboratory primates not killed, many are subjected to psychological trauma or physical pain and mutilation. According to sta-

*The total estimate of 100,000 to 110,000 primates required each year by laboratories during this period seems to conflict with the estimate of about 64,000 live primates traded internationally in 1979, the most recent year for which good data are available. The conflict probably arises from three sources. First, not all of the 100,000 to 110,000 laboratory primates were killed during the year, so not all had to be replaced by imports. Second, some of the replacements for killed animals came from captive breeding projects — perhaps 4,000 to 5,000 primates were captive-bred outside of the United States and a higher number within the country. Third, the data on the international trade in 1979, based on published import and export figures, is incomplete. At least two major consuming nations, France and the Soviet Union, have been remarkably dishonest about their primate imports. France acknowledges in 1979 the import of only 15 live primates; published records of *exports* to France, though, indicate a trade of at least 1,299 animals in 1979; France's actual need for laboratory primates at that time has been estimated at 4,000 animals. The Soviet Union officially lists the importation of only 9 live primates in 1979; published records of exports to the Soviet Union indicate a trade of at least 1,147 live primates; estimates of actual research needs place the number at 10,500 to 12,000, about half of which would have been imported (the other half captive-bred).

tistics published by the U.S. Department of Agriculture, some 2 percent of the primates used by American laboratories between 1978 and 1982 experienced "pain or distress." An additional 35 percent were given pain-relieving drugs to avoid the pain or distress they would otherwise have experienced during intrusive or mutilating research. What sorts of research cause "pain or distress"?

Item: over a ten-year period the University of Pennsylvania's Head Injury Laboratory experimentally produced head injuries in about 200 laboratory monkeys, first rhesus macaques, later baboons. A special helmet was rigidly cemented to a monkey's head; the helmet was then coupled to a pneumatic cylinder, which delivered precisely calibrated forces of acceleration-deceleration, producing a controlled level of brain injury. The helmet was then banged away from the animal's head by repeated blows with hammer and chisel, and the animal was tested behaviorally or examined surgically before being killed.

Critics of the experiments have charged cruelty and unprofessional behavior. Videotapes taken (stolen, to be more precise) from the labs show researchers neglecting to anesthesize animals in apparent pain, making jokes at the expense of injured animals, smoking during surgery on animals, performing surgery with an obviously dirty scalpel (dropped to the floor and picked up), and so on. Critics have charged, further, that knocking away the cemented helmet with hammer and chisel after the experimental injury caused not only unnecessary pain but additional injury, thus invalidating the research data (which were based on the concept of injuries produced in a carefully measured fashion). According to one account, Thomas Genarelli, the director of the laboratory, never mentioned the hammer-and-chisel removals in his published reports, arguing later that they caused no additional injury and were therefore "not germane to the results of the experiment."

Item: in his laboratory at the Institute for Behavorial Research in Silver Spring, Maryland, Edward Taub carried out "sensory deafferentation" studies on monkeys. By severing certain sensory nerves near the spinal cord, Taub disabled the limbs served by those nerves. He then subjected the animals to extended behavorial studies. In May of 1981, however, Taub made the mistake of taking into his laboratory as a volunteer an animal rights activist, Alex Pacheco, who secretly took photographs, gathered documents, and sought expert opinion on the laboratory conditions. By September of that year Pacheco was able to persuade Maryland authorities to act: The police of Montgomery County, Maryland, raided the lab, removed seventeen monkeys, and charged Taub and his assistant, Joseph Kunz, with fifteen counts of cruelty to animals.

Taub had severed the nerves of newborn and even fetal monkeys, and in one instance had experimentally blinded his newborn subjects by sewing their eyelids shut. Although the monkeys probably experienced no pain immediately after the nerve surgery, the eventual regrowth of nerves may have been painful. In any case, Taub was convicted by a Maryland District Court of cruelty to animals — not for his experiments, but for the quality of care given to his monkeys during the experiments. One scientist who had investigated Taub's laboratory noted in an affidavit that "many of the deafferented monkeys were mutilated, either by self-mutilation or by their nearest neighbors. Fingers had been completely bitten off. One monkey had only a palm left: of all his fingers, only bloody stumps remained." The self-mutilation may have been one response to the experience of deafferentation, but there was little evidence of veterinary care: "Forearms and biceps were often lacerated and gaping wounds were common, as nobody had taken the time to bandage the wounds. Antibiotics were not to be found. Of the medication available, all was outdated, some by as much as four years." Photographs taken of Taub's desk showed papers neatly stacked and a decorative monkey skull and monkey's foot ashtray neatly arranged. The office was clean and tidy. But the laboratories seemed extraordinarily dirty. Investigators found mice droppings "evident throughout the rooms"; a refrigerator full of rotting apples; dirty, plugged-up sinks; and dried blood "splattered in several places on restraining chairs." The monkeys' cages were "beyond filth: they contained remnants of ancient and rotting bandages, jagged wires broken from the cages themselves and left to threaten the occupant at every turn, crusted piles of feces which were used as perches by the hapless monkeys."

The cruelty conviction was overturned by a Maryland appeals court, which ruled that state anticruelty laws did not apply to federally funded research. Nevertheless, the National Institutes of Health (NIH), which had funded Taub's work to the tune of about $100,000 a year, conducted its own investigation and finally decided to terminate the grant. Several prominent scientists rallied to Taub's defense, apparently agreeing with his statement that the cruelty charges "were based on distortion of the facts and total misunderstanding by an untrained young man of the research I have been doing. The police raid is an interference with the spirit of free enquiry." While Taub appealed for the reinstatement of his NIH grant, the American Psychological Association donated $16,000, and in 1983, he was awarded a major grant by the Guggenheim Foundation to continue his life's work.

Item: at the University of Wisconsin in Madison, psychologists have used monkeys since the early 1950s for experiments in deprivation and depression. Harry F. Harlow, director of the project until 1970, carried out a series of classic studies on laboratory monkeys' needs for maternal contact that involved removing newborn rhesus macaques from their natural mothers and providing them with variously discomforting mechanical surrogates — "mothers" made of wire and cloth, "mothers" with ejecting devices, with jets of compressed air, with spikes, with violent rocking mechanisms, and with variable surface temperatures (from near freezing to near burning). Harlow's laboratory also produced clinical depression in monkeys by isolating them from their peers. This early method took too long, however, and eventually Harlow introduced a much more efficient technique for creating depression, which he called *pitting*. Individual monkeys would be placed at the bottom of small pits — actually inverted pyramids, with sloping steel sides — and left there in complete isolation, without even visual contact with humans or monkeys, for weeks. According to a 1972 progress report at the laboratory, the pitting technique produced "severe behavorial disturbances," possibly irreversible, that included "elevated levels of self-clasp and huddle and several diminished levels of locomotion, exploration and social activity of any kind" within a mere four to six weeks. Previously the laboratory had been able to produce such severe symptoms of despair only after six to twelve months of using more traditional techniques.

Stephen Suomi, a psychologist at the Wisconsin laboratory, claimed that Harlow had introduced the pitting technique and that after Harlow left in 1970 the pits were taken out. According to Suomi, "pitting . . . is a sledgehammer technique that failed scientifically because they could not eliminate individual differences in monkeys' reactions." Also the technique was "unnecessarily harsh," "unpleasant," and "distasteful," and gave Suomi nightmares. So pitting as a technique for producing depression was ended; laboratory researchers introduced such alternatives as depression-producing drugs, repeated separations and reunions with peers, and learned helplessness, in which the monkey is subjected to inescapable electric shock.

Items: at the University of Chicago several rhesus macaques were shot in the head with a power rifle from a distance of slightly more than one inch. At the University of Michigan, according to one source, seventy-two rhesus macaques, ten baboons, and three squirrel monkeys were struck in the abdomen with a cannon impactor accelerated to 70 miles per hour in a study of "blunt abdominal trauma." At the Medical University of South Carolina, possibly thirty to fifty monkeys a year for several years were experimentally crippled by the dropping of a heavy weight on their spines, as part of a spinal injury study. It has been reported that the monkeys were kept alive for two to three weeks after injury for observation and testing before they were killed. In another research project scientists immobilized eleven chimpanzees in restraining chairs with plaster casts fixed to their elbow and ankle joints; then, without any anesthesia, ten of the chimpanzees were smashed at the base of the skull by a mechanical device capable of delivering "impact forces up to 4,000 lbs. with satisfactory reliability." Those animals not sufficiently injured by the first blow were struck again until they were unconscious or at least "stunned" ("responded to stimuli slowly and in a dazed manner but did not lose reflex responses or muscular tone"). At intervals from two and a half hours to three days after injury, the animals still alive were killed and autopsied.

Scientists at the Letterman Army Institute of Research in San Francisco have tested the effects of laser beams on monkeys' eyes. At the U.S. Army Medical Research Institute of Chemical Defense in Aberdeen, Maryland, monkeys and other laboratory animals have been used in tests of chemical warfare agents. At the U.S. Army Medical Research Institute for Infectious Diseases in Fort Detrick, Maryland, rhesus macaques, crab-eating macaques, and squirrel monkeys have been experimentally infected with such potential biological warfare agents as staphylococcal enterotoxin B, shigella, and the organisms that cause Legionnaires' disease, Ebola fever, Rift Valley fever, Lassa fever, cholera, and Rocky Mountain spotted fever. Some of the research animals were directly "challenged" (to use the official terminology) — that is, innoculated — with the diseases. Others were placed in isolation chambers and sprayed with aerosol versions of the infectious agents.

In 1957 fifty-eight rhesus macaques were placed inside tubes at varying distances from Ground Zero during H-bomb tests in Nevada; many of the animals subsequently developed cancer. After those early tests approximately 3,000 monkeys were experimentally exposed to varying degrees of radiation (often supralethal doses) at the Armed Forces Radiobiology Research Institute, in Bethesda, Maryland, and at the School of Aerospace Medicine, Brooks Air Force Base, in San Antonio, Texas. Some of the research examined the long-term effects of radiation on monkeys' eyes: cataracts and blindness. But most of the research investigated how long monkeys could function effectively while they were dying of radiation sickness. The monkeys were first trained, using electric shock as a primary motivator, to carry out such tasks as running on a treadmill. They were then exposed to radiation, up to 200 times the standard lethal dose (a dose that will kill half of the exposed animals in sixty days), and required to carry out their tasks until they could no longer do so.

In 1977 Admiral Monroe, director of the Defense Nuclear Agency, defended some of the radiation experiments by stating that they were "essential to the medical support of the Department of Defense," and that "to the best of our knowledge, the animals experience no pain in the radiation experiments, though some of them die." However, it appears that most of the monkeys used in those experiments died, and that they did experience considerable pain. A 1966 report from the Armed Forces Radiobiology Research Institute described monkeys responding to radiation with extreme forward slumping, massive convulsions, spasticity, rolling of the eyeballs, stumbling, falling, vomiting, frequent shifts of position "as though in great discomfort," "wild . . . twisting, twirling and throwing of the body," "crawling and frantic pushing and scraping" against the sides of the cage, and "passive draping of head over the back while in an awkward crouched, clinging posture." The radiation experiments may not even have been very useful. A 1972 doctoral dissertation indicated that the purpose of the research was to provide data that would suggest how long humans in the military could function after they had been exposed to a nuclear bomb blast. Yet a comparison between the reactions of rhesus macaques and baboons shows that they reacted quite differently to radiation; if the reactions of one monkey species correlate poorly with the reactions of another monkey species, then "to extrapolate on the basis of observations from a single species related only through membership in the same order as are the Rhesus monkey and man, is completely indefensible." Also, humans exposed to the radiation of a bomb blast would have to respond to at least one other, major factor: knowledge of their fate.

TWO SORTS of people see no ethical complexity in such kinds of research: those who believe that pain and death inflicted on laboratory animals is a necessary evil, always justified, and those who believe that such research is simple cruelty, never justified. For the rest of us, however, the issues are anything but simple.

Yet along with the labyrinthine ethical issues that accompany the still extensive laboratory trade in caged primates rides a direct and obvious conservation issue. While primates constitute fewer than 1 percent of all research animals, they are the only laboratory animals still acquired mainly by capture from the wild. All other laboratory animals are produced almost entirely in breeding colonies. Thus laboratory research has contributed and still contributes to the decline of wild primate populations. It is part of the Deluge. The once numerous rhesus macaque serves as a major example. Whereas only three to four decades ago rhesus macaques were shipped out of India by the hundreds of thousands each year, today a mere 180,000 to 250,000 are left in all of India.

The cycle of trade begins in the tropical forest, with capture. And for the largest primates, gorillas, for instance, capture has often been a bloody affair. Sometimes animal merchants simply purchased orphaned gorilla infants directly from indigenous hunters who had eaten the parents. American adventurer Armand Denis, for example, acquired thirty gorillas in 1942 by purchasing infants and juveniles from M'Beti communal roundups. (Denis might have sold his gorillas for a small fortune; however, all thirty died before he could transport them out of Africa.) Denis argued reasonably that his captured animals would have been killed by their captors if he hadn't purchased them. But less scrupulous professional collectors have deliberately killed entire family groups in order to acquire the infants. Zoologist George Schaller cited an incident in which colonial African officials slaughtered sixty adult gorillas in order to acquire eleven infants for zoos; of the eleven taken, only one survived.

Probably the most common method of collecting orangutans was, likewise, simply to produce orphans: shoot the mother and pick up the infant. A less common but perhaps more productive method was to find a tree containing orangutans, typically a mother and her juveniles, and isolate the animals by cutting down all adjacent trees. The final tree could then be cut down; alternatively, the treed animals could be besieged until they clambered reluctantly, out of sheer hunger, to the ground. At that point they could be captured in nets and placed in boxes or cages. Until the mid-1960s the Gajo people of northern Sumatra regularly captured young orangutans with the latter method, selling them to animal dealers in the town of Kutacane for slightly more than a dollar per pound of live animal. All methods of collecting, though, severely traumatized the surviving animals. One authority described a group of eleven baby orangutans, all of whom had been gathered and offered for sale after their mothers were killed. On average the infants were one year old. Deprived of their mothers' milk, all were suffering from severe malnutrition. Several were worm-infested; some were infected with malaria. One had a broken arm, another had pneumonia. Despite intensive veterinary care, three of the eleven soon died.

While capturing great apes typically involves taking juveniles after killing the protective adults of a group, both adult and juvenile monkeys are taken with a broad variety of techniques.

Biologist Pekka Soini, who studied the Peruvian trade (centered on the Amazonian city of Iquitos), reported in 1972 that most monkeys there were caught by Indian or mestizo subsistence farmers living along the river. Nonprofessional trappers usually took infant or juvenile monkeys by shooting their mothers with a gun or a blowgun and poisoned darts. The mother would fall out of the tree with her offspring still clinging; the young monkey would be pulled away and then hand-reared until it could be sold to a trader. Sometimes adult monkeys could also be collected in this way, if the hunter used diluted poison or if the poison dart was immediately removed and the stunned monkey given an antidote.

But professional trappers in Peru were more efficient. They would seek out an area where monkeys were plentiful and build a camp. Then they would erect a feeding platform and load it with ripe bananas. After a few days, when the monkeys had become habituated to this easy source of food, the trappers used the bananas to bait their traps, often just cane or lath boxes with vertical doors that were automatically tripped when the bait was taken. Or the trap might be a larger cage capable of holding several monkeys; when they had crowded inside, the door was dropped or sprung shut with a cord yanked by one of the trappers hiding nearby. Sometimes the trappers placed an open container of *cacasa* (an enticing concoction of rum, brown sugar, and sliced bananas) on the feeding platform. Once the monkeys had drunk themselves into a stupor, the trappers could simply gather them up. And sometimes Peruvian trappers found monkeys feeding on a peninsula formed by a river's meander, which they would isolate by clearing vegetation and erecting a long, high, slack nylon net. The monkeys would be chased into the net; the trappers would then pick them up and cage or bag them. If necessary, the monkeys were doused with pails of water to quiet them down.

Once trapped, the monkeys were sold to river traders plying the Amazon in boats. The monkeys were crowded into cages made of wood, cane, or wire and transported to Iquitos. That trip sometimes lasted several weeks, during which time the monkeys were fed an inadequate diet of bananas and water. The ones that died yet had value in Iquitos: their skulls and teeth could be made into necklaces for the tourist trade. The ones that survived, albeit frequently "wounded, weak, or diseased," were passed on to middlemen on the waterfront, who then sold them to government-licensed animal dealers. In turn, the dealers either sent their monkeys to Lima, for likely export into the European market, or — after a ten-day quarantine — to Miami.

Iquitos was one of the two major Amazonian trade centers for live monkeys; the other was Leticia, Colombia. (The third major export center, Barranquilla, Colombia, which is north of the Amazon, specialized in non-Amazonian species, such as the cotton-top tamarin.) In 1953, when Colombian animal dealer Mike Tsalickis first came to the Leticia region, Indians would shoot adult monkeys for food and keep their infants and young juveniles as pets. Tsalickis would buy the young, hand-reared monkeys from the Indians for export. Soon, however, he decided to acquire monkeys of all ages, "in order to choose those best suited for research purposes." He designed wooden, then wire mesh, and finally stainless steel traps that prevented damage to the monkeys' tails during capture, and he provided the traps to his many subcontractors along the river.

When he began the business, Tsalickis's subcontractors brought their monkeys into Leticia. However, it took up to four days for them to reach town by canoe, during which time the monkeys would not be fed, given water, or protected from the sun; the animals often arrived "in pitiful condition." To lower the mortality rate, Tsalickis began sending his own power boat up the Amazon. By the early 1970s, he had acquired a small fleet of fast boats, which could pick up and deliver monkeys to Leticia in a single day. There the monkeys were sprayed with insecticide and dewormed. If healthy, they were taken to holding cages and separated by species. Unhealthy animals were treated with antibiotics and vitamins. During a three-week quarantine, the monkeys were given a diet similar to what they would receive once they reached their destination: Purina Monkey Chow, fruit, pablum, and milk.

It must be clear by now that each stage in the journey from forest to laboratory involves losses: some primates die during capture, some die during transportation to the point of export, and some die during quarantine at the point of export. Tsalickis claimed an overall mortality rate of about 8 percent during quarantine. At another South American quarantine station mortality rates were from 3 to 33 percent for "marmosets" (meaning the several marmoset and tamarin species), 15 percent for squirrel monkeys and capuchins, 20 percent for night monkeys, and 27 percent for common woolly monkeys.

After quarantine the surviving primates are provided with appropriate shipping documents (assuming the operation is legal), placed in shipping cages (such features as size and ventilation are usually defined by law and airline requlations), and loaded into the belly of an airplane. Usual destination, an industrialized nation in the temperate zone.

How many die in the airplane? Occasionally most or all. For example, in 1981 a shipment of 220 vervet monkeys and 100 baboons packed by an Ethiopian dealer, Workneh and Nadir, in partial fulfillment of a huge order to the Soviet Union, was seized at its first stop in Nairobi, Kenya, when inspectors discovered that the shipping crates for the vervet monkeys were less than half the size required by airline regulations. The shipping crates for the baboons were so flimsy that the baboons had actually broken out and were wandering about the hold of the airplane. Within a few hours after the animals were seized, 157 were dead; the rest were seriously ill and so were destroyed in Nairobi to prevent the spread of disease. A 1979 shipment of 402 vervet monkeys from Ethiopia, again packed by Workneh and Nadir, arrived in the United States with 135 already dead and another 74 dying — a mortality rate higher than 50 percent. Yet another 1979 shipment by Workneh and Nadir, of 60 vervet monkeys, arrived with 43 monkeys dead and another 14 dying — a mortality rate of 95 percent. A 1980 shipment of 20 silvered leaf monkeys ordered by the Minnesota Zoo from an Indonesian firm, C.V. Primates, arrived in New York with 4 dead. By the time that shipment got to Minnesota, another four had died, and the rest were dying from "malnutrition conformation"; a zoo veterinarian described them more bluntly as "nothing but skin and bone." A 1979 shipment of 625 crab-eating macaques from Indonesia, bound for the State Bacteriological Experimentation Unit of Stockholm, Sweden, arrived at the Amsterdam airport "tightly packed in small wooden cages with no food or water," according to a Dutch journalist. Altogether, 480 of the monkeys were dead; some "had obviously been killed by lack of fresh air, their eyes were hanging out of their sockets and they had bitten off their tongues." In yet another instance the Charles River Breeding Laboratories of Wilmington, Massachusetts, received in 1973 a shipment of 200 rhesus macaques from India. Because of plane failure, the shipment had been held up in extremely hot weather at a Turkish airport, so by the time the laboratory got its monkeys they were quite ill. So ill, in fact, that, according to a veterinarian at Charles River, it was economically inefficient to care for them. In his rather mechanical language, "It was decided that the input required for this group would be much greater than the output in usable animals, and, therefore, the entire group was discarded."

Obviously such examples are extreme. What are more typical losses during transport? The most thorough survey of shipping mortality, an examination of import records from the United States Center for Disease Control, found that of 28,558 primates entering the United States in 1978, some 5,206, or 18.2 percent, were dead on arrival or died within ninety days; of 22,276 primates entering in 1979, a total of 3,818, or 17.1 percent, were dead on arrival or died within ninety days.

Clearly the figures on the international trade in primates, based mostly as they are on official export or import figures, tell only part of the story. Sometimes we know only how many animals came out of the airplane, instead of how many were put in, and never do we know how many animals came out of the forest.

George Schaller has stated that for every gorilla successfully kept by a zoo, perhaps five more gorillas died. An Indonesian animal dealer, Charles Dasorno, estimated that 71 percent of Indonesian monkeys captured would be dead before reaching their foreign destinations: a ratio of almost three dead for every one alive. It is probably fair, indeed quite conservative, to double the standard trade statistics in order to begin visualizing the actual losses that the international trade inflicts on wild primate populations. In other words, if 200,000 to 400,000 caged primates were traded yearly in the 1950s, some 64,000 in 1979, and roughly 43,000 in 1983, probably twice that many, conservatively, were removed from the forest. And if to the losses of actual mortality (deaths occurring between capture and destination), we add the more subtle but genuine losses in such events as the disruption of primate family units and the selective trapping of young animals which have not yet begun their breeding careers, then the full impact of the live animal trade emerges in even clearer focus. It is not trivial.

DEMAND OUTSTRIPS SUPPLY, and the history of the international trade in live primates is the story of a stable group of consumer nations steadily drawing on the resources of a shifting group of supplier nations.* All told, more than a hundred nations, consumers and suppliers, have been actively involved in the trade.

The largest sale of primates in world history took place between India and the United States during the middle of this century. For

*In order of importance, the biggest consumers are: the United States, Japan, the United Kingdom, Taiwan, and Canada. Indirect evidence indicates that the Soviet Union and France should also be placed on that list of major importers.

decades India, with its large population of rhesus macaques, supplied those monkeys in almost completely unregulated fashion to various industrialized nations, particularly the United States. Yet elements of Indian cultural and religious tradition hold all forms of life sacred; and in 1955, yielding to pressure from its own animal protection groups, India banned all rhesus macaque exports. At that time rhesus macaques were considered essential to the production and testing of polio vaccine, so within four months of the ban India was persuaded once again to supply monkeys — this time formalizing its exchange with the United States through an agreement guaranteeing that the Indian rhesus macaques would be used only in medical research and polio vaccine production and that they would be used humanely. To ensure the appropriate use of its rhesus macaques, the United States created the U.S. National Advisory Committee on Rhesus Monkey Requirements and commenced a program of certification.

The certification program was probably as straightforward as most bureaucratic attempts to oversee and control complex events. Any research institution in the United States that wished to use rhesus macaques had to submit a formal application with a description of the proposed research. Each application would be considered by the Committee on Rhesus Monkey Requirements and the surgeon general's office; if the application was deemed to meet India's standards, a formal Certification of Need for Rhesus Monkeys would be signed by the committee's executive secretary and countersigned by the surgeon general. Proper signatures scrawled in proper places, the certificate of need asserted that "upon recommendation of the National Advisory Committee on Rhesus Monkey Requirements," the United States Public Health Service "certifies to the Government of India, (1) that these Rhesus Monkeys shall be used only for medical research or the production of antipoliomyelitis vaccine, and (2) that regular inspections shall be made as needed to assure humane treatment of these monkeys." The certificate was then returned to the requesting institution, which in turn forwarded it to one of two firms allowed to import rhesus macaques from India — Primate Imports of Port Washington, New York, and Primelabs of Farmingdale, New Jersey.

At some point, however, the certification program devolved into a rubber stamp operation. It appears that the National Advisory Committee did not review the requests on an individual basis, and eventually the committee died and went to bureaucratic heaven. Thus certificates of need were regularly passed on for review to other convenient review stations. Nonetheless, the paperwork machinery continued to grind; the certificates were still signed by someone identified as the executive secretary of the National Advisory Committee on Rhesus Monkey Requirements. And requesting institutions found that their applications needed only the briefest of research descriptions.*

During the early 1950s, as I mentioned previously, the United States was importing as many as 200,000 rhesus macaques yearly from India. Eventually India began instituting quotas, and by the mid-1970s it was allowing the export of only 20,000 per year. Meanwhile Indian conservationists noted an alarming decline in wild rhesus macaques. At about the same time in the United States, an organization known as the International Primate Protection League learned of the extensive use of rhesus macaques in painful and fatal radiation experiments at the Armed Forces Radiobiology Research Institute. In an effort to halt the experiments, Shirley McGreal, chairwoman of the International Primate Protection League, compiled a press release describing that research, as well as other American research with rhesus macaques, in vivid detail. She sent copies to every major newspaper in India — and got fast results. An editorial appearing in the *Times of India* on November 9, 1977, epitomized the common sentiment: "The details are gory enough and would shock even a half-wit but there is more to this non-research. The monkeys were obtained from this country under false pretenses — in the normal course they were to have been used for research aimed at benefitting humanity, and for preparing polio vaccine. The whole procedure was illegal because one of the conditions of export was a guarantee by the American companies to give the animals humane care and treatment." On December 3, 1977, Indian Prime Minister Shri Morarji Desai announced that as soon as the year's quotas expired, on March 31, 1978, India would export no more rhesus macaques.

That source was gone. For virtually all research, other primate

*Thus the National Institutes of Health was able to obtain a certificate to import 4,125 rhesus macaques for several different projects with a mere twenty-seven-word description; the University of Michigan obtained rhesus macaques for, among other projects, experimental addiction to cocaine and heroin and car-crash simulations by stating that their monkeys would "be used by various investigators at the University"; Walter Reed Hospital obtained monkeys for radiation experiments under the assertion that the animals were to be used for "studies in behavioral patterns of primates as well as immunological and parasitological investigation"; the U.S. Army Medical Material Agency, of Frederick, Maryland, obtained certificates of need for work that was, simply, "classified." In at least one instance, apparently, an official of the Oregon Primate Center casually requested a certificate after the monkeys had already been acquired. (Letter of June 20, 1976: "I'm enclosing a certificate of need for your signature. It covers the animals recently purchased by us. The monkeys are here and doing well.") And, it has been asserted, the Medical University of South Carolina obtained rhesus monkeys for its spine-crushing experiments with no certification at all.

species would do as well. But because rhesus macaques had been used in the original work on poliomyelitis vaccine, it seemed important for scientific consistency to continue using the same species for vaccine production and testing; indeed, in the United States and elsewhere, the vaccine production industry was required by law to continue testing vaccine batches with rhesus macaques. Thus India's sudden and irrevocable export ban produced reactions varying from mild alarm to near hysteria in the United States. As a scientist from Lederle Laboratories, the sole producer of the polio vaccine in the United States, stated to an NBC television camera in 1978: "There will be no polio vaccine in 1979 and there will be many crippled children and even deaths."

A business opportunity had been created, and even before India's ban, four Oregonians — a veterinarian and three businessmen — were ready. In 1976 the four had formed their own rhesus macaque import company, which they called MOL Enterprises. One of the company's principals was sent to lobby in Bangladesh, and after nearly seventy trips to that country he managed to acquire exclusive permission to import Bangladesh's rhesus macaques. The contract, signed in March of 1977 by Q. J. Ahmed, secretary to the Bangladesh Ministry of Agriculture, called for purchase of at least 18,500 rhesus macaques in the first three years, and up to 71,500 monkeys over ten years. It was not clear, however, that 71,500 rhesus macaques actually existed in Bangladesh. MOL claimed to have determined that the number was reasonable after it surveyed the rhesus population in Bangladesh. But Kenneth Green, an American scientist associated with the Smithsonian Institution, had surveyed the Bangladesh population of this species in 1976 and found "a paucity of Rhesus macaque populations in all habitats"; Green had petitioned the U.S. Fish and Wildlife Service to place that primate on the United States' list of endangered species. Nevertheless, once India banned the export of its rhesus macaques, MOL Enterprises' contract with Bangladesh suddenly became extraordinarily valuable, since it conferred a near monopoly on the world trade in a very important commodity.

MOL had already imported about a thousand rhesus macaques from Bangladesh, some of which were passed on to the controversial radiation experiments at the Armed Forces Radiobiology Research Institute, when Shirley McGreal of the International Primate Protection League, continuing the effort to halt radiation research on monkeys, sent press releases on the subject to Bangladesh newspapers. And within two weeks, by January 1979, the Bangladesh government had expelled MOL Enterprises and canceled their contract. The negative publicity had obviously embarrassed Bangladesh officials, but the cancellation was officially justified by a claim that MOL had violated its part of the bargain, including an agreement to create a breeding farm and a stipulation that the monkeys were to be used only for research "of benefit to the whole of humanity."

Bangladesh never reinstated its contract with MOL Enterprises, and the company slowly went down in a blaze of lawsuits. Bangladesh continued to export small numbers of monkeys in the 1980s, but with the collapse of the MOL Enterprises deal, the Asian trade shifted away from the Indian subcontinent to Southeast Asia, which began trading large numbers of other macaque species for laboratory use. Indonesia soon became the largest exporter of primates in the world, sending out more than 46,000 pig-tailed and long-tailed macaques between 1979 and 1981. The Philippines moved into second place, exporting almost 25,000 long-tailed macaques in the same period.

Even though it is a major primate region, Africa has never had the same significance as Asia in the international primate trade. Nonetheless, Kenya now ranks fourth in the world as an exporter.

From South America, an international trade in live monkeys has existed ever since the first Europeans arrived in the sixteenth century. Until comparatively recently, though, the South American trade was mainly for the pet market. South American squirrel monkeys were used in North American laboratories for behavioral experiments in the 1930s, but not until the 1950s were the export floodgates opened. Beginning then, hundreds of thousands of monkeys were captured in Amazonia and elsewhere and airmailed to North American pet shops, zoos, and laboratories. For some time, Peru dominated the South American primate trade. Between 1964 and 1973 the United States alone imported more than 360,000 monkeys from Peru, mostly from Iquitos. Colombia was the second major South American connection, by way of Barranquilla and Leticia. All told, the South American live monkey trade to the United States reached an annual volume of about 26,000 in 1961 and 1962, rose to a peak of 83,000 live primates in 1968, and gradually declined to about 58,000 by 1970. Around half a million New World monkeys, representing many different species, were imported by the United States during the 1960s.

Half a million is either a small or a large number, depending on one's perspective — and in this case, depending on the species. According to one source, about 80 percent of the monkeys exported from Iquitos were squirrel monkeys, common denizens of Amazonia. On the other hand, Iquitos also exported uncommon and even rare species. In fact, of the nineteen species inhabiting Peruvian Amazonia, all but two were at one time or another exported from Iquitos. Most likely the majority of monkeys exported from Leticia were also squirrel monkeys. In any event the South American trade depleted several species, and the profits poured into a few dealers' pockets. Fully considered, the trade probably took more than it gave, and Peru ceased exporting primates in 1973. Colombia ended its trade soon after.

In the last two or three decades many exporting nations have passed laws to control or end their trade in primates and also to protect endangered primate species. But given the powerful forces of world demand, an inconsistent patchwork of national laws has frequently only shifted the trade to other nations. The termination of major trade from Peru and Colombia, for example, merely cleared the way for Guyana to a small degree and Bolivia, which soon became the third largest primate exporter in the world.

Perhaps the most disturbing aspect of the international live animal trade is the trade in endangered species; and the shifting patterns of export have been accompanied by a substantial underground trade in protected and endangered species.

Brazil was never a major exporter of live monkeys. As early as 1967 the government banned any exploitation — hunting, capture, or trade — of its primates. Yet that protective legislation has been inadequately enforced, and during the 1960s and early 1970s, monkeys from Brazilian Amazonia were often captured and taken upriver to the Colombian city of Leticia. Leticia had been declared a free port in the late 1960s, which freed it from normal customs controls. At least for a time, the city massively exported illegal wildlife and wildlife products. When Colombia withdrew from the primate trade, the pipeline of smuggled wildlife and wildlife products from Brazil was simply reconnected to a Bolivian outlet.

In Asia, Singapore was for many years a major player in the illegal wildlife trade. Although Singapore's native wildlife was wiped out long ago, it was possible to purchase live animals whose official country of origin was Singapore, thus avoiding the export restrictions of other Asian nations. Gibbons and other primates from Thailand, Indonesia, and Malaysia would be smuggled into Singapore, on freighters, inside false gas tanks, and so on, for re-export. As late as 1975 a Singapore dealer, Mayfield Kennels and Zoos, advertised for sale such protected animals as orangutans and Sumatran rhinos.

Thailand provided a second outlet for the illegal Asian wildlife trade, exporting both animals smuggled from other countries and its own protected species. Thailand protected all its gibbons from export in 1961, and in 1975 it banned the hunting, trapping, and export of all primates except for approved educational or scientific purposes. Yet Thai wildlife laws still allowed individuals to possess two primates of one species as pets, which hindered prosecution of illegal dealers. Even in 1981 lar gibbons, long-tailed macaques, and slow lorises were openly sold in the markets of Bangkok.

Gibbons have been smuggled out of Thailand in a number of ingenious ways. Recently Japanese citizens have carried gibbons and other endangered primates out of Bangkok inside their personal luggage, bound, presumably, for the pet shops of Tokyo. In one instance a Japanese citizen was caught with eleven baby monkeys in his luggage, five of which had already died from suffocation. One notorious Thai dealer, the Bangkok Wildlife Company specialized in shipping illegal wildlife inside containers supposedly containing poisonous snakes. In 1971, for example, the company packed four gibbons inside snake crates in fulfillment of an order from the United States; at a stopover in London the crates were opened to reveal three dead gibbons and a fourth that was dying. In 1974 the Bangkok Wildlife Company mailed fifteen slow lorises in a crate labeled *Spitting Cobras* to a post office in San Carlos, California; five of the fifteen animals were dead when the crate was seized by the California Department of Fish and Game. Bangkok Wildlife and other dealers commonly overrode Thai export prohibitions by transporting their animals across the Mekong River to Vientiane, Laos, where the animals would be "legalized" with Laotian shipping documents and then sent back to Bangkok.

While Thailand has had difficulty policing its own wildlife trade, it has also had little help from importing nations in this task. In the United States the Comparative Oncology Laboratory at the University of California at Davis used gibbons for many years in its cancer studies; young gibbons would be inoculated with a leukemia-like virus that caused death within nine to fifteen months. The Davis laboratory, which at times had as many as fifty live gibbons, regularly imported the animals from Thailand in apparent violation of Thai law, which allowed exports for some scientific purposes but still required clearance from Thai Customs, a health certificate from the Thai Department of Livestock Development, and export permits from the Royal Thai Forest Department and the Department of Foreign Trade.

For example, on September 20, 1972, the Davis Laboratory received a shipment of eleven gibbons from the SEATO Medical Research Laboratory in Bangkok; the gibbons indeed were accompanied by an export permit from the Department of Foreign Trade, but no permit had been granted by the Forest Department, so the shipment was illegal. On December 31, 1973, Pimjai Birds and Wild Animals of Thailand sent to the Davis labs ten unweaned gibbons, perhaps one or two months old, accompanied by obviously inadequate documentation — only a Thai health certificate for "80 Mynah Birds" and "10 Heads, White-handed Gibbon." But the official record for that health certificate, held by the Department of Livestock Development in Bangkok, listed only "80 Mynah Birds," so it appears that the "10 Heads, White-handed Gibbon" was added later, making even that document illegal. The animals were sent via a Canadian dealer, Kenneth Clark of Ark Animal Exchange, Ontario, and by the time they finally arrived in California, on January 16, 1974, one was already dead (from a shotgun pellet lodged in his skull), while the rest were suffering from pneumonia, possibly as a result of their brief midwinter stay in Canada. Of the full shipment, only four ultimately lived long enough to die in the Davis experiments. On February 16, 1974, the Davis laboratory received another shipment of gibbons, this time six older animals, routed again through the Ark Animal Exchange, accompanied again only by a Thai health certificate.

Concern over the destructive effects of commercial hunting in the United States led to the passage of the Lacey Act of 1900, which prohibits interstate traffic in birds or mammals that have been killed or captured illegally in their state of origin. The Lacey Act was later amended to prohibit the importation of wildlife taken illegally from its country of origin. Thus the Davis Oncology Laboratory's importation of Thai gibbons appeared to violate the Lacey Act; in July 1974 the International Primate Protection League asked the U.S. Fish and Wildlife Service to investigate. Crucial to that investigation, apparently, was the issue of whether laboratory personnel at Davis had known that their gibbons had been exported in violation of Thai law. Some evidence suggests they had known. A 1970 memo to the University's Purchasing Department, sent by the director of the laboratory, Thomas Kawakami, called for a standing purchase order for twenty-four gibbons annually. In the memo Kawakami stated: "Gibbons are obtained primarily from Thailand but the Thai government refuses to release any gibbons at this time. Since they are extremely difficult to obtain, we are required to purchase the animals whenever they are available." And in a June 1974 interview, recorded and notarized as evidence in the case, a manager of the animal colony for the Davis labs flatly stated that since gibbons were protected in Thailand it was necessary to purchase them through the black market and pay bribes at the Bangkok airport to ship them out. The transshipment through Canada (which then did not protect the wildlife of other nations and so issued legal import permits) eased the gibbons' entry into the United States.

By March of 1975, however, the U.S. Fish and Wildlife Service had announced it was dropping the case, supposedly because the chief of the Wildlife Conservation Section of the Royal Thai Forest Department had not cooperated. When the chief protested that he had indeed cooperated and had sent the requested information to an appropriate official at the U.S. Embassy in Bangkok (who received but lost the letter, according to U.S. sources), the investigation was reopened — and finally terminated in 1977 for lack of "evidence of a prosecutable nature." Nevertheless the investigation led to the suspension of three customs officials in Thailand; and in 1980 the National Cancer Institute canceled its contract with the Davis laboratory, ending the gibbon research.

THE LACEY ACT was the first law in history to regulate wildlife imports according to the laws of exporting countries. In its 1960 convention in Warsaw the International Union for the Conservation of Nature called for a global version of the Lacey Act: an agreement that would help control the trade in endangered species through cooperation among the world's exporters and importers. By 1973 a final draft for such an international trade agreement was complete; at a conference in March of that year in Washington representatives of thirty-one nations signed the Convention on International Trade in Endangered Species of Wild Fauna and Flora — CITES for short.

The CITES agreement, which came into force on July 1, 1975, is not a law, it is a treaty. The signatory nations agree to abide by its terms through their own internal mechanisms. Stating that "wild flora and fauna in their many beautiful and varied forms are an irreplaceable part of the natural systems of the earth" and that "international cooperation is essential for the protection of certain species against over-exploitation through international trade," the treaty begins by listing endangered or threatened plants, and their derivatives, and animals in three appendices. The plants and animals in Appendix I are defined as immediately threatened with extinction, requiring both export and import permits for international trade between CITES nations; the permits can be issued only for noncommercial purposes and only if the removal of the plant or animal from its country of origin will not additionally threaten the species. Of the primates, currently forty-two species, subspecies, and species groups are listed in

Appendix I — including all of the endangered primates featured in this book. The plants and animals listed in Appendix II are those that could become endangered if their trade is not controlled. (It also includes species that are difficult to distinguish from Appendix I species.) Appendix II species require a permit for export but no additional permit for import, and the permits can be issued for commercial trade. For the primates, all species and subspecies not listed in Appendix I are given Appendix II status by default. Appendix III includes plants and animals which, although not endangered worldwide, may be vulnerable locally and are thus protected by individual member nations; such species must be shipped with a certificate of origin, and if they are sent from a nation that protects them, they must also have an export permit.

If CITES is a net cast across the global traffic of plants and animals in an attempt to halt the movement of endangered species, it is yet a wide-meshed net. Several animal-exporting nations with significant primate habitat have not yet joined. Until January 1, 1984, when the European Economic Community joined CITES as a bloc, at least one important importing nation, Belgium, had also not subscribed to the treaty.

Belgium, long noted for its nearly complete absence of control over the live animal trade, was for years a major laundering point for smuggled wildlife, impoverishing the ranks of CITES-protected animals while enriching a few already wealthy Belgian dealers. Protected wildlife from many parts of Asia and Africa would be exported through such easy outlets as Singapore, Hong Kong, and Bangkok in Asia, and Kinshasa (Zaire), Bujumbura (Burundi), and Kigali (Rwanda) in Africa, then airmailed to Zaventem, Belgium's international airport. If sent as freight, the animals would often be packed tightly in substandard containers, in higher numbers than officially declared, to cover the anticipated mortality in flight. But to avoid freight taxes Belgian dealers often simply ordered live animals such as parrots and monkeys to be shipped in containers identified as personal luggage. They hired African students to carry the luggage through customs and meet their patrons just outside the airport. Altogether, the Belgian wildlife and wildlife product trade was enormous — and enormously profitable. Two to three tons of ivory, much of it poached and smuggled from Zaire, Kenya, and Tanzania, and exported from Burundi (which has no elephants), would arrive in Belgium each week. In 1979 alone, Belgium officially imported 1.7 million pounds of ivory, tortoise and turtle shell, horns, antlers, and whalebone; 4.9 million pounds of whale meat, seal meat, and frogs' legs; and 7.6 million pounds of live animals (excluding fish) — much of it destined for a properly documented and profitable re-export.

Among the live animals were primates, of course, and at least two Belgian dealers, René Corten and George Munro, appeared to specialize in handling endangered primates. Corten acquired world notoriety in 1983 when he attempted to sell more than two dozen golden-headed lion tamarins (a nearly extinct species closely related to the nearly extinct golden lion tamarin) that had been stolen from a Brazilan preserve and illegally exported through Bolivia. Munro maintained a zoo that was open to the public; but a select clientele would be taken into the basement of his large house, where he kept in cages such protected primates as Indian rhesus macaques (for which he was asking $1,000 each in 1982), chimpanzees ($5,000 each), pygmy chimpanzees ($12,000 each), and the occasional gorilla ($30,000 each). Pygmy chimpanzees live only in Zaire, where they are entirely protected by law; nevertheless, in 1982 Munro possessed and then sold probably the world's largest collection of that species.

While CITES is weakened by the noncooperation of nonmembers, the treaty itself includes several important loopholes. For one thing, member nations can exempt themselves from controls over particular species. When Japan and France joined CITES they exempted themselves from controls on trade in some marine turtle species; Japan, Norway, and the Soviet Union exempted themselves from restrictions on trade in most whale products; France and Italy are exempt from limitations on the trade of some crocodile species' skins. Also CITES makes no provision for control over protected animals in transit. Thus, unless they have specified otherwise in their own customs laws, CITES nations cannot seize shipments of protected wildlife passing through their borders and bound for other destinations; and transit procedures are sometimes deliberately used to obscure a smuggled shipment's true country of origin.

Beyond the matter of loopholes, there is the issue of actual compliance. Of the eighty-six nations that by May 1984 had signed CITES, several abide only moderately by the spirit or letter of the agreement. Bolivia, a CITES member since April 10, 1979, may stand as the most shameful example, for Bolivia's recent dominance of the South American primate (and bird, reptile, and reptile skin) trade seems to have brought about an actual increase in illegal South American exports. In a 1985 report the secretariat of CITES noted that the Bolivian government appeared to be "strongly promoting" a traffic in wildlife smuggled out of neighboring countries. According to the secretariat's report, "Huge quantities of CITES specimens continue to be 'exported' from Bolivia. Most of these specimens have been taken illegally in other countries. . . . Forged documents are common, stolen permits are used . . . genuine permits cover shipments of illegally acquired or traded specimens. The scale of the illegal trade is vast, involving tens of thousands of specimens."

Bolivia placed a one-year moratorium on its entire wildlife trade as of May 1, 1985. But the ban was vigorously opposed both by importing nations and by an organization of Bolivian dealers known as the Association of Fauna Exporters of Santa Cruz. In anticipation of the end of the ban on May 1, 1986, nine members of the Santa Cruz group quickly filed applications to export more than 10,000 primates as well as 300,000 birds. More than twenty international conservation groups signed a letter asking Bolivia to make its ban permanent. But a Bolivian government official pleaded poverty; the wildlife trade constitutes "an important source of income" for the nation and "an irreplaceable means of subsistence for some of its most famished and poverty-stricken citizens."

Some CITES-nation customs officials are poorly prepared for the difficult task of distinguishing controlled from noncontrolled wildlife products (distinguishing, for instance, wolf from coyote skin, Asian from African ivory, one crocodile species from another); and since the format of CITES export and import permits has not yet been standardized, in some cases customs officials are unable to distinguish between legitimate and illegitimate permits. Some CITES nations allow tourists to transport many supposedly protected products, such as stuffed turtles, dried turtle shells, cat skins, and ivory products. Also, because enforcement of CITES is a responsibility of member nations, the treaty's terms are inconsistently enforced. In some countries, prosecutions are common and penalties are severe; in others, prosecutions are rare and penalties mild. The United States prosecuted 650 cases of wildlife import violations in 1978, for example, while in the same year the United Kingdom prosecuted only a handful. In the United States a person who attempted to smuggle twelve radiated tortoises from Madagascar was imprisoned for a year, but in Hong Kong, the Hong Kong Fur Factory was fined a mere $1,100 for illegally importing 319 cheetah skins, valued at $43,900, from Ethiopia.

Yet in spite of its several inadequacies, in spite of the inconsistencies of member nations, CITES has proven an extraordinarily important agreement, in some instances quite effectively regulating the huge international trade in wildlife and wildlife products. Combined with protective legislation in tropical countries, CITES has greatly diminished the trade in most of the endangered (Appendix I) primate species and has helped to regularize the continuing trade in threatened (Appendix II) primates.

I LEFT THE CAGES of cotton-top tamarins at the New England Regional Primate Center, picked up my umbrella — it was still rain ing — walked out into the parking lot, and began the drive back to my house on a little hill in Arlington, where dinner and family awaited.

We all have cages, I thought as I drove home, structures visible and invisible that we receive or construct and carry about for a while or forever. I've known schizophrenics living in the cages of their own distraught and disintegrating minds. I've known drug addicts and alcoholics living in the cages of their own overwhelming addictions. I've known fanatics living in the cages of neurotic projection, sybarites living in cages of desire, illiterates in cages of ignorance, and poor people in cages of poverty. I've also known rich people inhabiting cages of illogical abundance, and educated people in cages of intellectual triviality.

A week or two before I went to the New England Primate Center, a close relative came up from Washington for a visit. Along with his regular luggage he carried a bagful of hypodermic needles, insulin vials, and blood-testing equipment. He lives in the family cage, diabetes, an inheritable disease that also restricted my father's well-being, indeed, his life.

The question occurs to me now: how many monkeys would I place in a cage, if in so doing I could open my family's cage? No quick answer appears, but perhaps the question itself is inappropriate.

Research using primates has indeed contributed to our collective biomedical knowledge. And yet, caging a laboratory primate does not necessarily open the cage of a person; not always does the death of a laboratory primate reduce the dying of humans; the pain inflicted on a laboratory primate does not inevitably alter the complex equation of human pain. Some research done with primates could as effectively be carried out with other animals or with human subjects or through computer simulations. And much research does not actually produce new knowledge. Some is simply repetitive, duplicating for no overwhelming purpose experiments carried out elsewhere. Some of it is trivial. Some of it produces no clear results. The problem is (on the other hand) that the usefulness of a research project becomes a good deal more obvious at its end than its beginning.

Many caged primates are used for research to improve human

fertility, which is already creating its own vast cage of human limitation. The world human population doubles in size every forty years, and the human population of the impoverished, less developed nations doubles every thirty-three years on average. We are thus moving into a century where the sheer reality of population growth may overcome most advances in medical knowledge, a world of basic shortages and limits in which too many are caged by their competition over too little. Persistent starvation in Africa today, perpetual poverty in Asia today predict a future in which human numbers have overwhelmed the carrying capacity of the planet, where any knowledge acquired from experimenting with caged primates will remain a flimsy luxury of the wealthy few.

It is sometimes suggested that the critics of laboratory research involving primates are sentimental, antiscientific fools; nineteenth-century antivivisectionists stumbling myopically against twentieth-century science; people who "love monkeys more than people." That sort of nonsense entirely clouds the issue. Many people believe that the use of any animal in laboratory research raises ethically complex issues. If complex to start with, however, the ethics of using animals as laboratory subjects becomes more so with primates, given their closeness to us in the zoological hierarchy of soma and psyche. Proponents of research with primates argue that their closeness to humans makes them essential as research subjects. But I believe that because primates are so close to us, they deserve to be protected by particularly exacting ethical standards. I also believe that our society can for the most part afford to make biomedical research with primates part of the Ark, through captive breeding, rather than part of the Deluge.

Twelve

ZOOS

I TOOK MY FAMILY to the San Diego Zoo to see douc langurs, leaf-eating monkeys from eastern Cambodia, Laos, and Vietnam. I had made arrangements beforehand to see not only the two male doucs on display but also ten others of this species kept in off-exhibit parts of the zoo, where they are encouraged to breed. As we waited at the zoo security office for our private tour, I looked out a window and read a sign: "Modern zoos are slowing the extinction rate." Nice — yet I had come in part to discover whether that blithe piece of optimism contained any truth whatsoever.

Our guide, Gale Foley, was a blond, athletic-looking man dressed in a khaki zoo uniform. Gale took us to the front of the long line waiting at the entrance, through the turnstile, past crowds of people, down a macadam walkway and into a building with glass cages and monkeys inside.

There, in the first cage, on a rack of wooden climbing bars, sat two male douc langurs. They sat upright, knees up, tails hanging down, and turned their heads to gaze at us. My children, delighted to see monkeys, chirped appropriately — "Look! Look! Monkeys!" — but I think I was even more pleased to see these particular monkeys, so rare in captivity. I recognized them immediately from pictures I had seen and descriptions I had read. I went as close to the glass as I could get without pressing my nose against it; one of the doucs made a few smooth hops down the climbing apparatus until he was close to the glass, then looked right back at me.

Their names, I learned, were Ashley and Nguyen. (San Diego has named some of its douc langurs after characters in *Gone with the Wind* and given the rest Vietnamese names.) Ashley and Nguyen had bright yellow-skinned faces surrounded by fringes of white hair and wispy white beards. Their eyes were shaped like big tear drops and set at an angle toward the nose. Their fur was brilliantly colored — gray around the body with a crown of black on the head, bands of maroon and white around the neck, broad white cuffs on the arms below the elbows, black on the legs from hip to knee, and maroon from knee to ankle. Their hands and feet were black, and their long, loose tails were white. But it was their faces that held my attention. I had expected to see bare yellow facial skin, but I hadn't expected it to be so thick and opaque — indeed leathery, and wrinkled around the nose. Gale reminded me of something I had read but forgotten: their yellow faces will darken, or "tan," with exposure to sunlight and turn shades of orange.*

We then piled into a truck, and Gale drove us slowly through the zoo, through crowds of strolling people, to an isolated concrete building with wire fencing. There we looked into a cage holding three more doucs, a male, a female, and their young male offspring. Gale explained that doucs at San Diego have frequently been plagued by vomiting, probably caused by psychological stress. In an attempt to reduce stress, the three breeding groups, including this one, are kept away from the exhibits, away from staring people and the commotion of other animals.

We drove farther, passed through a gate, continued up a hill, and parked the truck before a series of long, wire-mesh, partitioned cages. One of those partitioned cages held the zoo's breeding collection of ruffed lemurs, and as we arrived, a couple hundred lemurs set up a loud, echoing chorus of alarm cries. The cries, sudden and loud, alarmed my son; I remembered then the echoing chorus of brown lemurs in the trees at Andasibe one evening a few months earlier. We entered a second long enclosure and stood before two interior cages containing two groups of douc langurs, a total of seven animals. About the first thing I noticed was that two of the adult males sat before us, at eye level, with their legs spread wide open, displaying erect penises. The penises were bright red, two or three inches long, thick as pencils, and I imagined — Gale confirmed it — that they were so boldly presented for our benefit.

Erections. One supposes that the erection of a human male indicates

*Two subspecies of douc langur exist. The better-known *Pygathrix nemaeus nemaeus* is easily distinguished by its yellow to orange face. The second, lesser-known subspecies, *Pygathrix nemaeus nigripes,* has a black-skinned face. There are no black-skinned doucs in captivity.

only sexual arousal, a simple flag of state hoisted by the brain. But at least with some other primate species, the state is not always so simple. Male black-and-white colobus monkeys sometimes threaten enemies by displaying erect penises, as do the males of a few other primate species, including, as I just discovered, douc langurs. Indeed — to continue the comparison with human behavior — an erect penis for some monkey species serves much the same function as a human finger rudely poked out a car window on the uncivilized streets of Boston.

Anyhow, there they were, seven douc langurs in two adjacent cages. I noticed once again the yellow and leathery quality of their faces; those impossibly large, almond eyes (eyes of Keene children in a five-and-dime art gallery); the bright patches and bands of color on their fur; the black, long-fingered hands; the long, hanging tails. One of the males was clearly the largest animal there. I spotted several females and two or three actively moving youngsters in the group, and recognized that here were two reconstructed douc langur families, caged refugees in diaspora. The San Diego Zoo managers were hoping to encourage normal sexual and parental behavior by imitating normal family groups.

Douc langurs are most probably declining in the wild, although no one knows how precipitously. The Dresden Museum once had a specimen of an adult male, shot on the Chinese island of Hainan (Gulf of Tonkin) in the early 1890s, but doucs are extinct there now. Some were captured in Thailand in the early 1970s, but none have been sighted there more recently. In Vietnam, Dao Van Tien, dean of the Faculty of Biology at Hanoi University, reports that the yellow-faced subspecies currently inhabits inland mountain forests over a wide area; the black-faced subspecies is "not rare," although its full range is smaller. Western experts, however, describe the status of this dramatic and colorful leaf eater as between "threatened somewhat" and "in serious danger of extinction."

During the recent war years in Indochina, douc langurs were hunted for food regularly and widely. Since the end of the war, subsistence hunting has probably declined — at least in Vietnam. Food is still scarce, but few civilians now possess guns. However, douc langurs depend entirely upon trees for food, so their future fully parallels the future of forests in eastern Cambodia, Laos, and Vietnam.

We know little about the forests of Cambodia. An inventory carried out by the U.S. Agency for International Development from 1960 to 1962 indicated that forests covered nearly three fourths of the country; but a more accurate survey in 1965 counted roughly half that amount. The effect of shifting cultivation on forests in Laos is better documented. As many as 3 million of Laos's 3.7 million people survive through subsistence farming; roughly 1 million practice shifting cultivation in hilly forest land. In northern Laos cultivation of corn and opium by hill tribes has exhausted soils to the point that grass grows where forest used to. Altogether, shifting cultivation may be entirely destroying almost 1,200 square miles of Laotian forest each year, while decades of "unduly wasteful" logging practices in wet lowland forests and highland teak forests have added to the toll.

Vietnam's population has tripled since 1930, yet during that same period the nation's rice production has merely doubled. The end of the war in 1975 left the country in shambles: one tenth of the full population dead or wounded, massive resettlement and political realignment, nearly a million refugees fleeing the country, and a new regime pursuing rigid policies of collectivization that so far have not worked very well. In spite of massive injections of Soviet aid (about $1 billion per year), since the end of the war Vietnam's gross national product has risen by only 2 percent yearly, while its per capita income ranks among the lowest in the world, below $200. Shifting cultivation remains, as it has been for centuries, a way of life for many Vietnamese. In the central highlands of the south, 5 million people belonging to perhaps sixty seminomadic tribes practice slash-and-burn farming. In the southern lowlands extensive tracts of secondary forest bear witness to the long tradition of subsistence agriculture, while large portions of forests within the Mekong Delta have been permanently converted to support settled farming.

Yet the forest destruction wrought during recent years of war far surpassed the more usual forms of exploitation, such as shifting cultivation and logging, and will probably continue to do so for some time. No war has improved the ecology of its battlefields, but in Vietnam the battlefields were, in large part, forests. The combatants included large numbers of elusive, highly mobile guerrillas who depended on the forest for cover. Thus Vietnam hosted perhaps the first war since Hannibal's time to include as a major military strategy the destruction of an ecosystem.

During the period of American involvement, B-52 bombers dropped more than seven million tons of bombs, nearly three times the total force of explosives used by the United States in all of World War II and Korea combined. Standard 500-pound and 750-pound bombs riddled selected areas in the south of Vietnam, creating a moonscape of craters thirty feet deep and forty-five feet across. More than two and a half million such craters were blasted out in 1967 alone, "filling now," according to a journalist's report in 1985, "with clay-yellow water in the flat plains of the south, or laterite red in the hilly areas around Hue and Da Nang."

Also, in the early 1960s the United States began using herbicides to defoliate enemy cover and destroy cropland in hostile areas. Before that decade was over, roughly one third of southern Vietnam's swampy mangrove forests and 16 percent of the inland forests had been defoliated, while altogether during the period of American involvement around four million acres of forest and cropland were sprayed with twenty million gallons of herbicide.

When the defoliation program first began, a spokesman for the U.S. Department of Defense insisted that it would not have serious ecological consequences. However, the herbicides — Agent Orange, Agent White, and Agent Blue — had never before been used in such concentrations.* Agent Orange, for instance, was applied at concentrations well over ten times that of standard domestic use.

Fred Tschirley, a plant scientist sent to Vietnam in 1968 by the U.S. Department of State to examine the results of defoliation firsthand, confirmed the general belief that such chemicals would not have a lasting effect on Vietnam's forests. "Some trees would be killed," he wrote, "and the canopy would be less dense temporarily. But within several years the canopy would again be closed, and even a careful observer would be hard pressed to circumscribe the treated area." However, Tschirley went on to warn against the consequences of repeated defoliation: "A second application during the period of recovery would have a wholly different effect." In fact many areas of Vietnam (one rough estimate suggests 20 to 25 percent of the entire area treated) were sprayed twice or more. Tschirley believed that after the forest canopy had been removed with a first spraying, the second treatment would kill more of the higher trees and also reach the lower vegetation — younger trees, saplings, and seedlings. Tschirley was able to visit a site in Vietnam, at Katum, that had been sprayed twice. Comparing Katum to other sites subject to only one spraying, he noted that the difference was obvious. He concluded that with the defoliation of lower vegetation and the loss of saplings and seedlings, "natural reseeding may be a problem": fast-growing bamboo might quickly dominate previously forested areas and inhibit the return of the original forest indefinitely.

*Agent Orange accounted for about half the herbicides used in Vietnam. Agent White was the preferred herbicide near Saigon; it decomposed more quickly and thus was considered less of a threat to untargeted crops in instances of accidental, windblown dispersion. Agent Blue, which includes organic arsenic, was specifically used on mountain rice crops considered to belong to the enemy.

Vietnam's swampy mangrove forests were much more vulnerable to defoliation than other forest types, and a single spraying was enough to kill most of the trees in those formations. Tschirley estimated that mangrove forests could completely regenerate in about twenty years, but his estimate may have been optimistic, since it was based on the theory that seeds would be immediately dispersed thoughout affected areas. A 1982 report indicates that the mangrove forests are returning, although some tree species have had to be planted by hand. The inland forests, according to the same report, "seem to have retained seedlings capable of regenerating as long as the clearing, originally the result of herbicide destruction, is not maintained by cultivation." On the other hand, observers who visited defoliated areas in 1985 tell us that "dead trees have been cut down for firewood, and defoliated areas now look like abandoned farms, overgrown with low brush and weed."

What were the direct toxic effects of Agents Orange, White, and Blue? Tschirley was most concerned about Agent Blue, which contains organic arsenic. Although inorganic arsenic is extremely toxic to mammals, Tschirley noted that the organic form of arsenic in Agent Blue was short-lived and had "low mammalian toxicity." Most scientific field studies of primates have taken place in the last three decades, when the various Indochinese wars effectively prevented major studies of douc langurs or surveys of their habitat. So we really don't know whether this primate, along with the eleven other primate species of Vietnam, was significantly poisoned by toxins sprayed on the leaves it eats. We do know that bombing and defoliation degraded and destroyed extensive areas of its habitat.

Douc langurs in the wild are protected in only one Ark, Vietnam's Cue Phuong National Park, which consists of about forty square miles of mountain forest in the north. The area was not affected by wartime herbicide spraying and is said to contain many "enormous" trees and considerable wildlife. That wildlife is only moderately affected by local people hunting with bow and arrow. Outside of their native habitat, doucs are held in significant numbers and bred only at three zoos — the Cologne Zoo of West Germany, a privately owned zoo in Rome, and the San Diego Zoo. Thus my family and I were standing before an important Ark of the species, albeit an Ark smaller than my living room.

A zoo employee came into the enclosure carrying a clipboard. When I asked Gale what they fed the doucs, he directed the question to her. She told me sweet potatoes, carrots, bananas, spinach, green beans, oranges, celery, and apples.

Gale sat back against the wire fence of the enclosure, smoking a cigarette, and described some of the problems with the breeding project. Doucs are sensitive animals, easily susceptible to stress. They often get lung mites, which used to be a serious problem, but a treatment has recently been discovered. Most troubling are the genetic constraints. All but two of the douc langurs in North America are related to each other through a single prolific, wild-caught male, Jack, so the North American population is quickly becoming inbred. Jack also seems to have a chromosomal abnormality, which has resulted in a 50 percent abortion rate for his conceptions. Thus, although San Diego has witnessed a total of twenty-eight douc langur births, its breeding colony is barely holding its own. New genetic strains are critical for the continued survival of this group. The zoo had recently acquired two wild-caught females from a collector in Italy, but one died in childbirth. A wild-caught male was due to arrive at San Diego from the Yokohama Zoo within a couple of months — perhaps he will become a significant breeding male. It might be possible to borrow some doucs from the Cologne Zoo to introduce new genetic strains onto this Ark. But the director of the Cologne breeding project is reluctant to subject her animals to the stresses of travel and a new environment. "Without new bloodlines," Gale concluded, "I'm not too optimistic."

IN 1975, conservationist Gerald Durrell called for the creation of Zoo Banks. These would be much like zoos, might even be zoos. But unlike a typical zoo, a Zoo Bank, as part or most of its mission, would breed and maintain endangered species. Captive breeding has already demonstrated its potential as a conservation measure, but in the past, it has been used in desperation, as a hastily planned last-ditch effort to save a species that would otherwise disappear. Durrell envisioned many captive breeding institutions to preserve endangered species permanently and methodically. Instead of last resorts, they would serve as standard links in the chain of conservation.

A captive breeding program begun early, when a species was first recognized as threatened but still substantially existed in the wild, would mean that founders could be captured without disastrously reducing the wild population. In the case of the California condor, for example, a serious captive breeding program begun in the 1950s, when about sixty condors still remained, would have threatened the wild population much less than the controversial $25 million dollar program begun in the 1980s, when fewer than twenty wild condors were left. Furthermore, captive breeding begun sooner rather than later and methodically carried through would have a better chance of succeeding. Perhaps the greatest problem with the condor program, for instance, has to do with genetic representation. Right now only five known family lines of California condor exist; not long ago, there were nine. The fewer family lines, the greater the inbreeding and loss of genetic variability. As a species declines toward oblivion, the genetic opportunity declines as well.*

Durrell himself established the first zoological institution in the world devoted solely to the captive breeding of endangered species: the Jersey Zoo, located on the British island of Jersey. Meanwhile, many traditional zoos are considering their own potential as Zoo Banks. Are traditional zoos ready?

A recent report on a major North American zoo, the Atlanta Zoo in Georgia, is revealing. The animals there are confined in small steel cages or kept at the bottoms of concrete pits. A gorilla named Willie B. has been locked inside the same small cage for over twenty years, without a single outing. A pair of monkeys on loan from another zoo have disappeared. Four cheetahs are also missing. Twinkles the elephant somehow died in the back of a pickup truck used by a traveling circus. Two bears, very generously loaned by the zoo's veterinarian to a tourist attraction, were subsequently shot because of "unruly behavior." The zoo is old and run-down. Atlanta recently announced plans to spend $21 million on renovating the zoo, but then abruptly dropped the matter because it didn't have the money.

The pattern is familiar. Zoos like the Atlanta Zoo are relics of the sensibility of another era. They are mere display cases, museums with live exhibits. And if such zoos are parsimonious in caring for their animals, they are yet profligate in using them. Between 1926 and 1962, to cite one example from a multitude, the National Zoo in Washington found it necessary to import giraffes from Africa four times to keep its giraffe exhibit stocked. Even a decade ago, the world's zoos were maintaining self-sustaining populations of only 26 out of a total of 275 rare mammal species in captivity.

Historically, zoos have not succeeded well in breeding animals in

*Very small and isolated populations of breeding animals face two genetic problems: inbreeding and loss of genetic variability. Inbreeding, the mating of closely related animals, usually leads to increased mortality and declining fitness and fertility over generations. Genetic variability refers to the variance in inherited qualities among unrelated individuals within a species: some individuals will be shorter than others, more tolerant of heat, less susceptible to disease A, more susceptible to disease B, and so on. The variance in inherited qualities provides insurance that sudden changes in an environment — for instance, the appearance of a new disease — although it may wipe out individuals will probably not wipe out the species. With very small populations of breeding animals, the genetic variability may be low to begin with.

captivity. Why? For one thing, zoo directors have sought to exhibit wide varieties of species rather than to collect enough animals of one species to make breeding possible. Exchanges and loans of animals among zoos for breeding purposes were possible but uncommon: zoos tended to be competitive rather than cooperative. Also, since many animals still could be collected from the wild, zoos were not economically motivated to breed their own.

But zoos need not follow the patterns of the past. Modern zoos can be, and often are, designed and maintained with a concern for the welfare of the animals. Zoos can display animals living within satisfying imitations of their natural habitats. Zoos can enlighten. Zoos can also contribute to conservation in a distinctly tangible way by breeding rare or endangered species. Captive breeding in zoos can itself serve as an important sort of public education. Breeding programs that maintain animals in seminatural conditions can increase public awareness of and sympathy for vanishing species, and they can publicize the need to preserve a species' natural habitat. At the same time, zoos can provide a more specialized education. Captive breeding projects can serve as zoological laboratories, where scientists learn about species' physical and psychological needs and reproductive biology through unusually close observation. Such learning can, in turn, improve attempts to preserve the animal in its wild habitat. Beyond education, captive breeding in zoos does two good things. It enables zoos to maintain their own populations rather than buy additional wild-caught animals, and it provides insurance against extinction in the wild.

We already have models, captive breeding projects that have saved three major mammalian species — the Arabian oryx, Père David's deer, and Przewalski's horse — from imminent extinction. The Arabian oryx, a small, long-horned desert antelope of the Arabian Peninsula, hunted to near extinction by sportsmen with modern weapons, was saved by a last-ditch breeding project at the Phoenix Maytag Zoo in Arizona. Père David's deer, eliminated from its original range in northeastern China by habitat destruction during the Shang Dynasty (1766–1122 B.C.), was preserved within managed herds kept in royal game parks. Later, after foreign soldiers ate most of the remaining deer during the Boxer Rebellion of 1900, this animal was saved by captive breeding in Europe. Today the world herd of Père David's deer numbers about eleven hundred, perhaps enough to return some of the species to its original wild habitat. Przewalski's horse, the world's only surviving wild horse, may still exist in small numbers within its original remote habitat in southwestern Mongolia. But the precarious and uncertain state of the species' existence in the wild is balanced by a stable captive population in Europe, begun in 1900 when England's Duke of Bedford paid a German animal collector to acquire a few breeding pairs.

Captive breeding may seem a simple enough enterprise (place Male A next to Female B), and we might still wonder why zoos have only lately attempted it in any determined fashion. In truth, however, it is extremely difficult to get most zoo animals to breed. Knowing an animal's physiological requirements (which vary tremendously from one species to another) is crucial, but it is merely the beginning. For most species, breeding is intimately associated with very complex environmental and social events in the wild.

No zoo can fully reconstruct a wild environment, but animals in captivity will not breed without certain environmental cues. They need a specific amount of space before they will contemplate procreation. Animals also need to feel familiar and safe in their surroundings. Since many mammals mark their home territories with scents, even cleaning a cage may disturb them. Some animals must be separated from others of their own species. Some will be stressed by exposure to unfamiliar humans — even a short visit by a new veterinarian or maintenance person may have drastic effects.

For many animals the breeding cycle is triggered by the presence of particular environmental "furniture." For instance, many bird species have evolved to lay and incubate eggs only in highly specialized nests. When the Bronx Zoo created simulated nesting habitats for its birds in 1964, species that had not bred for fifty years while living in wire-mesh cages began nesting and laying eggs. Tree-dwelling animals may need branches in their cages, while many other species require nests, caves, or similar areas where they can retreat and feel safe. Tree shrews need more than one nest in order to breed at all, since in natural conditions the parents live in a separate nest from their infants.

Among environmental factors, perhaps most important is the animal's natural climate — including humidity and temperature, seasonal variations in rainfall, daily light cycles, and yearly changes in light cycles. For lemurs the beginning of the breeding season is associated with the start of a rainy season. At least with mouse lemurs, the seasonal increase in day length directly triggers breeding. (In captivity their breeding can be artificially accelerated by sneakily shortening the yearly cycle of day length.) The Hawaiian goose will begin to lay eggs only when day length diminishes to 8.8 hours in a day.

Yet environmental simulations are not enough. In many cases animals will not reproduce unless they are living within a normal — or normal enough — social situation. Captive flamingos are stimulated to breed only when they congregate in large flocks. Cheetahs mate successfully only if the males are kept separate from the females except when the female is quite ready (in heat). When this knowledge was applied to captive groups, the number of cheetahs born in captivity rose from none in 1965 to 102 by 1975. Mouse lemurs too will breed only if the male and female are kept apart except for the few days when the female is fertile; in addition, the female must enter the male's cage, not vice versa. The insecure male can perform only in the security of his own home.

Some animals are monogamous; others are polygamous, polyandrous, or just plain promiscuous. And with some mammals, a single male in a group will mate with all the females after intense male-to-male fighting. But fighting among males, even if it is a necessary prelude to mating, makes normal zoo keeping difficult. A classic example: in 1925 officials at the London Zoo built an artificial cliff face, which they named Monkey Hill and populated with hamadryas baboons from northeastern Africa. Male hamadryas baboons are large, powerfully built animals, graced with flowing leonine manes and shoulder capes of gray fur. The zoo management, feeling that the be-maned and be-caped males would attract more visitors to the zoo than the smaller, less dramatic females, chose to populate the exhibit with only six females and about a hundred males. But the hamadryas baboons' primary social unit is the single-male family. Each adult male typically possesses and actively herds — with such enticements as threats and biting — a harem of several adult females. Not surprisingly, then, the many adult male baboons in the Monkey Hill exhibit fought continuously and violently over the few females. A male acquiring even brief control over a female would immediately mount her, try to keep her in his possession by biting her, and then be forced to fight other males for possession. Even after a female was killed, the males would fight over her dead body until it was removed by zoo keepers. Within a brief period almost thirty baboons from the exhibit were dead, and eventually over half the animals had been killed. For some years after the exhibit was closed down, the story of London Zoo's Monkey Hill was considered evidence of the vicious aggressiveness and intense sexuality of baboons and, by pseudological extension, evidence of the highly aggressive and sexual nature of all Old World monkeys. We now know that the Monkey Hill exhibit was primarily an illustration of zoo-keeping ignorance.

Few zoos have repeated the London Zoo's mistake — crowding animals together in a fashion so contrary to their normal social inclinations that violence is provoked. But many zoos have done the reverse — isolating animals to suppress violence. For purposes of breeding, that may also be a mistake. It has been demonstrated, for example, that threats and fighting among male rhesus macaques beneficially elevate their levels of male sex hormone (androgen). When the males are not allowed to compete or express aggression, they tend not to be sexually inclined. And the female rhesus macaques similarly appear to lose sexual interest if the males are not properly aggressive among themselves.

North American zoos and aquariums, which now confine more than 47,000 mammals, 33,000 birds, 19,000 reptiles, 4,500 amphibians, 159,000 fish, and 79,000 invertebrates, are a potential Ark of some significance. Recently the American Association of Zoological Parks and Aquariums, an organization that includes most North American zoos and similar institutions as members, has established a Species Survival Plan to breed rare and endangered species methodically and to maintain them over many generations. North American zoos might be able to support stable breeding groups for perhaps one hundred to two hundred species. But so far the American Association of Zoos has concentrated its efforts on about thirty rare or endangered species and subspecies, including the following primates: ruffed lemur, black lemur, golden lion tamarin, lion-tailed macaque, western lowland gorilla, and orangutan.

The Species Survival Plan begins at a bureaucratic level. A particular species will be chosen on the basis of need (how threatened is the species? how common or uncommon is it?) and likelihood of success (how many are in captivity? have they bred before?).* Once the Amer-

*Selecting species for inclusion in the Survival Plan, incidentally, is itself a major task. Consider, for example, a problem that appeared a few years ago: choosing which rhinoceros species to preserve. All five species of rhinoceros were and are endangered, but the southern subspecies of the African white rhino was temporarily "secure," with a world population of almost 3,000 on the increase. By contrast, only about 1,000 northern white African rhinos remained in the wild, and their numbers were rapidly declining. Some 14,000 to 24,000 African black rhinos still survived in the wild, but that population was declining precipitously. Of the other species, 2,000 Indian, 200 Sumatran, and perhaps 50 Javan rhinos remained. American zoos had already decided they had room for only about 300 rhinoceroses altogether. How many species should be preserved? It might have been reasonable to preserve 100 individuals from each of three species. But which three? At first glance, it might have seemed appropriate to include any group except the "secure" southern subspecies of the African white rhino. However, world zoos already possessed over 500 of this kind and few of any other — the southern African white rhinos were already there. To complicate matters further, many zoos had only enough room for one pair of southern white rhinos, but the species generally reproduces only when lolling about in larger herds. Other rhino species would reproduce in simple pairs. A solution? Perhaps the beginning of one was to move all of the southern white rhinos out of American zoos and place them together in larger herds on private land somewhere, thus opening up enough rhino space in zoos to accommodate other, more immediately endangered types. And that is precisely what the American Association of Zoos began attempting.

ican Association of Zoos has laboriously chosen a species for inclusion in the Survival Plan, member zoos already possessing animals of the species are solicited. Those that agree to participate in the program elect a ten-member Propagation Committee, which selects a coordinator and studbook keeper for the project and begins documenting a breeding plan.

To begin, a thorough genealogy of all animals of that species in zoos is necessary. (For many zoo animals, studbooks have been kept for years. But the American Association of Zoos has recently started something called the International Species Inventory System, a large computerized database containing genetic and other information about zoo animals. Though this inventory currently includes records mostly from American zoos, the plan is eventually to gather records on captive animals from around the world. At the moment the database contains basic information — sex, parentage, place of birth, blood chemistry, and so on — for about 53,000 mammals and birds.) Genealogies in hand, the Propagation Committee members are now ready to begin serious planning for their particular species.

First, the committee must decide which animals of the species should be bred and which not allowed to breed further. Already inbred animals or hybrids of different subspecies, for instance, will probably not be included in the breeding group. Second, the committee must decide how much each individual animal should reproduce in order to equalize *founder representation*.

The founders of a zoo population are the original wild-caught animals. It is presumed that the founders are unrelated to each other and that they represent a reasonably well-distributed genetic sample of the full species in the wild. If all existing zoo animals of a species are founders — all born in the wild — then founder representation is already equal. No problem. Usually, however, many animals in zoos are captive-born descendants of founders. In a group of captive-born animals, many or most may be descended from a few prolific founders. In other words, some founding animals have already reproduced more than others and are thus more fully represented in the population's gene pool. To maximize the genetic variability of the breeding group, the Propagation Committee must determine which animals are descended from the underrepresented founders and encourage them to breed more fully. At the same time, descendants of overrepresented founders must be limited in their breeding.

The committee must also decide on the ultimate size of the species' *maintenance population*, the generations after the founding generation. Typically, that size will be a compromise between what is needed ideally to maintain full genetic variability over time (fifty managed individuals for a few generations; five hundred for indefinite generations) and how much space is available.*

In most cases the maintenance population will be distributed among several zoos. Unhappily, though, the collective capacity of zoos to hold breeding members of a species is terribly limited. The average North American zoo covers less than 55 acres of land. Combined, North American zoos possess about 20,000 acres, or some 31 square miles — an area smaller than the New York City borough of Brooklyn. Most of that land is already used for existing animal exhibits that do not involve breeding and for parking, concessions, and so forth, and is not actually available. Of course, captive groups of many species can be confined to relatively small areas. And several North American zoos — such as the Los Angeles Zoo, the San Francisco Zoo, the New York Zoo, and the National Zoo in Washington — are acquiring additional land solely for breeding. The land used for breeding groups of animals can be further expanded through the donation or lending of private lands. Two exotic game ranches in Texas, for instance, have voluntarily set aside land for the breeding of Grévy's zebra and the scimitar-horned oryx. African rhinos may soon breed on private land in Texas. Yet altogether, the land available for captive breeding by North American zoos is minuscule, and in effect all endangered species are competing for it. Ultimately, the limited carrying capacity of North American zoos means that the maintenance population for most animals under the Species Survival Plan will include far fewer than five hundred individuals.

Once inbred and other undesirable animals have been excluded, breeding has begun, founder representation has been equalized, and the captive group has reached its maintenance size, the next phase of the Species Survival Plan begins. At this point the full captive population must be managed in a fashion that assures continuing maximum genetic diversity. Over time the full captive group, generation after generation, must be regulated to provide balanced family size,

*A small founding population can present a reasonably broad genetic sample of the wild species' full gene pool. But to pass that broad genetic sample on, each founding animal should be equally represented in subsequent generations — that is, founders should have equal numbers of offspring that, in turn, breed equally. Even if the founder representation is equal, in later generations the original genetic sample will still tend to narrow: in a small population, a random loss of particular gene forms over generations means that the inherited variability of individuals diminishes. To counter this loss of genetic variability, it is necessary to have a large maintenance population. If they are carefully managed to ensure equal genetic representation, fifty animals will pass on a stable level of genetic variability for a few generations; five hundred animals will pass on a stable level of genetic variability indefinitely.

balanced sex ratios, and an overall stable population size. Older and otherwise nonbreeding animals will probably be removed to assure that the limited space is maximally used by the breeding group.

The Species Survival Plan for the Siberian tiger provides a good example of how North American zoos might try to preserve an endangered species. Although all eight subspecies of tiger are nearly extinct in the wild (from loss of habitat and hunting), American zoos have decided they have room only for 500 tigers, or 250 individuals each of two subspecies. The Siberian tiger is one of the lucky two.

American zoos already possess around 250 Siberian tigers, descended from seventeen founding animals. However, five founders are distinctly overrepresented in the genealogy of the present group, while eight others are underrepresented. Another four founders are proportionately represented. So founder representation must be equalized by expanding the breeding of some animals and limiting the breeding of others. Fortunately, the Soviet Union and China are lending some Siberian tigers to expand the genetic base of the American population. These animals will be thoroughly interbred with the American group and thus will become additional founders.

The present group of Siberian tigers is poorly distributed in terms of sex ratio and age structure; proportionately there are too few young just reaching breeding age, which means that the present population size could soon drop, reducing the genetic sample we now have. So to equalize sex ratio and stabilize the age distribution, zoo managers will probably have to encourage the breeding of some animals and, simultaneously, remove others. Once a maintenance population of 250 animals has been achieved, with equal founder representation, balanced sex ratio, and a stable age distribution, then theoretically each animal will be allowed to produce two litters in a lifetime. From the two litters all but two animals, one male, one female, will be removed — probably killed (or, to use the preferred expression, "euthanized"). Animals over a certain age (between thirteen and fifteen) will also be removed.

Considered coldly, as a semimathematical proposition, the Species Survival Plan for the Siberian tiger, this largest and most powerful of all tigers, is disturbing indeed. But the Siberian tiger might not survive much longer in its original wild habitat. Captive breeding may be this animal's final Ark.

The ultimate goal of captive breeding is commonly thought to be the return of endangered animals to the wild, but in fact the great majority of captive breeding projects will probably never reach that goal. All zoo and medical research collections began with wild-caught animals. Now that many of those animals are rare, difficult to obtain, and very often illegal to capture in the wild, zoos and research centers are forced to breed their own animals. One may or may not decry the perpetual captivity of animals in zoos and research centers, but the fact is that maintaining captive populations indefinitely both ensures against future extinction of species and — equally important — reduces present pressures to collect more animals from their wilderness homes. And breeding projects that do not currently include plans for reintroduction still provide the potential for return in the future. Should the entire wild population disappear, the captive population remains as a hope for the species.

In spite of the many difficulties, captive breeding projects have begun to reintroduce several endangered bird species. The Hawaiian goose, or néné, once existing in large numbers, was reduced by hunting, habitat destruction, and the predations of introduced animals in Hawaii to a population of about forty-two birds in 1935; captive breeding and reintroduction have returned the wild population to around six hundred currently. The North American whooping crane, brought down to a 1937 population of about eighteen by habitat destruction and wanton hunting, is now nurtured within protected sanctuaries and by captive breeding at the Patuxent Wildlife Research Center in Maryland; some of the captive population at Patuxent provide eggs for the slowly expanding wild population. The masked bobwhite is being reintroduced. The peregrine falcon, eliminated in the eastern United States during the 1960s largely because of DDT and other toxins, is being returned to that area. Similarly, a captive breeding and reintroduction project may help preserve the endangered American bald eagle. In Europe several conservation groups are breeding and attempting to reintroduce the European eagle owl into parts of its former wild range. Captive breeding and subsequent release have also helped preserve several kinds of waterfowl: the Aleutian Canada goose, the marbled teal, the brown teal, the Hawaiian duck, the Mexican duck, the New Zealand scaup, the white-winged wood duck, and the North American wood duck.

Reintroduction so far has succeeded mostly with endangered birds, but a few mammals and reptiles have also been returned to their original wild habitats. A few Arabian oryxes, bred in captivity, have been returned to parts of their former range in the Arabian Peninsula. American bison bred in the Bronx Zoo were reintroduced between 1907 and 1917 into some of their original range lands in the American

West. European bison bred in Swedish zoos since the 1930s have been returned to wild portions of their former habitat in Poland and the Soviet Union. Wolves bred in German zoos have been reintroduced into the semiwild lands of the Bavarian National Park. Mexican crocodiles bred in the Atlanta Zoo have been returned to appropriate areas of Mexico.

Of course, captive animals can be returned to the wild only if a wild exists. In most cases, species are taken to the edge of extinction in the first place because their natural habitats are being destroyed. Returning these animals to a dwindling habitat seems particularly futile, like saving a drowning man only to cast him back into the sea. In some instances, captive-bred animals have been introduced to wilderness areas that were not their original homes — the black buck, the Himalayan tahr, the aoudad, the axis deer, and the Bali myna, for example. But in many cases, captive breeding has become a waiting game. We are waiting for the possibility that at some time in the future humans will save or protect, restore or recreate some particular lost Eden.

After we had watched the douc langurs for some time and taken a few photographs, Gale Foley transported us by truck back into the public part of the zoo. It was late in the morning by then, a hot and sunny Sunday, and the zoo was very crowded. Gale confessed that he likes the zoo best when it is closed: "The most exciting time is when the zoo closes. Suddenly it's very quiet, and the animals come alive. Especially the hoofed animals. As soon as everybody has left, they perk right up. Their ears start to twitch. They can hear the wind and falling leaves again."

At the moment, however, the zoo contained far more people than animals, and it wasn't quiet. We drove very slowly, carefully, along a wide macadam path through thick crowds of people and came to a great walled and fenced exhibit that, according to its sign, contained orangutans and a Sumatran siamang. Gale parked the truck, and we got out and looked. I saw briefly a flash of black fur as the siamang flew from one log to another within his artificial forest. On the other side of the macadam path was a large cluster of cages with many monkeys inside. My wife touched my arm and pointed over there, where a cage contained perhaps half a dozen very familiar-looking black monkeys with white manes and tufted tails: lion-tailed macaques. I was happy to see them, having recently met their relatives in southern India.

Gale led us down the macadam path to the ape exhibits. We looked over a wall into an impressively realistic environment of rocks, grass, trees, and caves, where a few dark-furred pygmy chimpanzees were sitting, standing, and moving around, astonishingly humanlike in form and even posture. Next door to the pygmy chimpanzees, a few gorillas hung out in their own enclosure, and we watched one walk back and forth over rocks with a heavy, hulking manner. Later we walked back up the hill to the enclosure containing the siamang and the orangutans. The siamang had disappeared from sight, but the orangutans were there, sitting motionless and huddled, like giant potatoes covered with rough, red fur.

My children were becoming tired and hungry, so — after saying good-bye to Gale — we walked over to a food stand and bought lunches for what seemed like a lot of money. The sun was becoming oppressive, so I walked over to the zoo shop and bought straw hats for all of us. My son fell asleep, and we sat on a bench under a tree near the monkey cages. I looked over at the monkeys, mostly obscured by people standing in front of their cages, and briefly imagined the zoo as a microcosm, an absurd prototype of earth near the start of the Big Bang, or a vastly collapsed world in which all the primates — humans, apes, monkeys, and prosimians — were gathered together in one big heaving huddle. The microcosm, however, was strangely antigeographical, since South American, African, and Asian species had been placed next to each other in what seemed to be an entirely random arrangement. Next — my daughter was falling asleep by then — I thought of the species I had visited in my tour around the world and tried to recall which ones were well represented in zoos and which were also successfully breeding in captivity.

Until recent prohibitions on trade in endangered animals, some of Madagascar's prosimians were represented well enough in zoos. Germany's Saarbrücken Zoo, for instance, held twenty lemurs of six different species and ten subspecies in 1965. Several zoos found these animals easy enough to care for and quite long-lived in captivity. A lemur in the London Zoo lived to an old age of twenty-seven years, and the Cairo Zoo had a twenty-two-year-old. For many years an interested European by the name of Georges Basilewsky kept records on all the captive lemurs outside of Madagascar and attempted to preserve them through his own private captive breeding project. Although Basilewsky provided about sixty zoos with lemurs, most of the zoos did not seriously try to breed them.

Today, however, many European zoos are responsibly and successfully breeding prosimians from Madagascar. In the Netherlands, the Rotterdam Zoo has long maintained a colony of lesser mouse lemurs. First started with a few wild-caught animals in 1968 at University College, London, the Rotterdam group now includes around twenty animals, including some from a fourth generation in captivity. Like many lemur and loris species, lesser mouse lemurs are intensely seasonal little creatures; their body weight, endrocrine activity, behavior, and sexual inclinations vary markedly during the year. (The male's scrotum increases up to eight times in size as day length increases.) Thus the Rotterdam Zoo maintains these prosimians in several small cages within a single closed room, where artificial daylight and temperature simulate the Madagascar seasons. Within their cages the animals have nesting boxes with circular entrances that imitate their natural home in Madagascar (tree hollows with little entrance holes). In the wild, lesser mouse lemurs sleep in sexually segregated groups, typically one or a pair of males together and two to fifteen females together. At Rotterdam, however, the animals are kept singly in their cages, and females are placed in the home cages of males during estrus.

The American Association of Zoos is organizing Species Survival Plans for the black lemur, best represented at the St. Louis Zoo in Missouri, and the ruffed lemur. So far the most extensive captive breeding project in the world for a Madagascar prosimian involves two subspecies of the ruffed lemur — the red ruffed lemur and the black-and-white ruffed lemur — both represented mainly in separate collections at the San Diego Zoo in California and the Duke University Primate Center in North Carolina.

San Diego's breeding group, which we saw in the cages next to the douc langur breeding enclosure, had produced as of mid-1984 some 87 captive-born red ruffed lemurs and 153 black-and-white ruffed lemurs, all descended from a very few imported founders. The San Diego Zoo has from time to time expanded the genetic base of its groups through exchanges with Duke. Still, physical anomalies, possibly as a result of inbreeding, have begun appearing in the black-and-whites. Some are born with scoliosis, or curvature of the spine, and some are born with funnel chest, a concave bend in the breast plate. If these or other anomalies are the result of inbreeding, the San Diego project may still hope to breed them out in later generations. New genetic strains may yet be imported from Duke and elsewhere, even possibly from newly captured wild animals.

Including its ruffed lemur group, the Duke University Primate Center now keeps more than 600 prosimians of about twenty species, making it the largest and most significant prosimian breeding colony in the world. Not surprisingly, it is also the world's most prolific, producing altogether about 110 new animals every year. Eighty percent of its current group is captive-born, and some species are now represented by a ninth captive generation. Duke keeps some of its smaller prosimians within environments that artificially approximate temperature and day-length variations over a typical Madagascar year. Other prosimians, which have become generally acclimated to the North Carolina seasons, are kept in outdoor cages with heated winter sleeping quarters. (The occasional new arrival from Madagascar spends a year and a half in an artificial climate in a closed building to become acclimated to North Carolina's seasonal sequence of temperature and day length, which is generally the reverse of Madagascar's.) Still other lemurs, members of five especially hardy species, are allowed to range freely over eleven acres of fenced forest.

The two lemur species I visited in Madagascar have never endured captivity very well. Eight indris were acquired by the Paris Ménagerie in 1939, but all died within a month. The indri's shy temperament, specialized diet, innate territoriality, and apparent sensitivity to environmental changes combine to make this animal difficult to keep in captivity, much less to breed. As for the aye-aye, after the discovery of ten small, scattered aye-aye groups in the wild in 1957, a few individuals were taken into captivity, to zoos in London, Amsterdam, and Berlin. The Amsterdam aye-aye lived for twenty-three years. One pair was kept in a Madagascar zoo for "breeding studies." A zoo assistant was employed full time to gather beetle larvae for these two animals. The breeding studies must not have been very successful, though, since this single captive pair never reproduced. At the moment, I understand, three aye-ayes are kept in a Parisian zoo.

The Artis Zoo of Amsterdam may have been the first zoo in the world to exhibit a lion-tailed macaque, in 1839. The first American zoo exhibiting a lion-tail was the Philadelphia Zoo, which began its collection in 1899 and hosted the first captive birth in 1932. Right now North American zoos hold about 225 lion-tailed macaques of a total world captive population of about 400. Five North American zoos — at Baltimore, St. Louis, San Diego, Seattle, and Winnipeg — have major collections. Major collections of the species can also be found at the West Berlin Zoo in Germany and the New Delhi Zoo in India.

Although several zoos successfully bred lion-tailed macaques in the past, the breeding was often carried out very carelessly. A few years ago, for instance, the Woodland Park Zoo in Seattle noted that its collection of this highly endangered species was becoming inbred. To reduce inbreeding, the zoo imported a male from the Toronto Zoo, which was not a close relative of any of the Woodland Park animals. The male did breed successfully, but it was later discovered that he and most of the Woodland Park animals were descended from the same wild-caught founder. In other words, the new animal was contributing to a narrowing of founder representation.

The American Association of Zoos has begun a Species Survival Plan for lion-tailed macaques. Right now this project involves creating a reliable genealogy for the captive population and determining the overall carrying capacity of North American zoos. Probably the maximum size of the captive group will be around 250 carefully managed individuals. But the size of the group will be finally determined by the space offered by zoos. At least two zoos, at San Diego and Baltimore, are constructing large (one-acre) off-exhibit breeding enclosures solely for the species. The Mesker Park Zoo of Evansville, Indiana, is constructing a $200,000 exhibit area to house its nine lion-tails. But future breeding of the lion-tailed macaque will have to contend with limited founder representation. Almost half of the current North American population of 225 animals is related to only three fecund founders: Goliath of Winnipeg, Caspar of Seattle, and another male currently in the Baltimore Zoo. So future breeding in North America will include the removal of animals from breeding groups, and exchanges with European zoos to broaden the genetic base of the North American group.

Since the remaining wild population is small and fragmented, it might be conceivable to reintroduce lion-tails to their wild habitat. Even a minor intermixing of captive and wild groups would increase the genetic diversity of both. However, as with other species, reintroduction of captive lion-tailed macaques into their original habitat could devastate the remaining wild members with diseases to which they have acquired no resistance.

The orangutan is a tropical animal, adapted over millions of years to a perennially warm or hot climate. In natural circumstances it spends most of its time high in the trees, sleeping, eating, or in rare moments looking down at potential threats far below. It is used to great heights and great space. Confinement in a cold cell in a cold climate at ground level is probably the most abnormal set of circumstances one could dream up for this animal. Not surprisingly, most early zoo orangutans soon died. Even now orangutans in zoos tend to sleep a lot and become obese. Barbara Harrisson, the person who first called attention to the decline of wild orangutans, gave her impression of the orang quarters at Munich's Hellabrunn Zoo it the late 1950s: "Their cage had a concrete floor and sloping shelves to climb on to, round the sides; a low wooden table and chairs, where they sat — the keeper with them — oblivious of what they were supposed to do. They had grown out of cute babyhood and were at an age where they should have exercised their bodies over long hours to become strong and vigorous animals. When they should have climbed, swung, investigated, built nests. Their hair was sparse and worn."

Orangutans, whether wild or captive, must be active. Confinement and boredom turn these intelligent, introverted animals into lethargic dullards. So a decent captive environment must be at least minimally entertaining. It might include a view of the outside. Since orangutans in their natural circumstances are intensely arboreal — much more so than the other great apes — a good captive environment must also include some climbing apparatus, ideally a real tree. Whatever cage furniture is provided, however, ought to be very solid, for orangutans are superb reverse engineers. They are intelligent, observant, and powerful, and they love to dismantle things. A rope for swinging will be quickly chewed to pieces. Wooden shelves will be torn and chewed away. An orangutan can easily remove nuts and bolts that have not been thoroughly tightened with a wrench. In addition to solid permanent furniture, these animals should have various materials such as twigs, wood shavings, and cardboard boxes that can be torn, gathered, built into nests, and so on.

Captive orangutans require a varied diet that approximates their highly varied natural diet. They are susceptible to tuberculosis, colds, and other human-carried ailments and must be inoculated and otherwise protected from infected humans. They should be protected from indiscriminate feeding by zoo visitors. They require a humidity above 60 percent and a warm shelter from wind and rain when the outside temperature is below 77 degrees Fahrenheit. Captive orangutans also have some very important psychological needs. The infants cannot stand, walk, or crawl, and need either an adult to cling to or some clingable substitute. Much more important, infants need the emotional nourishment a mother or mother substitute provides. If an adult female orang is not available, they will (and must) develop a powerful attachment to whatever human caretaker happens to be around. Ideally, of course, infants and young orangutans should remain with their mothers or mother substitutes for several years. Hand-reared infants should frequently spend time with others of their kind in the first few months, since so much of their normal behavior, including mothering and other important social interaction, is learned.

In 1928 three zoos — in Berlin, Nuremberg, and Philadelphia — hosted the first captive births of orangutans. However, all three infants soon died of rickets because their mothers were producing inadequate milk (probably because the mothers' diets were inadequate). In the years to follow, several zoos, including those of Düsseldorf, Dresden, Havana, Moscow, Rome, and St. Louis, successfully hosted captive births and raised the young, often with supplementary bottle feeding. Food shortages and the bombing of major European cities during World War II killed much of the captive population at that time, but by the early 1960s about 250 orangutans were living in zoos around the world. Apparently zoos had learned by then how to keep their orangutans healthy: one pair in the Philadephia Zoo had survived thirty-five years of captivity; and zoos in London, New York, and Birmingham, Alabama, had kept orangs for twenty years. The Rotterdam Zoo maintained a large group of orangutans until a 1965 smallpox epidemic almost entirely destroyed it.

By 1974 zoos and other institutions throughout the world had a total of about 625 orangutans. Only one other endangered species, Père David's deer, was better represented in captivity. Captive births had become common; more than a third of the 1974 population had been born in captivity. Yet some observers felt that such success was more apparent than real. Although about thirty-five captive-born females in the 1974 group had reached sexual maturity, not one had produced offspring: no second generation had appeared. The entire captive group seemed in danger of ending at the first generation. Why? It appeared that the unnatural circumstances of captivity were producing a generation of psychologically abnormal orangutans — physically mature animals who were, to repeat the jargon of one expert, "behaviorally deficient for adequate procreative activity." Probably the common practice, convenient for zoo management, of removing infants from their mothers at birth had a major impact on the later urge for "procreative activity." In the wild, adult male and female orangutans come together only for mating. So perhaps the common closeness of zoo life, the casual contiguous confinement of males and females, served to dull a sexual appetite that is, even in the most stimulating of circumstances, periodic.

Over time many zoos improved their care of the red ape, giving special attention to maternal matters. An example: in the midafternoon of October 12, 1977, a ten-year-old orang female at the Toronto Zoo gave birth to a 4.4-pound male. The mother was part of a larger social group, five Sumatran orangutans, only one of them male, that had lived together in the same exhibit space for a long time. Although this was the mother's first birth, she had already had some practice in motherhood, having handled and pretend-nursed another infant born two years previously. At the miraculous appearance of her own infant, though, the mother seemed disturbed. She neither handled him nor attempted to clean him. Since he appeared to be suffocating from the remains of placenta around his face, zoo attendants quickly removed, cleaned, and revived the infant. The mother showed no further interest in her baby when he was later returned to the cage, so five hours after birth, the infant was given baby formula in a bottle. Hand rearing had begun.

The infant orangutan was never placed in an incubator. For the first three days he was warmed by a heating pad, and thereafter by blankets at room temperature. A rolled-up sheepskin provided some tactile substitute for a mother. When he began eating the wool, the sheepskin was replaced with a rolled-up towel. But a zoo attendant served as the important mother surrogate. The infant slept in the attendant's bed at night (usually on her stomach), and during the day she frequently carried him around. Within three days the infant was actively clinging to the attendant and staring fondly into her face during feeding.

Meanwhile the mother had been returned to her home cage with the larger social group. On the eighth day after birth, it was observed that she had begun lactating, and she actually began nursing a two-year-old in her social group. She was once again separated from her group and, on the twelfth day, shown the infant. The mother reached out, touched the infant's head with a finger, then sniffed her finger and showed no further interest. After six hours the infant was again taken away. The next day, though, the mother was given a dose of tranquilizer (hidden in her figs), while the infant was gently rubbed with a piece of her dung to make him smell familiar. When the infant was placed before his mother that day, she immediately picked him up, took him to a far corner of the cage, constructed a comfortable nest of wood shavings, lay on her back, and clasped the baby to her breast. Within an hour he had begun nursing. The father was introduced to the mother and infant at six weeks, but, predictably perhaps, he showed absolutely no interest. After several months the mother and infant were placed back in the home cage with the original social group.

As of late 1982 the world captive population of orangutans, living in an Ark of about 170 zoos and similar institutions, included approximately 840 animals of three sorts: the Sumatran subspecies, the

Bornean subspecies, and hybrids. The world captive population now appears to be self-sustaining. A viable second generation has appeared, and currently some thirty-five to forty orangutan infants are born every year in zoos. The overall captive population is descended from a large number of founding animals, so there is no problem with founder representation. Some wild-born animals are still taken into captivity (illegally, and confiscated by authorities in Borneo and Sumatra), so new genetic strains could expand the captive genealogy if it should become necessary.

The most serious problem today has to do with hybrids. Although the two subspecies, Bornean and Sumatran, can be distinguished visually and by chromosome testing, many zoo managers have not known enough to make the distinction and so have inadvertently (sometimes deliberately) crossed the subspecies, creating hybrid offspring. About ten hybrids are born yearly, accounting for almost a third of all captive births. There is nothing intrinsically wrong with hybrid animals, of course, but since they do not represent a naturally occurring type, most zoos consider them of little interest.

So many orangutans now live in captivity that zoos will soon have to limit their reproduction and begin working toward a stable maintenance population. What will be done with the surplus animals, particularly the hybrids? A number of small, low-budget zoos are still willing to take young hybrid orangutans, but who wants to see these intelligent apes placed in substandard exhibits with small, cramped cages? Male orangutans are more common in captivity than females, so males are in less demand. What will be done with the surplus males? Euthanasia is considered an acceptable option for surplus zoo animals of other sorts, but probably not for orangutans. They are too intelligent; they resemble humans too much. Nonbreeding older animals will have to be placed somewhere, away from actively breeding groups. But who will provide the space? The American Association of Zoos, which in 1983 began organizing a Species Survival Plan for the orangutan, must now confront such problems.

The first gorilla exhibited in captivity in Europe was a young female, named Jenny by her captors, who survived a few months in 1855 as a caged exhibit in a traveling menagerie. The Falkenstein Gorilla, bought by the Berlin Aquarium from a Dr. Falkenstein in 1876, survived sixteen months in captivity. As of 1980 a total of 472 gorillas, including no mountain gorillas and only around two dozen of the eastern lowland subspecies, were held in some 115 zoos and similar institutions.

Although their record is improving, generally zoos have been part of the problem, not the solution. Between July 1, 1970, and December 31, 1976, for example, the zoos of only three nations produced more gorillas from captive births than they took from the wild: the United States (with +55), Spain (+4), and Switzerland (+5). All other nations with gorillas in zoos contributed to a net loss — most notably, West Germany (−12); Holland (−11); and Japan, Czechoslovakia, Denmark, and Poland (−9). If we consider the number of gorilla deaths in zoos during that period, then the United States' contribution shrinks to +5. And if we presume, very conservatively, that for each wild-caught gorilla imported into the United States during that period, an additional wild gorilla was killed, then the United States also was a net consumer. In actuality, probably only Switzerland, which took no gorillas from the wild, increased the species' world population during that period.

In the past, gorillas were nearly always kept in scandalous conditions in zoos, often in the solitary confinement of a concrete and steel cage. And with a few exceptions, zoos are not yet fully prepared to provide for the psychological needs of these animals: their need for companionship within a family group; their need for privacy, variety, and stimulation. Captivity should include something to climb on, a tree or something similar, and material for nest building. Food should be provided in semiprepared form, so that the gorillas can entertain themselves by manipulating their food. In their natural habitat gorillas love to sunbathe. In captivity they should be given an outdoor enclosure and some opportunity to sun themselves undisturbed. Most important, however, gorillas need to live within a natural family group.

The Columbus Zoo, in Ohio, was the first zoo in the world to host a captive gorilla birth. On December 22, 1956, a gorilla named Christina deposited an infant, still viable within its unbroken amniotic sac, on the floor of her cage. Christina appeared dazed and confused by the strange arrival, so her infant was taken from the cage and successfully hand reared by humans. The world's second captive birth for the species occurred at the Basel Zoo in Switzerland. A female named Achilles bore an infant late at night on September 22, 1959. A keeper found Achilles the next morning gently holding the infant to her chest, but Achilles didn't know how to nurse it. Zoo keepers removed it, and Ernst Lang, the zoo's director, raised the infant in his own home. Achilles bore another baby in 1961, which she successfully nursed and raised.

The Lincoln Park Zoo of Chicago has managed one of the best individual captive breeding projects so far for gorillas in North America. Lincoln Park's gorilla exhibit began in the 1920s with the arrival from Africa of a young western lowland gorilla male, who was christened Bushman. Bushman spent his entire captive life alone in a cage. In 1948 the zoo acquired a second western lowland male, who was also kept alone. Although this second male is still alive, he has never adjusted to the company of other gorillas: he remains alone. Lincoln Park's breeding program began in the mid-1960s when the zoo purchased several young males and females. The zoo built large new quarters, paired their young animals, and hoped eventually to establish family groups.

The zoo's first gorilla birth occurred in 1970. The mother neglected her infant, however, and it was taken away to be hand reared in the zoo's nursery. Over the next few years other infants were born, and in several cases mothers neglected their infants. But as the size of the family groups increased, it appeared that young females were learning to mother by observing mothers who did successfully nurse and care for their babies. In one instance a female, herself captive-born, neglected her female infant, so the infant was hand reared for a year. When she was at last introduced to a family group, one of the mature females "adopted" her, and she eventually integrated well into the family group.

The genealogy of Lincoln Park's gorillas was expanded by loans from and exchanges with other institutions. The zoo traded one of their successful breeding males, Kisoro, for two young males from Howlett's Zoo Park in England. Kisoro went on to become one of the major studs in England. The Milwaukee County Zoo lent a female for breeding in 1978, and the Rotterdam Zoo lent a young male in the same year. By 1984 Lincoln Park had a thriving collection of twenty-three western lowland gorillas, mostly divided into three family groups.

Very recently the American Association of Zoos began a Species Survival Plan for the western lowland gorilla. Forty-seven zoos holding well over 200 gorillas agreed to participate. Those institutions elected ten representatives, who met as a group in February 1984 to discuss captive breeding issues. For years zoos have bred western lowland gorillas through informal breeding loans and exchanges. Now, however, the American Association of Zoos' captive breeding project will manage the full North American population with an eye toward long-term survival.

Thirteen

WORDS

I WENT TO SEE KOKO, the talking gorilla. I drove to the top of a mountain in northern California, turned left, turned right, stopped at a gate. I drove down a worn road, arrived at a clearing. To one side a modest house, two cars. To the other side a large steel-frame and wire-mesh cage attached to a house trailer. Near the cage stood two people. Inside the cage, looking out, a large, heavy, dark, and furry form, now looking at me: deep glowering eyes, a heavy brow, massive arms and wide powerful shoulders, a look in the eyes at once vaguely intelligent and remote. One of the people standing next to the cage spoke to the gorilla. "You have a visitor, Michael. Yes, someone has come to see you."

That was Michael, Koko's companion, a young male gorilla weighing perhaps 250 or 300 pounds. Michael stared at me, unmoving, sullen. Having a vague idea of gorilla etiquette, I made a point of not staring at Michael. I glanced at him from time to time. Suddenly Michael leapt to the side of the cage and with an open palm slapped the steel frame, slamming the entire cage into vibration, a large atonal gong, then leapt back to his original position and resumed his original hunkered posture.

He watched me directly. I watched him obliquely, with occasional glances. I was struck both by the primitiveness of his expression, an uncompromising glowering glare, and by his humanlike form — in face, equally in body. His legs were short, thick, perhaps awkward; his arms long, thick, and muscled like great, powerful legs. He leaned, hunched forward, balancing himself with those great arms, knuckles to the earth. Yet his bent body still carried an unmistakable human resemblance, particularly in the upper torso, the chest, and shoulders. The chest was as wide and powerful as any I had ever seen, more obviously powerful than one would expect even in an extraordinary human — an Olympic weightlifter, a well-toned sumo wrestler. Without question, Michael was a very handsome animal.

Inside the cage were a few gorilla toys, a large rope net and hammock, an old truck tire hung by a thick rope from the top of the cage. Three sides of the wire-mesh enclosure were open to the air and sun; the fourth was closed and shaded by the side of the trailer. Presently, from around the corner of the trailer appeared a rather slight, attractive woman with long, light blond hair, whom I recognized as Francine Patterson, the psychologist who has worked with Koko for several years. At the same time, Koko boisterously made her appearance inside the cage, having entered from a small door at the side of the trailer. Koko was much smaller than Michael. While Michael's nose was long and straight at the bridge, Koko's nose seemed short and upturned, imparting an undignified, squashed appearance to the middle of her face. She moved quickly, and instead of Michael's great, dignified reserve, she displayed an open curiosity toward her visitors.

I noticed Koko making some gestures with her hands — she has been trained to communicate in American Sign Language, the sign language taught in North American schools for the deaf — that included a reference to her knee. Patterson responded to Koko with both hand signs and speech: "No, Koko, he doesn't want to play the knee game." But Koko insisted. She repeated the gestures. "No, Koko," Patterson said and signed, laughing. "He doesn't want to put his knee against the fence. He's not interested in the knee game." Then Patterson explained to me that the "knee game" consists of having the unsuspecting visitor place a knee against the side of the enclosure. Koko runs along the fence and then smacks it with her hand or foot, jolting the visitor's knee. It is more practical joke than game, but Koko seems to find it amusing.

After more signed exchanges, Koko finally asked (according to Patterson's spoken translation) if I had any gold teeth. I had recently had a gold crown placed on a cracked molar, so I moved close to the cage, pulled my mouth open wide, and displayed the back molar as fully as possible. Koko peered in with a deep seriousness and, breathing heavily, contemplated this wonder for two, three, even four minutes. At last, the examination over, Koko and I were left looking at each other very closely through the wire mesh. Then, marvelously, Koko began blowing air through her parted lips — a gentle, affectionate gesture. Responding, I softly blew air through my lips. We looked at each other eye to eye, with Michael yet standing back, yet glowering.

PLATO SAID that all living matter contains one or more of three souls: nutritive, sensitive, and rational. Plants possess merely a nutritive soul. Animals operate with both nutritive and sensitive souls. But only humans are lucky enough to be animated by all three.

Other philosophers and opinion makers, according to historian Keith Thomas, have defined human uniqueness differently. Some have pointed to the religious impulse as a distinguishing human quality. Some have felt that humans were elevated above the animals by virtue of modesty and control of the sexual impulse. Aristotle considered man a political animal; elsewhere he noted that only men laughed, found their hair to grow gray with age, and were unable to wiggle their ears. Martin Luther considered, and a few centuries later Pope Leo XIII agreed, that man was distinguished from the beasts by property ownership. Cambridge cleric George Abbot noted in 1600 that humans were marked by, among other qualities, the inability to swim. Eighteenth-century author Oliver Goldsmith asserted that humans were distinguished by their lack of pleasure in eating. For Thomas Willis, man was a laughing animal. For Benjamin Franklin, man was a tool-making animal. Edmund Burke saw a religious animal. James Boswell described man as an animal that cooks its food. And toward the end of the eighteenth century, Uvedale Price, a student of aesthetics, absurdly suggested that only humans had noses: "Man is, I believe, the only animal that has a marked projection in the middle of the face."

Even more than the distinction between human and animal, the unique position of the most humanlike of animals, the primates, has provoked speculation for centuries. Aristotle included in his fourth century B.C. encyclopedia, *Historia Animalium,* a description of the ape, monkey, and baboon as animals that "share the properties of man and the quadrupeds." His study of apes is remarkable for its point-by-point comparison with human form:

> Apes are hairy on the back in keeping with their quadrupedal nature, and hairy on the belly in keeping with their human form — for, as was said above, this characteristic is reversed in man and the quadruped — only that the hair is coarse, so that the ape is thickly coated both on the belly and on the back. Its face resembles that of man in many respects; in other words, it has similar nostrils and ears, and teeth like those of

> man, both front teeth and molars. Further, whereas quadrupeds in general are not furnished with lashes on one of the two eyelids, this creature has them on both, only very thinly set, especially the under ones; in fact they are very insignificant indeed. And we must bear in mind that all other quadrupeds have no under eyelash at all.
>
> The ape has also in its chest two teats upon poorly developed breasts. It has also arms like man, only covered with hair, and it bends these legs like man, with the convexities of both limbs facing one another. In addition, it has hands and fingers and nails like man, only that all these parts are somewhat more beast-like in appearance. . . . The genitals of the female resemble those of the female in the human species; those of the male are more like those of a dog than those of a man.

A few centuries later the Roman court physician Galen encouraged his students to dissect apes in order to perfect their understanding of human anatomy. In the thirteenth century Marco Polo returned from his Eastern explorations and reported on the "long-tailed monkies" and "apes, so formed and of such a size as to have the appearance of men." In a 1625 edition of *Purchas his Pilgrimes,* the English explorer Andrew Battell (who was captured by the Portuguese in 1559 and imprisoned in a western African outpost for several years) described as "two kinds of Monsters" what were probably gorillas and chimpanzees:

> The greatest of these two Monsters is called, *Pongo,* in their Language: and the lesser is called, *Engeco.* This *Pongo* is in all proportion like a man, but that he is more like a Giant in stature, than a man: for he is very tall, and hath a mans face, hollow eyed, with long haire upon his browes. His face and eares are without haire, and his hands also. His bodie is full of haire, but not very thicke, and it is of a dunnish colour. He differeth not from a man, but in his legs, for they have no calfe. . . . They cannot speake, and have no understanding more than a beast.

But with these and a very few other exceptions, Europeans did not actually see any of the great apes until the expanding explorations of the seventeenth century. In 1641 a Dutch physician and anatomist, Nicolas Tulp, described his dissection of an ape, probably a chimpanzee, that had recently been acquired from Angola by the Prince of Orange. Early in 1698 another chimpanzee was taken from Angola and brought to England. Perhaps the first ape ever to appear on that island, it lived for only a few weeks. Upon its death Edward Tyson, an English physician with an interest in anatomy, took the body home and dissected it. With the help of draftsman William Cowper, Tyson prepared an extensive illustrated description of the animal, which he presented to the Royal Society in June 1698 and then published as a book, *Orang-outang, sive Homo Sylvestrius: or, the Anatomy of a Pygmie Compared with that of a Monkey, an Ape, and a Man,* in the following year. Tyson's anatomy of a chimpanzee (which he had misnamed *orang-outang*) deeply affected not only his small group of scientific peers but also contemporary philosophers, poets, and the general public. Tyson described his subject as an animal with a profound similarity to humans; his calling it a *Homo Sylvestrius* and a *pygmie* emphasized that similarity unmistakably for the general public.

By 1758 the Swedish botanist Carl von Linné (popularly known as Linnaeus) had knighted Tyson's discovery with the sword of scientific nomenclature, pronouncing that lemurs, monkeys, apes, bats, and humans, by virtue of undeniable anatomical continuities, belonged to a single biological category, the primates. A century later Charles Darwin's theory of evolution affirmed the rightness of Linnaeus's general system by suggesting that anatomical similarity resulted from common descent.

Today, with the exception of the bats, Linnaeus's original grouping of primates remains the scientific standard. Recent discoveries have only confirmed the great anatomical affinity of humans and apes. Geneticists have found the chromosomes of chimpanzees and humans to be very similar, with only a 1 or 2 percent variance, indicating that chimpanzees are more closely related to humans than, for instance, zebras to horses. Comparisons of the structure of ape and human brains seem to indicate that human brains are somewhat more complex but show no obvious qualitative differences.

Perhaps the clearest distinction is in the size of the brain. Human brains are bigger. Within the normal range, human brain size has no obvious relationship to intelligence (Lord Byron's brain was 2,238 cubic centimeters while the brain of Anatole France measured around 1,000 cubic centimeters — both were sufficiently smart), although it is certain that below a critical point brain size does limit intelligence. In any case, the smallest human brain ever measured had a volume of about 790 cubic centimeters, not much larger than the largest gorilla brain ever measured, about 690 cubic centimeters. The comparison implies that humans are more intelligent than apes (as we already suspect), but it leaves open the possibility that the mentality of the great apes is only quantitatively rather than qualitatively different from that of humans. The several dozen skulls and skeleton parts from humanlike apes and pre-*Homo sapiens* lines that have been dug up periodically during the last several decades have furthered our sense of continuity. *Homo habilis* skulls (dated through studies of radioactivity at between 2.2 million and 1.6 million years old) range in cranial capacity from 500 to 800 cubic centimeters, while *Homo erectus* skulls found in Africa, China, Indonesia, and Australia (dated from 1.6 million to 0.4 million years) exhibit cranial capacities of 900 to 1,200 cubic centimeters.

Many dissectors have penetrated brain with scalpel, but no one has touched a mind. Mental reality is the flimsiest, most obscure of constructs — too obscure even for scientific scrutiny until the late nineteenth century, when Sigmund Freud began inventing various physical analogies and "mechanisms" to describe the mind's intricacies. And if the issue of anatomical continuity between man and animal has been resolved, matters of psychological and intellectual continuity remain far more abstruse. With various transmutations, Plato's attribution of a rational soul to humankind alone has long informed both popular and scientific belief. A few individuals might assert, as God did to Adam in Milton's *Paradise Lost,* that animals "reason not contemptibly" but for the most part, reasoning Europeans adhered to John Locke's dictum: "Beasts abstract not."

Probably most people have subscribed to the idea of a fixed and unassailable boundary, an uncrossable Rubicon, found somewhere in the psychological or intellectual realm of existence, that forever separated man and animal. As naturalist William Bingley expressed the concept in his *Memoirs of British Quadrupeds* (1809): "How slender so ever it may sometimes appear, the barrier which separates men from brutes is fixed and immutable." Or, as the English physician William Lambe wrote in 1815: to ignore the great human physical similarity to monkeys, apes, and baboons was to demonstrate "misplaced pride and an ignorant apprehension," but in his rationality, man was yet "distinguished from the whole tribe of animals by a boundary which cannot be passed."

The issue of belief or disbelief in an absolute boundary, a Rubicon separating the mentality of humans and animals, remains a profoundly significant philosophical (and emotional) schism in the natural sciences. All biologically-oriented scientists now agree that Darwin's *On the Origin of Species* (1859) described somewhat accurately a great chain of genetic relationship between all living beings, a clear biological continuity. But not all fully subscribe to Darwin's later assertion, in *The Descent of Man* (1871), that "there is no fundamental difference between man and the higher mammals in their mental faculties."

Part of the problem lies in how one thinks of *continuity.* After all, many continuities are not really perfectly continuous; they are rough, continuous in general but discontinuous in particular. But more important, scientists toward the end of the nineteenth century began discovering how the basic forces of instinct (fixed and inherited behavior) and conditioned learning (behavior acquired as a response to reward and punishment) could explain even very remarkable animal behavior.*

In the mid-1850s, Lewis Henry Morgan, American lawyer, politician, entrepreneur, and anthropologist traveled by rail through forty miles of rugged wilderness south of Lake Superior that contained, in his estimation, a "beaver district, more remarkable, perhaps, than any other of equal extent to be found in any part of North America." Morgan saw streams "bordered continuously with beaver meadows, formed by overflows by means of . . . dams, which had destroyed the timber upon the adjacent lands. Fallen trees, excavated canals, lodges, and burrows, filled up the measure of their works."

Awed by the complexity of such constructions, Morgan began studying the beavers in that region. In 1861 he explored the Red River Settlement portion of Hudson's Bay Territory, and the following year the Missouri River to the base of the Rocky Mountains. This unprecedented study of the American beaver concluded with the publication of a classic account, *The American Beaver and His Works* (1868). It included detailed descriptions, surveys and measurements, photographs of beaver constructions, and the results of hundreds of beaver dissections and concluded with a strong personal testament to the intelligence of beavers. "A beaver canal," Morgan wrote, "could only be conceived by a lengthy and even complicated process of reasoning. After the conception had been developed and executed in one place, the selection of a line for a canal in another would involve several distinct considerations, such as the character of the ground to be excavated, its surface elevation above the level of the pond, and the supply of hard wood near its necessary terminus." Indeed, Morgan believed that the animal must not only be considered rational, but conscious as well. For when the beaver paused and considered his labors, "evidently to see whether it is right, and whether anything else is needed, he shows himself capable of holding his thoughts before his beaver mind; in other words, he is conscious of his mental pro-

*In the laboratory, rewards and punishments are carefully controlled by the experimenter or trainer; in the wild, an animal's conditioned learning may occur in response to the rewards or punishments of a parent, or through random, trial-and-error explorations of the environment.

cesses." Obviously beavers were inferior in intelligence to humans. Yet, Morgan believed, they and what he called the other "mutes" possessed "a thinking self-conscious principle, the same in kind that man possesses, but feebler in degree."

Morgan may have believed he was echoing Darwin's assertion of mental continuity, but not all of his readers were convinced. Darwin himself, when Morgan visited him in 1871, expressed firm skepticism regarding thinking beavers. Though Morgan had scoffed at the idea that mere instinct could explain the architectural virtuosity of beavers, Darwin believed that in fact instinct could well account for such impressive behavior. Darwin's inclination was probably correct. Modern studies of beavers in laboratories suggest that much of their engineering tendencies derive from an instinctual antipathy to the sound of running water. Beavers will flee from tape recordings of the sound. If given sticks and mud, however, they will blindly begin plastering dams against a tape recorder's speaker. Beavers probably do not consider thoughtfully what they are doing. They simply do it. They do it because they are programmed to by inherited instinct.

While Morgan, perhaps foolishly, attributed to beavers a humanlike intelligence when instinct would do, other people were fooled by the conditioned learning of a well-trained horse. The horse was owned by Wilhelm von Osten, a Berlin schoolteacher of aristocratic descent who believed that some animals were capable of intelligence rivaling that of humans. In 1888, von Osten fitted his stables with several accoutrements of the teaching profession — blackboards, flashcards, an abacus, and so on — and proceeded to train a stallion he named, with great expectations, Clever Hans. After years of the best education von Osten could provide, Clever Hans became world famous; the brilliant horse could solve mathematical problems, distinguish between left and right as well as up and down, and correctly select colors — all for the simple reward of a carrot. Of course, Clever Hans couldn't speak. He simply responded to questions with gestures. He gave numerical answers by tapping, somewhat ponderously yet precisely, with his front hoof. He indicated yes, no, and directional answers with various motions of his head. He answered questions about color by selecting the appropriate piece of colored cloth from those spread before him on the stable floor.

Most people who witnessed a performance in von Osten's stables were entirely convinced of Clever Hans's humanlike intelligence. But von Osten, wishing to convince even the most skeptical, finally petitioned Emperor Wilhelm II for a formal investigation of the phenomenon. So it was that a commission of thirteen highly respectable men from a variety of relevant fields — including a psychologist, a zoologist, a circus trainer, a veterinarian, and a politician — arrived at the stables one September morning in 1904 to observe that remarkable horse first-hand. Clever Hans did not disappoint. He seemed to answer complex verbal questions, solve difficult problems in arithmetic, and so on. After several days of such examinations, the commission reported that they could find no evidence of trickery.

Finally Oskar Pfungst, a bright young student from the Berlin Psychological Institute, came to the stables and began considering Clever Hans in the light of double-blind testing. The principle of double-blind is to present a test in such a way that the experimenter does not know the answers, thus avoiding the possibility of accidentally cueing or guiding the subject. In the stables Pfungst had von Osten ask simple arithmetic questions of Clever Hans by holding up number cards in front of the horse. Pfungst asked von Osten to look at some of the cards and not to look at others before he held them up. The results were immediate and definite: whenever von Osten had seen the face of the card, Clever Hans was able to give the right answer, but whenever von Osten had not seen the card face, Clever Hans's responses were confused and entirely random. Although no one doubted von Osten's good faith, it was apparent that somehow he had been accidentally giving the answers to his horse. He was cueing in some very subtle way.

Eventually it was revealed that von Osten controlled his horse's responses with cues of expression and posture so subtle that only the horse noticed them. In the responses that involved head motion (yes, no, up, down, and so on), von Osten would unconsciously anticipate Hans's responses with his own body postures. Clever Hans simply followed the motions he saw his trainer anticipate. In the mathematical questions that required hoof counting, von Osten would lower his head slightly when he expected Hans to begin tapping and elevate his head and brow slightly, in relief and satisfaction, when he knew Hans had reached the desired number.

Oskar Pfungst became famous for his solution to the problem of Clever Hans. Wilhelm von Osten was devastated. The horse was sold. And the story of Clever Hans remains today both as a classic example of the immense degree to which a highly perceptive but nonthinking animal can be trained and as a warning of the possible pitfalls in attributing humanlike intelligence to an animal.

Perhaps the most palpable expression of mental reality is language. And while the Rubicon of rationality that supposedly distinguished humans from animals remained elusive, most people were convinced by the Rubicon of language. The evidence seemed certain: all people but no animals use language. Yet if any animal could talk, it would be the animal closest to humans, the ape.

Not until 1931 did two optimistic Americans first try to teach an ape to talk. In that year Winthrop and Luella Kellogg began raising an infant chimpanzee named Gua alongside their own infant son. The Kelloggs provided near identical treatment for the two babies; both were bathed, bottle fed, diapered, spoken to, and coached in spoken English. After about sixteen months, the Kelloggs' own baby had begun to speak, but the chimpanzee had not. Although the Kelloggs believed that Gua understood about a hundred words, she could not articulate one. A decade after that experiment, Keith and Catherine Hayes raised a chimpanzee named Vicki in their home, making a concerted effort to teach her spoken English. After six years Vicki understood a large number of spoken words, but she could only and with great effort roughly articulate four: "mama," "papa," "up," and "cup."

Both of these early experiments seemed to show conclusively that chimpanzees, and by logical extension probably all the great apes, were incapable of learning language. In the 1960s, however, two research psychologists at the University of Nevada, Allen and Beatrice Gardner, carefully examined a film made by the Hayeses showing little Vicki attempting to say her four words. The Gardners noticed that Vicki accompanied her laborious speech with very expressive gesturing and concluded that the Hayeses' experiment had demonstrated only that chimpanzees were incapable of *vocalizing* human speech.

So in June of 1966 the Gardners acquired a young, wild-caught female chimpanzee which they named Washoe (honoring Washoe County, home of the University of Nevada). The Gardners themselves learned American Sign Language, or Ameslan, a gestural language for the deaf consisting of a large number of well-defined signs and motions (some obviously iconic or representational, many not), made with the hands, arms, and other parts of the body, that function as words or concepts.* Like the Kelloggs and the Hayeses before them,

*Ameslan is not a translation of English; it functions as a language in its own right, and it is not necessary to know English before learning Ameslan. Many deaf people supplement their gestural Ameslan vocabulary with a second form of signed communication, finger-spelling, in which English words are spelled out letter by letter, using the fingers to shape the letters. Since finger-spelling requires a knowledge of written English, however, none of the ape language experiments have introduced it.

the Gardners recognized the importance of providing their chimpanzee with an emotionally supportive and intellectually stimulating environment. They hugged Washoe, played with her, tickled her, fed her, bathed her, brushed her teeth, and signed to her in Ameslan whenever appropriate. The Gardners and their assistants attempted to use only Ameslan, and no vocal speech at all, around Washoe. They used several techniques to teach and encourage Washoe to sign, and by early 1969 Washoe seemed to have an active vocabulary of thirty definite signs; by 1971 she had acquired eighty-five signs.* The signs represented not only objects (such as "toothbrush") and actions (such as "open"), but abstract qualities and concepts (such as "sweet," "funny," and "sorry"). Later Washoe learned to use pronouns ("I-me" and "you") as well as various grammatical connectors ("and," "for," "with," and "to").

Many wild animals communicate with other members of their species, and even other species, in remarkable and rather sophisticated ways — with postures, gestures, facial expressions, vocalizations, and so on. But animal communication is different from human language. One simple distinction involves the matter of flexibility: animal communication is *closed,* while human language is *open.* That is, animals may learn to communicate with a "vocabulary" in which one signal, perhaps a certain cry, corresponds to one event, perhaps the appearance of a flying predator, while other signals correspond to other events. Animals may use such signals in impressively precise ways, but they do not, it seems, spontaneously combine their signals to produce higher levels of complexity. Human language, by contrast, begins with a vocabulary of individual signals — words — which can be combined and recombined endlessly to produce a very, very large number of meanings.

When laboratory-trained apes acquire large vocabularies, even when those vocabularies parallel human vocabularies, the apes are not "talking," not using language in a humanlike fashion, so long as they reproduce in a rote fashion only single elements of vocabulary. Washoe's acquisition of eighty-five words by itself was still not different in kind from a parrot's acquisition of three or four words, which it can only repeat mechanically and without variation. Crackled articulations such as "Polly wanna peanut," or "Sandwich please," or "Give Polly butter" may have some meaning both for the parrot and the human observer. But only if Polly, trained to emit the above three

*The Gardners described a sign as part of Washoe's active vocabulary only if she used it spontaneously, that is, not in immediate imitation, every day for fifteen consecutive days.

lines, were suddenly and spontaneously to combine elements of each and say, "Please give Polly peanut-butter sandwich," might we believe that Polly is able to use language as a flexible, open system — a system that combines individual vocabulary elements through a systematic grammar to produce new meaning. In short, one crucial test of whether apes use language as humans do is not the measure of vocabulary, but the measure of spontaneous and appropriate vocabulary combinations.

How well did Washoe combine her vocabulary? The Gardners reported that once Washoe had acquired a vocabulary of about eight or ten signs, she began producing sensible combinations, many of them obviously original — combinations that the Gardners had not previously made. Washoe signed "gimme tickle" when she wanted to be tickled; later she learned to be more specific: "Roger you tickle," or "Greg you peek-a-boo." When she wanted to go outdoors, Washoe would stand near the doorway and sign, "You me out," or "You me go out," or "You me go out hurry." She signed "open food drink" to refer to the refrigerator (although the Gardners had always signed it "cold box"), "go sweet" to indicate she wanted to be carried out to the raspberry bush, "listen dog" at the sound of a barking dog beyond her line of vision. She learned to sign "dirty" to refer to her own feces and combined it into "dirty good" to indicate her toilet. Eventually she began using "dirty" as a way of insulting people she didn't like.

In 1971 Francine Patterson, then a graduate student in psychology at Stanford University, attended a public lecture by Allen and Beatrice Gardner in which they presented some of the results of their pioneering language work with Washoe. Inspired by that lecture, Patterson decided to do her thesis work on the topic of ape language acquisition. By July 1972 she had persuaded the San Francisco Zoo to allow her to work with an infant gorilla named Hanabi-Ko — Koko for short.

No scientist had attempted to teach language to a gorilla before. Many considered chimpanzees more intelligent, and most believed that chimpanzees were more tractable. Gorillas were known to be introverted and very self-possessed, while chimpanzees were extroverted, eager to please, and seemed to enjoy working with humans. Scientists were also cautious about working with an animal that grows to be so much bigger than themselves. Even chimpanzees, which reach an adult weight of about 120 pounds in males, 90 pounds in females, quickly surpass humans in strength.

Patterson decided to follow the Gardners' example and try to teach

Koko Ameslan. Unlike the Gardners, she decided to accompany sign language with simultaneous spoken translations. At that time Koko was still a zoo exhibit, and Patterson reasoned that it would be impossible to isolate her from the sounds of human speech. Of course, Koko could never be expected to speak back, but Patterson believed she might learn to understand spoken English.

Patterson began by trying to teach Koko the sign for "drink": she would speak the word, make the sign, then hand Koko her bottle. (In Ameslan, "drink" is indicated by closing the hand into a hitchhiker's fist, thumb extended, and placing the thumb to the lips.) At first Koko showed little interest in such weird human behavior, preferring to grab the bottle or to play games, such as closing her eyes and spinning around, or spinning around with a blanket over her head. Soon Patterson and her assistants began a more formal regimen of training. They would hold up Koko's bottle, then give it to her if she signed "drink." If she didn't, they would prompt her with the sign, "What's this?" If she still didn't sign appropriately, they would mold her hand into the sign for "drink" and then give her the bottle.

Within a couple of weeks Koko appeared to sign her first word, "food," by placing all the fingers of one hand, palm down, to her mouth. The gesture seemed quite deliberate. Quickly Koko learned other Ameslan words: for "drink," "more," "out," "come-gimme," "up," "toothbrush," and "that." A week later she produced her first two-sign combination, "Gimme food." Eleven days later she expressed another two-sign combination, "food-drink," to describe her normal food, which was a combination of cereal and milk. By the end of the project's second month, Patterson had recorded sixteen different two-word combinations.

Koko had difficulty making some Ameslan signs because of anatomical limitations. Her thumb is smaller and farther away from her fingers than a human's thumb; she cannot touch her little finger with her thumb. Also her motor control is not as precise as a person's, so some of Koko's signs are gorilla variants of human Ameslan.

Koko (or was it Koko's teachers?) seemed to have a problem with motivation. She quickly learned the signs for "swing" and "berry," but she was very slow to learn the sign for one of her least favorite foods, "egg." Patterson noticed, moreover, that although Koko quickly learned new signs, she almost as quickly became bored with performing them simply for the satisfaction of her trainers. So Patterson decided to alter her criteria for determining the acquisition of vocabulary. At first she had followed the Gardners' example: if Koko spontaneously and appropriately used a sign for fifteen consecutive days, it would be listed as part of her active vocabulary. But to allow some flexibility in the training and testing schedules, Patterson began counting a sign as part of Koko's vocabulary if at least two observers noted its appearance for fifteen days out of thirty. After eighteen months of training Koko possessed an active vocabulary of 22 signs. After three years and three months, Koko had used 236 signs, of which Patterson accepted 78 as entirely valid according to her criteria.

Koko has described herself as a "stubborn devil." Actually, Koko's willfulness proved to be an obstacle. When Patterson attempted to assess Koko's development and gather data in a reasonable and responsible manner, Koko would sometimes produce unreasonable and irresponsible responses. Once Patterson tried to get Koko to sign "shell," first by displaying a shell. Koko did not respond. "Forgot?" queried Patterson. No answer. At last Patterson sent Koko to her room, closed the door, and said out loud, "Well, I'll just take these goodies [food treats] to Michael." Koko, who by then seemed to understand both spoken English and Ameslan, immediately appeared at her door and signed "shell." The same day Patterson attempted to elicit the sign for "rock." She asked Koko, "What is the sign for 'rock'?" Koko responded with several wild gestures, using her two fists, but never with the appropriate sign (striking the back of one hand with the other fist). At last Patterson said, "I won't give you your night dish unless you say 'rock.' " Immediately Koko signed a perfect "rock."

At times Koko's responses defied the literal and pedestrian methods of typical scientific data collection. When asked to make the sign for "drink" one day (hand in a hitchhiker's fist with thumb brought to the lips), Koko consistently refused, signing only "sip," "thirsty sip," "apple sip," and so on. Finally, after an extended and frustrating session, the assistant working with Koko pleaded: "Koko, please please sign 'drink' for me." Koko grinned, leaned back, and produced a perfect "drink" sign — except that she brought her hitchhiker's thumb up to her ear. Realizing that "drink" was the second word Koko had learned, and that she had used it properly on thousands of previous occasions, Patterson concluded that Koko had just produced a marvelous joke. Yet if one rigidly considered Koko's actions to see if she had produced the appropriate sign, one would have to conclude (as at least one critic of the project does) that Koko had simply made an error.

By 1984, after twelve years of training, Koko displayed an active vocabulary of over five hundred signs, which she was combining into sentences commonly three to six words long. Koko's companion, Michael, who was born in the wild and bought by Patterson in 1976, had by then acquired a vocabulary about half that size.

Several critics of the ape language studies remain unconvinced. One major line of criticism attributes Koko's remarkable productions to imitation. The idea is that Koko has learned, without any serious language fluency, simply to return with minor variations the words and phrases that her human companions are signing. If a trainer says and signs, "Koko, you want out?" and Koko responds with "Koko want out," then perhaps Koko understands nothing about language beyond a few vague associations between objects or events and certain words — not much more impressive than a dog wagging its tail when its master says, "Out?" Yet Patterson insists that she has consistently and carefully evaluated the levels of imitation and spontaneity in Koko's signing. If we are to believe Patterson, an average of 41 percent of Koko's signings are spontaneous.

A variation on the imitation theme imputes unconscious cueing to Koko's trainers — the Clever Hans effect. This sort of criticism is effective partly because of the resonance of the Clever Hans story. A scientist, or any intelligent person for that matter, can only applaud the unmasking of fuzzy thinking through responsible skepticism and sound scientific method. The story of Clever Hans is a good story; but does it entirely apply to the ape language studies? Some critics believe it does, yet in all fairness we ought to recognize some differences between the clever horse and Koko.

First, Clever Hans was responding to questions whose answers were extremely easy to cue. The mathematical questions required only that Clever Hans respond to a cue for starting and stopping: the horse began tapping his foot when von Osten tilted his head forward; he stopped when von Osten stood erect. By contrast, the cueing necessary to make apes use language would have to be very complex. Second, Clever Hans was unmasked by a double-blind experiment, in which von Osten was "blinded" to the questions. But Patterson has carried out double-blind tests, at least on vocabulary, and Koko performs well on them. Third — and most intriguing — Koko and Michael have sometimes been observed signing to each other, and Koko often signs to herself, both when people are in the same room but not communicating with her and when people are out of the room but peeking in. Once Koko examined herself in a mirror and signed, "Eye, teeth, lip, pimple." Sometimes when she leafs through the pages of a favorite magazine, looking at the pictures, she seems to comment to herself with signs. While playing with dolls and toys, Koko sometimes signs to herself, although she apparently does not like being watched when she does. Once Patterson noticed Koko signing "kiss" to her toy alligator. Once Koko picked up a toy horse, signed "listen," and breathed noisily against it. Another time Koko was observed playing with two gorilla dolls, one blue and one pink. She signed "bad bad" to the pink doll and "kiss" to the blue one. Finally, she pushed the two dolls together in a mock wrestle-play session and concluded: "Good gorilla good good."

Imitation and the Clever Hans effect represent one major line of criticism. A second focuses on the matter of word order. Noting that Koko and Michael's word combinations are very poorly sequenced, some critics contend that word order is the crucial indication of true language. Word order, after all, is syntax, indicating grammatical relationship. Unless an ape's spontaneous sequences of words are in an appropriate order, it can be argued that the animal is randomly tossing out several vaguely appropriate pieces of vocabulary without understanding their combined meaning, their crucial grammatical relationship.

Michael once said of a bird: "Bird good cat chase eat red trouble cat eat bird." We might conclude, as his trainers did, that Michael had recently seen a cat eat a bird (he was never told that cats eat birds), but is he using language in the way humans do, deliberately producing meaning through syntax? Certainly both Koko and Michael, as well as other language-using apes, are notoriously poor grammarians. And yet (to rush to their defense), is the poor grammar their fault or their trainers'? From the beginning of the experiment, Koko and Michael were surrounded by anxious humans only too eager to reward word combinations that made even slight sense, whether or not those combinations were grammatically sound.* Moreover, at least part of the roughness of Koko's transcribed verbal productions may result from the limitations of Ameslan, which is ordinarily a laconic language. The number of distinctly recognizable signs is very small, a few thousand words altogether, compared with the hundreds of thousands of words in English. Even very fluent human signers take about twice as long as speakers of English to express the same ideas, so Ameslan users are strongly motivated to be brief. Commonly, Ameslan signers use no articles, conjunctions, or auxiliary verbs, and only a few temporal expressions.

Critics also scoff at Patterson's retreat from rigorous scientific debate into the more lenient arena of popular opinion, thus finding an audience only too eager to believe in talking animals. One critic dismisses Patterson's book-length account of the study as a "trade book." "En-

*In the one ape language experiment that emphasized rigidly correct syntax — Ann and David Premack's extensive and well-documented work with a chimpanzee named Sarah — the ape did seem to achieve considerable grammatical precision.

tertaining," scolds another. "The whole thing is ridiculous," complains a third: "If Penny Patterson tried to publish in a scientific atmosphere, then she would be laughed out of court." Others are disturbed by the anecdotal quality of much of her reporting. As still another critic expresses it: "I personally feel that she has gone way beyond the data in her claims for that gorilla — *way* beyond the data."

Some critics, at least in unguarded moments or in preparation for more reasoned arguments, seem to be saying that the ape language experiments have failed because they can't succeed: apes cannot learn to use language because they cannot. After all, some modern linguists believe the essential structure of language to be part of the human genetic inheritance. Although the culturally transmitted surface elements vary from language to language, these linguists see a common deep structure, apparent proof that the essence of language behavior is passed not through culture but through a code lingering in the biochemical double helix. There is no reason to believe, so the argument proceeds, that animals pass on an equivalent genetic inheritance, since even the great apes do not use anything seriously resembling human symbolic language in the wild. As linguist Noam Chomsky summarizes the belief: "Acquisition of even the barest rudiments of language is quite beyond the capacities of an otherwise intelligent ape."

Even more provoking than the problem of talking versus not-talking apes is the problem of what they might be saying. For if we are to believe the reports of Patterson and her group, Koko and Michael argue, insult, joke, make moral judgments, play with language, and are aware of time, past and future, other animals, and themselves. Language is the primary means by which humans convince each other that they can think, that they are conscious; if animals can talk about some of the same abstractions we do, even primitively, then perhaps they too can think and are conscious. But if the matter of animal language is controversial, the issue of animal thinking and consciousness is much more so. Scientists are properly wary of the kinds of slovenly presumption that led Lewis Henry Morgan to insist that beavers can think, that allowed Wilhelm von Osten to delude himself and others as to the mental capacities of a handsome horse. Further, the discipline of behaviorism has allowed psychologists and animal scientists to keep a clean intellectual house for nearly a century now, ignoring the very abstruse concepts of thinking and consciousness (both in humans and animals) by focusing on learning and behavior. Some scientists believe that animals simply do not think and are not — in any way comparable to humans — aware or conscious. Others believe that animals feel, may think, and may be aware or even conscious, but since we hardly know what thinking and consciousness are, such abstract issues are better left to the philosophers of this world.*

And yet what are we to say about the following exchange between Koko and Barbara Hiller, one of her trainers? Koko was making a nest of white towels and signing the word "red." Barbara Hiller said and signed: "You know better, Koko. What color is it?" But Koko insisted "red," repeating herself three times with increasing emphasis. At last, grinning broadly, Koko picked up from the nest of white towels a tiny speck of red lint, displaying it ostentatiously. Was Koko making a joke? The following day she repeated the same behavior with Patterson.

How are we to regard an exchange between Michael and assistant Ellen Strong that occurred in May 1978, just after Michael had ripped Strong's jacket? "Who ripped my jacket?" Strong demanded. Michael signed "Koko." Strong repeated the question, and Michael signed "Penny" (Francine Patterson's nickname). After a third repeat of the question, Michael at last seemed to admit guilt: "Mike." Had Michael been lying? Was Koko lying when — after Patterson caught her eating a crayon — Koko signed "lip" while drawing the crayon across her upper lip, then her lower lip, as if applying lipstick? Was Koko lying when — after Patterson caught her trying to break a window screen with a stolen chopstick — she signed "smoke mouth" and placed the stick in her mouth as if smoking a cigarette?

And what does one make of an animal that seems to know how to insult people?

Hiller: "What do you say when you really want to insult people?"

Koko: "Dirty."

Hiller: "OK, can you think of another one?"

Koko: "Sorry, gorilla polite."

Hiller: "It's OK to tell me."

Koko: "Toilet."

Koko seems to sense time. One of her trainers, Maureen Sheehan, had begun a new schedule. On her fourth visit, though, Sheehan was required to stay late because her replacement was late. When she stayed past her normal time, Koko reminded her: "Time bye you." "What?" asked Sheehan. "Time bye good bye," responded Koko. Another time Koko roundly scolded her trainer Debi Wilkinson for being twenty minutes late.

*For a more thorough discussion of the controversy, see Donald R. Griffin, *Animal Thinking* (1984).

When Wilkinson entered the trailer, Koko remained in a corner, and stared with an angry expression. "Koko, are you mad at me?" Wilkinson asked.

"Toiletface Debi," Koko responded.

"Toiletface, why?"

After some further exchange which concluded with Koko sticking out her tongue, Koko at last signed: "Time time Debi time."

Wilkinson both spoke and signed: "Are you mad because I am twenty minutes late? There was very bad traffic Koko. I am sorry."

Koko: "Bad."

Wilkinson: "You're right, but I wasn't late on purpose."

Koko: "Pick pick pick."

Wilkinson: "Yes, you are picking on me. Let's play now."

Koko: "Owe cookie Debi time."

Wilkinson: "I have no cookies."

Koko: "Me me me apple."

After Wilkinson gave Koko a treat and further apologies, she asked: "Friends again?" Koko made what Wilkinson later described as "her happy purring sound," and signed: "Chase time."

Both Koko and Michael seem thoroughly aware of other animals. Koko hates birds, but she is fond of many other animals, including several pets she has had — rabbits, frogs, and cats. When she was told that her pet kitten had been killed by an automobile, Koko reportedly cried. "She was quiet for a very short time. Then she let out a high, hooting call, *whoo, whoo, whoo,* the cry she used to sound as a baby when Patterson left her alone at night in her cage."

Koko sometimes observes herself in a mirror: chewing gum, curling her tongue, combing her hair, brushing and flossing her teeth. She has amused herself by decorating her head with hats, wigs, scarves, and makeup while looking in the mirror. When asked who she sees in the mirror, Koko has responded: "Think that me animal gorilla animal Koko-love." Asked whether she is animal or human, she has said: "Fine animal gorilla." Asked to describe the difference between herself and a human, Koko has signed "head," and then beat her head with her open hands very hard. Asked to expand on the difference between gorillas and humans, Koko has indicated the sign for "blanket" against her stomach, then stroked and pulled the fur on her stomach. Asked to name what she shares in common with humans, Koko has responded with "eye," and then "love."

On December 8, 1978, Maureen Sheehan showed Koko four pictures of animal skeletons and asked her to pick the skeleton of a gorilla. Koko did. Was the gorilla alive or dead? Koko signed: "Dead drapes." (The drapes in Koko's trailer are closed in late afternoon, a sign of approaching bedtime.)

Sheehan pursued the question further, speaking and signing: "Let's make sure, is this gorilla alive or dead?"

"Dead good bye."

"How do gorillas feel when they die — happy, sad, afraid?"

"Sleep."

What does death mean to Koko? And how did she acquire the concept? When Sheehan once asked Koko if she thought Francine Patterson would die, Koko fidgeted for several seconds and then signed, "Damn!" Another time Sheehan asked, "Where do gorillas go when they die?"

Koko: "Comfortable hole bye."

"When do gorillas die?"

"Trouble old."

WHETHER OR NOT Koko has actually crossed the Rubicon of language or thinking, one is tempted to ask her: "Koko, what do you think about extinction?" Extinction is a sophisticated concept, though, and even a very well educated Koko could not be expected to understand it well. But we can.

Gorillas have been killed for food and for symbolic purposes. Gorillas have been decimated by people gathering live specimens for zoos and research institutes, dead specimens for museums, and severed heads and hands for the amusement of tourists. In many parts of Africa gorillas are threatened by the destruction of their habitat. While the entire world population of wild gorillas has dropped to a few dozen thousand and is probably still declining, the human population numbers over 5 billion, *increasing* by 90 million every year. And if we might fit the entire species, the few dozen thousand gorillas left, into a small town, many other endangered primate species could be crammed into a stadium, an auditorium, or even a single room.

Primates evolved as tree dwellers. With a few exceptions, they cannot live without a forest habitat. Yet the world's exploding human populations encroach directly on the tropical forests. By the year 2000 roughly half of West Africa's forests, a third of central tropical Asian forests, a quarter of southern Asian forests, and significant portions of Central and South American forests will be gone. As the forests go, so go the primates.

One can value the world's primates from several perspectives. Ecologically they are inseparable from their forest homes: they are necessary parts of extremely complex and delicate ecosystems, part of an intricate chain of life 75 million years in the making. Scientifically they are of tremendous value. Anthropologists, sociologists, and psychologists believe they can learn much about the bases of human behavior by studying primate behavior. Medical research depends deeply upon this particular group of animals, which shares with us so much physiology and biochemistry. From a moral perspective the primates may deserve our stewardship no more than the millions of other forms of life that we now, for the first time in the history of the planet, are able to do with or dispose of as we wish. And yet there is something special about the primates.

Writing in 1896, H. O. Forbes described shooting a gibbon in Malaysia: "Falling on its back with a thud to the ground, it raised itself on its elbows, passed its long tapered fingers over the wound, gave a woeful look at them, at his slayer, and fell back at full length — dead." Forbes's Malaysian guide cried out: *"Saperti orang!"* — So like a man! And without weak sentimentality or foolish anthropomorphism, we might recall that phrase: *So like a man!* For if the primates deserve our special attention, a special protection from the Deluge of our time, they deserve it because they are so close to us: our nearest biological relatives. In their faces, bodies, and gestures we see the reflection — sometimes clear, sometimes distorted — of our own. Thus we find them beautiful and fascinating, or peculiar and comical. Their perceptions, their vision, their sense of smell, hearing, taste, and touch are like ours. They communicate in complex ways, and some may even be capable of learning language. Some, such as the chimpanzee, make and use crude tools. Primates learn, and they can acquire learning in groups and pass it through generations. They feel pain, care for their young, care for each other. They are some of the largest, often the most dramatic, and usually the most complex and intelligent of the forest dwellers. By such criteria alone the primates command our particular interest and sympathy.

One has a feeling of inevitability about biological events. They are large and exceedingly complex, and they extend in time beyond our own small vision — apparently beyond our control. When faced with the decline and disappearance of life forms, extinctions of species, one's first impulse is to turn away, to place one more notch in the great stick of accounting and then learn to live and take pleasure in a diminishing world. And quietly, perhaps with an undefined and unspoken measure of sadness, to pass on to our children and their children a biological deficit far more profound than the economic deficits we have come to tolerate. Yet biological events are not inevitable; the course of biological history is by no means beyond human control — as we now have proven only too well on the negative side. It is surprisingly easy now, with today's levels of human population and our century's technology, to extinguish some species by choice. It is even easy to extinguish a species without intending to. Given our great power for negation, one might consider whether we possess any powers of positive design. I believe we do.

The people of San Francisco recently spent nearly $50 million to restore a piece of their historic heritage, the cable car system. The people of the United States spent a good deal more than that to restore their great symbol of personal freedom and national conscience, the Statue of Liberty. Few will deny that preserving the symbol of a city or a nation is important. Yet cable cars and statues are inert objects. They can be preserved or reconstructed at any time. The San Francisco cable cars could as easily be rebuilt a thousand years from now. The Statue of Liberty could entirely disintegrate, fall into a bubbling underwater shambles of concrete, copper, and steel, yet still be resurrected from photographs or drawings at any time in the future. But the world's endangered primates, once lost, will never be regained. No sum of money, no remotely foreseeable future technology will ever return them to this planet or this universe. And when they finally disappear, species by species, only the physically desperate, the intellectually dead, and the spiritually degenerate among us will remain untouched.

To preserve endangered species for all time is beyond our meager power. But of one thing I am certain: small solutions right now can save the many threatened primate species for a while, perhaps for a century, long enough to give our children and their children a chance to preserve them in better ways. We can build an Ark — of parks, of preserved forests, even of cages — to carry the primates that far, and I have written this book, these words, to explain why we ought to and how we can. Comparatively small amounts of money — contributions to the Primate Division of the World Wildlife Fund, for example — can still today make a world of difference for some of the endangered primates I have come to know. I have no money, only these words, but I hope that my small contribution might, too, make a difference, so that my own children may mature on a planet that still carries all of these remarkable, beautiful animals.

Notes
Works Cited
Index

Notes

1 / WORLDS

My general introduction to and definition of the primates in this chapter begins with material from Eimerl and DeVore, 1965, pp. 9–16; and Kavanagh, 1983A, pp. 16–23. Background material on the muriqui is from IUCN, 1982B; Da Fonseca, 1985; and De Assumpçao, 1983. General statements on size and appearance are mostly from Grzimek and others, 1968, p. 354; and Napier and Napier, 1967, pp. 66–68. Both Da Fonseca and De Assumpçao provide recent reports on ecology and behavior, which I have used. I have reported De Assumpçao's descriptions of vocalizations. Population figures are cited in Wolfheim, 1983, p. 258; the estimate on current populations is from Milton and Lucca, 1984; and Mittermeier, 1987.

Much of the information on the decline of the Atlantic forests is based upon Mittermeier, 1982, 1987; and Mittermeier and others, 1982. The quote ("dying forest formation") is from Mittermeier, 1982, p. 18. The effects of habitat loss on muriquis are described generally in several sources, including Thornback and Jenkins, 1982, pp. 181, 182; Da Fonseca, 1985; and Wolfheim, 1983, p. 258. The effects of hunting on muriquis are described in a number of sources, including Mittermeier and others; Thornback and Jenkins, pp. 181, 182; and Wolfheim, p. 258. Information about various protected areas for the muriqui is mostly based on Mittermeier and others; Thornback and Jenkins, p. 182; and Wolfheim, pp. 258, 259. Also Hatton, Smart, and Thomson, 1984; and Padua, Magnanini, and Mittermeier, 1974. Mittermeier and others describes four areas where the muriqui is known to exist; Da Fonseca mentions a fifth, Fazenda Esmeralda. Milton and Lucca, 1984, describe the remaining population in Fazenda Barreiro Rico. Mittermeier, 1987, mentions ten isolated forest pockets. See also Mittermeier, Valle, and Coimbra-Filho, 1985.

Background information on the lion tamarins can be found in Bourne, 1974, pp. 71–75; Coimbra-Filho and Mittermeier, 1977, 1983; Dietz, 1984; Kavanagh, 1983A, pp. 81, 82; Mallinson, 1984A and B; Thornback and Jenkin, 1982, pp. 137–147; and Wolfheim, 1983, pp. 159–165; also IUCN, 1982A. The quotation from Magellan's chronicler is cited in Coimbra-Filho and Mittermeier, 1977, p. 61. Specific information on the size and anatomy of lion tamarins is mainly from Napier and Napier, 1967, pp. 197–199. Information on the number of golden lion tamarins remaining is mostly from

Thornback and Jenkins, pp. 137, 141, 145; as well as Wolfheim, pp. 159–165; and conversations with Andrew Baker and Russell Mittermeier.

Specific information on habitat loss as it affects the golden lion tamarin is cited in Wolfheim, pp. 160, 161. The information on Atlantic forest sanctuaries for lion tamarins is from Mittermeier and others, 1982. Other specific information on the status of lion tamarin reserves is mostly from Mallinson, 1984B; Mittermeier and others; and Thornback and Jenkins, pp. 137–147. Information on the captive breeding and reintroduction of golden lion tamarins is from Dietz, 1984; Kleiman, 1977, 1984; Kleiman, Ballou, and Evans, 1982; and Kleiman and Jones, 1977. Mallinson, 1984B, provides an update on the Rio de Janeiro Primate Center. The quotation in this section is from Dietz, p. 6. References to human population growth in this and all other chapters are based upon statistics in "1988 World Population Data Sheet," 1988.

2 / FORESTS

In preparing this chapter I have particularly relied on Myers, 1984B. Norman Myers is one of a few top experts on tropical forests and world deforestation; this book is probably the best single popular account of the subject. I have concentrated on this source also for reasons other than the author's expertise: the dimensions of deforestation are described with many rough estimates, and I wanted to cite estimates from one or two sources (I also refer to Myers's more detailed study of 1980) to achieve some internal consistency.

The definitions of various tropical forests, based on temperature and rainfall rates, are from Myers, 1984B, pp. 36–49. For an alternative definition, see Committee, 1982, pp. 25–40. The description of multilayered plenitude includes material from Myers, pp. 22–25, 70–73, 77, 78, and 255; also from Arditti, 1966; and Caufield, 1985, pp. 48–50, 61. *Diversity and dispersion* includes material from Myers, pp. 50–64, 68–72; Caufield, pp. 59–61; and Committee, p. 39. The suggestion that 40 percent of the world's 5 to 10 million plant and animal species live in the tropical forests is cited in many places, including Myers; the more radical idea that tropical forests contain as much as 90 percent of 30 million species has been expressed by Brent Blackwelder, vice president of the Environmental Policy Institute, noted in Peterson, 1987. *Mutualism* includes material from Myers, pp. 68–88. Myers thoroughly discusses durian and bats, Brazil nuts, and figs and wasps, pp. 82–86. *Fragility* includes material from Myers, pp. 79–82 (seeds), 75–79 (nutrient cycle); and Sanchez, 1981 (soils). See Committee, pp. 40–45, for a more technical discussion of oxisol and ultisol chemistry, including the problems of low cation exchange and phosphorus sorption. I have mostly depended on Herrera and others, 1981, for information on experimental work in Venezuela. The speed of leaf decay is noted in Myers, p. 34. Committee, pp. 59–61, includes a detailed discussion of tropical forest nutrient cycles, which I have relied on.

The material on deforestation for ranchland is mostly based on Myers, 1984B, pp. 127–134; and 1981. Material on ranching in the Amazon is mostly based on Myers, 1980A, pp. 123–125. King, 1984, is one main source on the protein value of wild meat, the wild origins of today's domesticated animals, and future possible domesticates. Myers, 1984B, is another. Pirie, 1967, also provides some information on this subject. Most of the suggestions for alternative approaches to ranching are from Myers, 1984B, pp. 139–142; see also Sanchez, 1981.

The rates of forest loss cited in the "science fiction/science fact" section are based upon Myers, 1984B, pp. 175–179. Another important study, sponsored by the United Nations Food and Agricultural Organization (FAO), indicates a deforestation rate of 30,000 square miles (76,000 sq. km) yearly, compared to Myers's estimates of nearly 80,000 square miles (200,000 sq. km). However, the FAO figures are actually biased toward total loss of forests (conversion to permanent farmland, for example). The FAO estimate of total deforestation per year is, in fact, not much different from Myers's estimate of permanent yearly loss (36,000 square miles). For a discussion of this issue, see Myers, 1984B, pp. 197, 198; and 1985. An extinction of up to a million species is cited in Wilson, 1980, and the quotation from Wilson is from the same source. See also Myers, 1979, 1980B, 1983, and 1984A.

3 / PIECES

Background information on the bearded sakis, particularly the southern subspecies, can be found in Johns, 1985; Johns and Ayres, 1987; and Van Roosmalen, Mittermeier, and Milton, 1981; also Kavanagh, 1983A, pp. 104–105; and Wolfheim, 1983, pp. 303–309. My general information on the hunting of primates in Brazilian Amazonia is largely based on Mittermeier and Coimbra-Filho, 1977; and Rylands and Mittermeier, 1982. Rudran and Eisenberg, 1982, provide some additional information on hunting in Venezuela. The description of an Indian blowpipe is from Bates, 1863, vol. 2, p. 320, as is Bates's reference to the two hundred Tucana Indians. Bates's quoted description of spider monkey meat is from vol. 2, p. 119. See Smith, 1976, for a more thorough description of primates as bait for cat hunting in Amazonia. More information on how hunting affects bearded saki species can be found in Mittermeier and Coimbra-Filho, 1977; Rylands and Mittermeier, 1982; Thornback and Jenkins, 1983, pp. 159–162; and Wolfheim, 1983, pp. 304, 307, 308.

Information on habitat loss in Amazonia, both generally and as it affects the southern bearded saki, can be found in Johns, 1985; Johns and Ayres, 1987; Rylands and Mittermeier, 1982, 1983; Thornback and Jenkins, 1982, pp. 161, 162; and Wolfheim, 1983, pp. 304, 307. Caufield, 1985, pp. 3–31, is my main source on the Tucuruí project and Amazonian dams in general. Orens, 1987, provides most of my information on the Grande Carajas Program, and I've quoted him on the potential fate for "remaining native forests." The material on Amazonian parks and reserves is largely based on Barrett, 1980; Padua, Magnanini, and Mittermeier, 1974; Rylands, 1985; Rylands and Mittermeier; also Caufield. A list of protected and unprotected primates in Amazonia is in Rylands. For more information on Indian reserves, see Barrett, 1980; and Caufield, pp. 11–16. The "American journalist" is Caufield, who quotes Lutzenberger, p. 235. The "best experts" are Johns and Ayres.

4 / HUNGER

Material on Old World leaf-eating monkeys is based upon Kavanagh, 1983A, pp. 117–174; and Eimerl and DeVore, 1965, pp. 35–58. General descriptions of the four black-and-white colobus species are mainly from Dandelot, 1971. For further anatomical information, see Napier and Napier, 1967, pp. 123–127 (the anatomical measurements are imperfect, since the authors define *Colobus polykomos* according to a different scheme of classification). Much of my information on black-and-white colobus behavior is based upon studies of the eastern species. The statement of typical group size is from Struhsaker, 1969, p. 22. The description of social relationships is mostly based on Schenkel and Schenkel-Hulliger, 1967; the quotation ("chewing-and-lip-smacking") is from that source, p. 188, as is the concept of a "grooming ceremony" and most of the description of the social intricacies of grooming. Struhsaker, p. 51, relates information on grooming frequencies. Kavanagh, 1983A, p. 120, summarizes territory size at 27 acres (15 hectares), while Struhsaker, p. 168, gives a fuller summary of various estimates of guereza territory size. The description of a dominant male's lookout activities is based on Schenkel and Schenkel-Hulliger, while the account of the jumping-roaring display is based on accounts in Marler, 1969; and Schenkel and Schenkel-Hulliger. Both sources also describe, in moderately different ways, the sequences of territorial conflict; I have attempted to combine their observations in a reasonable fashion. Wolfheim, 1983, p. 768, describes the western species as "threatened substantially." Other information is from Grzimek, 1968, vol. 10, pp. 464–469.

Information on the black-and-white colobus fur trade is largely from Oates, 1977; Wilcox, 1951; Mittermeier, 1973; and, to lesser degrees, Asibey, 1974; Bachrach, 1953; Happold, 1972; Jones, 1970; Wilkinson, 1974; and "Pope Gets Monkey Rug," *IPPL Newsletter* 12 (Dec. 1985): 11. The statistics on game meat consumption in West Africa are from Asibey, 1974. The first quotation in that section ("starved of meat") is from Jeffrey, 1970, p. 240. Asibey's list of foods is from Asibey, p. 33. Asibey, 1974, mentions the decline of monkeys in Ivory Coast. Comments on the decline in Ghanian wildlife are based on Asibey, 1978; Booth, 1956; and Jeffrey, 1970. The "heavily poached" quotation is from Asibey, 1978. See also Robinson, 1971. The information (including the quoted remark) on hunting in Sierra Leone is from Lowes, 1970; and from Mack and Mittermeier, 1984A. General information on deforestation by tropical forest farmers is almost wholly based on Myers, 1984B, pp. 143–162. Myers describes 200 million forest farmers, a number he revises to 250 million in Myers, 1985. Committee, 1982, p. 96, puts the number at 240 million. For figures on deforestation across the western black-and-white's strip of forest in West Africa, see Myers, 1980A (Sierra Leone, p. 164; Liberia, pp. 160–162; Ivory Coast, pp. 165–169; Ghana, pp. 155, 156); and from Harding, 1984 (Sierra Leone); and Asibey, 1978 (Ghana).

For a count of parks and reserves harboring the western black-and-white colobus in West Africa, see Wolfheim, 1983, pp. 448, 449. Information on Tiwai Island comes mostly from conversations with Anne Todd. Information on Outamba and Kilimi is largely from Lowes, 1970; and Harding, 1984; I have also referred to Teleki, 1980; and McGiffin, 1985, to update some of the information on Outamba-Kilimi. Information on the Susu is from Thayer,

1983. Further information on parks and sanctuaries elsewhere in West Africa can be found in Asibey, 1978, pp. 56–59. For information on the Tai National Park in Ivory Coast, see Caufield, 1985, pp. 236, 237; and Dosso, Guillaumet, and Hadley, 1981. "Ivory Coast Poaching Ring," 1980, gives some specifics about poaching in Tai Park. Asibey, 1972, 1978, discusses some of the general history on legal wildlife protection in Ghana. See Jeffrey, 1970, 1975A, for more information on the forest zone reserves of Ghana.

5 / FOOD

This chapter is substantially based on material from Myers, 1984B, pp. 189–205. Pirie, 1967, is the source of the N. W. Pirie quotation. King, 1984, provides information on human protein needs; much of the information on the winged bean is from Myers, 1984B, pp. 192–193. King, 1984, is the main source of information on the importance of wild plants in supporting domestic crops and the decline of wild cultivars; Myers, 1984B, is another important source on this subject; Committee, 1982, p. 83, provides some further information. The description of pesticide use and problems in modern agriculture is based on Begley, Lubenow, and Miller, 1986. The subsequent material on natural alternatives to chemical pesticides is almost wholly from Oldfield, 1981; and Myers, 1984B, pp. 199–201. The quotation from a symposium for botanists is taken from Alvim, 1977, p. 347. Suggestions for improvements on uncontrolled slash-and-burn farming are based on Dickinson, 1972; Myers, 1984B, pp. 162–169; and Sanchez, 1981; also Alvim, 1977; Greenland, 1975; and Myers, 1985. Information on Mayan farming and recent experiments is from Chen, 1987.

6 / ISLANDS

The introductory material on apes is from Eimerl and DeVore, 1965, pp. 61–84; and Kavanagh, 1983A, pp. 176–178. The brief discussion of human evolution is based on Kavanagh, 1983A, pp. 203–207; and Rensberger, 1984. The background material on gorillas is from Fossey, 1983; Kavanagh, 1983A, pp. 191–196; and Schaller, 1963, 1964; also, IUCN, 1982A; and Napier and Napier, 1967, pp. 160–167. The quotations attributed to Savage and Owen can be found in Schaller, 1964, pp. 4, 5. The Du Chaillu quotations are from Du Chaillu, 1861, pp. 60, 70, 71. The quotation of Schaller is from Schaller, 1964, p. 10. Census estimates are derived from three sources: Butynski, 1985; IUCN, 1982A; and Weber and Vedder, 1983. See also Goodall, 1978; Groom, 1973; Harcourt, 1977; and Harcourt and Groom, 1972.

Most of the material on the hunting of gorillas comes from IUCN, 1982A; Cousins, 1978C; Denis, 1963; Du Chaillu, 1861; Fossey, 1983; Merfield, 1954; Merfield and Miller, 1956; and Schaller, 1964. Cousins provides the list of West African tribes that hunt or have hunted gorillas. Denis relates the story of the M'Betis' hunt (the expression "from time immemorial" is from that source, p. 185). Merfield and Miller, and Merfield tell of the hunting practices of the Mendjim Mey and the Yaounde tribes; the quotations in the account of the Yaounde gorilla roundup are from Merfield and Miller, pp. 74, 75. Information on the dietary habits of the Fang is from Sabater Pi and Groves,

1972; the quoted "almost human" is from that source, p. 240. The Du Chaillu quotation is from Du Chaillu, p. 71. The quoted descriptions of Powell-Cotton and his museum are from Merfield and Miller, p. 68. Hall and Hall, 1964, give some background on the stuffed mountain gorilla at Harvard.

The section on gorillas in parks and preserves is mostly based on material from Curry-Lindahl, 1964; Fossey, 1983; Schaller, 1964; Weber and Vedder, 1983; also material in IUCN, 1982A; Goodall and Groves, 1977; Wolfheim, 1983, pp. 689, 691, 692; and Myers, 1980A, pp. 165–167. Butynski, 1985, is the source of all my information on mountain gorillas in Uganda's Impenetrable Forest. Comments and information specifically about Dian Fossey and her death are based on Battiata, 1986A and 1986B. "Dian Fossey Asked for It" is briefly described and quoted in *IPPL Newsletter* 13 (Apr. 1986): 8. The *Boston Globe* quotation is from Lessem, 1986. The remarks of a Rwandan journalist were quoted in Battiata, 1986A. Information and the single quotation regarding the Mountain Gorilla Project are from personal communications with Amy Vedder.

MacArthur and Wilson's work, including their theory of island biogeography, has been summarized many times in many places. I have most fully relied on MacArthur, 1972; MacArthur and Wilson, 1967; and Soulé, Wilcox, and Holtby, 1979. The data on the California Channel Islands are from MacArthur. The quotation from Wilson is in Lessem, 1985. Information on Barro Colorado Island is based on Ehrlich and Ehrlich, 1981, p. 267; and Lovejoy, 1980. MacArthur, in a discussion on the theory of island biogeography, refers to the experiment on four small mangrove islands in Florida, as well as to the surveys of the California Channel Islands. See also Diamond, 1975, 1980; Liu and Godt, 1983; Lovejoy, 1982; and Wilcox, 1980. Some ideas briefly touched on in the final section are from Lusigi, 1981; also Coe, 1980.

7 / HEALTH

General information on prosimians is from Eimerl and DeVore, 1965, pp. 9–16; Grzimek and others, 1968, pp. 270–277; Kavanagh, 1983A, pp. 26–67; Sussman, Richard, and Ravelojaona, 1985; and Tattersall, 1982, pp. 1–35. Tattersall, pp. 1–18, includes historical material on the discovery of prosimians; the Keeling quotation is from that source, p. 4; the De Flacourt quotation is from Tattersall's translated version, pp. 1, 2. General information on the indri is from Petter and Peyrieras, 1974; Pollock, 1975; also Kavanagh, 1983A, pp. 51, 52. The quoted descriptions of indri calls are from Petter and Peyrieras, p. 42. General information on the aye-aye is from Kavanagh, 1983A, pp. 52–54; also Bourne, 1974, pp. 34, 35; Caras, 1966, pp. 58–61; Constable and others, 1985; Goodwin and others, 1978; Grzimek and others, 1968, pp. 270–277, 295–297; Richard, 1982; Simon and Geroudet, 1970, pp. 184–186; Tattersall, 1982, pp. 111, 112, 307, 308; Walker, 1979; and Wolfheim, 1983, pp. 133–136. Descriptions of aye-aye calls are from Grzimek and others, p. 297.

My main sources on hunting in Madagascar are Richard, 1982; Richard and Sussman, 1975; Sussman, Richard, and Ravelojaona, 1985; Wolfheim, 1983, p. 121; Grzimek and others, 1968, p. 297; and Tattersall, 1982, p. 335.

Most of the general information about Madagascar and its forests is from Heseltine, 1971; Richard, 1982; Tattersall, 1982; and Jolly, 1987. The figures on endemic species are from Richard, 1982; Tattersall, 1982; and Jolly, 1987. For specific figures on deforestation, see Sussman, Richard, and Ravelojaona, 1985; and Myers, 1980A, p. 162; also Jolly; Richard; Richard and Sussman; and Tattersall. Specific information on the loss of indri habitat will be found in Goodwin and others, 1978; and Wolfheim, 1983, p. 121. Information on the aye-aye's loss of habitat can be found in Goodwin and others; Tattersall, pp. 112, 113; and Wolfheim, pp. 133–136.

For further information on the Ark in Madagascar, see Jolly, 1987; also Richard, 1983; Richard and Sussman, 1975; Tattersall, 1982, pp. 337–340; and Simon and Geroudet, 1970, p. 193. Additional information about the indris in reserves can be found in Goodwin and others, 1978; and Wolfheim, 1983, p. 121. For information about the aye-aye reserves, see Goodwin and others; and Wolfheim, 1983, pp. 134–136. The story of Elizabeth Bomford's visit to Nosy Mangabe is told in Bomford, 1981. The quotation is from that source.

The most important sources for the section on tropical forest pharmaceuticals are Humphreys, 1982; and Kreig, 1964. The material on alkaloids is based on Humphreys. The story of curare is mostly from Kreig, pp. 223–241. Oldfield, 1981, adds information about tetrodotoxin; Humphreys, about ouabain. The story of reserpine is mostly from Kreig, pp. 317–339; Myers, 1984B, p. 216, provides a little more on the extent of its modern use. The story of quinine comes from Kreig, pp. 165–206; and to a lesser degree from Oldfield; some minor information on quinine in Java is based on Taylor, 1945; additional information about the disease is based on "Malaria," 1986. Information about anticancer compounds from the rosy periwinkle is from Oldfield. Caufield, 1985, p. 220, describes the improved survival rates of some cancer patients after these drugs were isolated; Myers, 1984B, p. 212, mentions the current $90 million value to the international pharmaceutical industry; on p. 213 he cites the number of potential anticancer plants in tropical forests. Oldfield mentions the Indian mandrake and *Brucea* as two new anticancer possibilities. Myers, p. 218, describes the use of the nut of the greenheart tree in Guyana as a contraceptive. The story of diosgenin is told in several places; my main source is Oldfield. Myers, pp. 209, 223, lists the number of native plants used by various traditional forest-dwelling peoples as medicines. Myers, p. 210, places a $20 billion yearly estimate on the medicinal value of tropical plants and animals.

8 / WATER

The reader may find that some of my words in this chapter sound strangely familiar. My experience in India during a severe drought powerfully reminded me of T. S. Eliot's *The Waste Land* and *The Four Quartets* ("Burnt Norton" and "East Coker"). The surprising parallels between my experience and Eliot's vision were hard to suppress; instead of trying, I incorporated some of his language and phraseology, in the form of several brief echos and the near quotation of four lines: "East Coker," line 2 and "Burnt Norton," lines 35, 38, and 39. Much of my information on lion-tailed macaques is taken from Sugiyama, 1968; and Green and Minkowski, 1977; also Karr, 1973. The descriptions and size estimates of the lion-tailed macaque in the two main sources differ slightly; I have presumed Sugiyama's to be more accurate. The quotations are from Sugiyama, pp. 284, 287. General comments on macaques, including the references to the toque and Barbary macaques, are largely based on Kavanagh, 1983A, pp. 167–171; also Lindburg, 1980. The 1975 population estimate is from Green and Minkowski. My current estimate is based on a personal communication from Dr. Steven Green, combined with Karanth, 1985.

Information on hunting in India and, more particularly, the hunting of lion-tailed macaques is based on material in Goodwin and others, 1978; Green and Minkowski, 1977; Krishnan, 1971; and Wolfheim, 1983, pp. 516–520; also Hutton, 1949. The quotations are from Krishnan, p. 534. Major sources on the loss of lion-tailed macaque habitat are Green and Minkowski; Karanth; Krishnan; also Mohnot, 1978; and Wolfheim, 1983, p. 517. The quotations attributed to Krishnan are from Krishnan, p. 532. An estimate of remaining habitat size is cited in Wolfheim, 1983, p. 516. Groombridge, 1984; and Karanth describe newly discovered habitat.

Comments on the recent status of Kalakad are based upon a personal communication with Steven Green. The quoted "Western scientist" who recently visited the sanctuary is Dr. Green. Johnsingh, 1984, provides information on the Pachayar Dam scheme. A small amount of additional information comes from Viswanathan, 1986.

The Fairlie quotation is from Fairlie, 1979. A rainstorm in Borneo is mentioned in Myers, 1984B, pp. 32, 33; and absorption rates of a forest, pp. 261–264. The loss of forests in the Ganges watershed and subsequent destruction downstream is mentioned in several places. I've used material from Caufield, 1985, pp. 68–70; Myers, 1984B, pp. 62–267, and 1985; and Postel, 1984. The erosion information is based on figures in Myers, 1984B, pp. 272, 273; and Brown, 1984. The quotation of a "Brazilian agronomist" is from Alvim, 1977, p. 349. The discussion of possible changes in Amazonian hydrological cycles is based on material in Barrett, 1980; Dickinson, 1981; and Henderson-Sellers, 1977. Dickinson, p. 425, comments on the uncertainty of this idea: "Early beliefs that forests increased rainfall have been superseded in the hydrological community by a consensus that the land use changes . . . have no effect on rainfall. Meanwhile, there have been a considerable number of computer studies over the last decade by meteorologists showing significant changes in regional climate, including rainfall, due to surface changes. There is, however, no real consistency between these two viewpoints." Information on the albedo effect is based on Dickinson, 1981; and Henderson-Sellers. The quotation and some information on the greenhouse effect is from Begley and Cohn, 1986. The role of carbon dioxide in the greenhouse effect is partly from Begley and Cohn, but is mostly based on Dickinson; Henderson-Sellers; Flavin, 1988; also Myers, 1984B, pp. 283–293, and 1985. My estimate that tropical forest destruction is adding another fifth or two-fifths or even more is from Dickinson, who cites past estimates of the contribution of forest burning to atmospheric carbon dioxide as varying from 17 percent of the fossil fuel contribution to 1.5 to 3.0 times that, i.e., 25 to 51 percent. Myers, 1984B, p. 286, says "between two-fifths and four-fifths as much."

9 / CHILDREN

Much of the general information on gibbons is from Chivers, 1977; and Kavanagh, 1983A, pp. 178–183; also Napier and Napier, 1967, pp. 172–178. The opening quotation is from W. C. Linnaeus Martin, 1841, as quoted by Chivers, pp. 539, 540. Details of siamang size are from Grzimek and others, 1968, p. 471. The story of the discovery of the bilou is mostly from Miller, 1903, pp. 1, 2, 70, 71. Details of bilou behavior are almost entirely from Tenaza, 1975; and Tenaza and Hamilton, 1971. The quotation describing a group conflict is from Tenaza, p. 72. The quotation describing a lar gibbon song is from Kavanagh, pp. 178, 179.

General information on the hunting of gibbons can be found in Wolfheim, 1983, pp. 649–682. Information more specific to the bilou (on hunting and deforestation) is from that source; also Caufield, 1985, pp. 94–96; Tenaza, 1975, 1987; Tenaza and Hamilton, 1971; and Tenaza and Mitchell, 1985. Goodwin and others, 1978; Tenaza and Mitchell; and Tenaza, 1987, provide an update on the status of logging in Mentawai; the statistics on rattan trade are from Tenaza and Hamilton. Summaries of habitat loss for the several other gibbon species are mostly based on figures in Wolfheim, 1983, pp. 649–682, although the "greatly reduced" description of Burmese habitat is from Chivers, 1977, p. 582. The "current rates of deforestation" estimate is from Chivers; he anticipates the destruction of most forest outside the 4 percent that is legally protected. Since he later modifies that estimate — "There are good chances of forest remaining outside protected areas in 10–15 years"— I have taken the liberty of tempering his estimate of 96 percent destruction with my own "nine tenths."

General estimates on total protected habitat for gibbons will be found in Chivers, 1977; for additional information on protected gibbon habitat, see Wolfheim, 1983, pp. 649–682. Information on Teitei Batti is from Goodwin and others, 1978; and Wolfheim, p. 659. Caufield, 1985, p. 96, describes a WWF/SI conservation plan for Siberut. Tenaza and Mitchell, 1985, mildly contradict information from some of the above sources, and I have presumed their report to be the most accurate.

10 / WOOD

The information on orangutans is from Kavanagh, 1983A, pp. 184–190; Grzimek and others, 1968, pp. 503–520; MacKinnon, 1971, 1973; Rijksen, 1982; and Rodman, 1977; also Kevles, 1976, pp. 111–144; and Wolfheim, 1983, pp. 720–730. The de Bondt quotation is from Yerkes and Yerkes, 1929, p. 12. The Beeckman quotation is from Harrisson, 1963, p. xii. The description of a long call is from MacKinnon, 1971, p. 170. The Rijksen quotation is from Rijksen, p. 323. "Several times that number": Goodwin and others, 1978, indicates 10,000 to 30,000; Rijksen estimates from 5,000 to 15,000 in Sumatra and around six times that in Borneo.

Further information on hunting of orangutans can be found in Harrisson; Rijksen and Rijksen-Graatsma, 1975; Wolfheim, 1983, pp. 723–726; Jones, 1982; and MacKinnon, 1977. Aken and Kavanagh, 1983; and Kavanagh, 1983B, provide additional information on hunting practices in Sarawak.

Primary sources on the trade in live orangutans are Harrisson; Rijksen and Rijksen-Graatsma, 1975; Wolfheim, 1983, pp. 723–726; Jones, 1982; and MacKinnon, 1977. The information on local orangutans kept as pets is from Rijksen and Rijksen-Graatsma; the quotation can be found on p. 65. The story of von Goens's work is from Harrisson, pp. 112–116. Most of the statistical figures on the orang trade are taken from Wolfheim, pp. 723–726. "As late as 1977 one expert" is MacKinnon, 1977.

I relied especially on Myers, 1984B, in preparing my discussion of tropical forest logging. The discussion of deforestation for fuelwood is almost entirely based upon that source, pp. 115–126; the section on commercial wood is from pp. 91–109. The remarks of "one Southeast Asian forestry official" were first quoted by Myers, p. 100. For further information on the potential for logging within secondary forests, see Myers, 1984B, pp. 106–108, 110–114; Myers, 1985; Caufield, 1985; and Postel, 1984. On habitat loss and orangutans, Myers, 1980A, pp. 68–76, provides a good summary of deforestation in Indonesia as a whole, which I have taken advantage of. Myers, 1980A, tells the story of Indonesian logging; however, my particular facts on the effects of logging in southeastern Borneo are based on Kartawinata and others, 1981. Other general information about orang habitat is from Rijksen, 1982; and Wolfheim, 1983, pp. 720–730. Information about Sumatra is mostly based on reports by Borner, 1976; and MacKinnon, 1973; the first quotation is from MacKinnon, p. 242. The Hornaday quotation regarding Borneo is from Hornaday, 1885, p. 335. The story of a rain forest fire in southeastern Borneo, including quotations, is from Webster, 1984. Information on the situation in Malaysian Borneo comes partly from IUCN, 1982A; and MacKinnon, 1971; as well as from Myers, 1980A, pp. 84–88; Aken and Kavanagh, 1983; and Kavanagh, 1983.

The general information about parks and reserves for orangutans in Borneo and Sumatra is mostly based on Rijksen, 1982; also Wolfheim, 1983, pp. 724–726. Information on conservation areas in Sarawak is from Aken and Kavanagh, 1983; and Kavanagh, 1983; the quoted remarks regarding orangutans in the Lanjak-Entimau sanctuary are from Aken and Kavanagh, 1983. Information about protected areas in Sabah is largely based on MacKinnon, 1971; also IUCN, 1982A. For more information on early rehabilitation projects for orangs, see MacKinnon, 1971; Grzimek and others, 1968, pp. 518, 519; also Harrisson, 1963; Rijksen and Rijksen-Graatsma, 1975. For further criticism of the rehabilitation centers, see MacKinnon, 1977; and Rijksen, 1982.

11 / CAGES

Most of the background information on cotton-top tamarins comes from Neyman, 1977; Kavanagh, 1983A, pp. 78–81; Grzimek and others, 1968, pp. 378–395; and Napier and Napier, 1967, pp. 303, 304; also Camacho and Defler, 1985; and Cerquera, 1985. Thornback and Jenkins, 1982, p. 127, describe it as "endangered"; Wolfheim, 1983, p. 767, lists it (*Saguinus oedipus*) as "severely threatened." Wolfheim describes as two subspecies of the *Saguinus oedipus* (which she calls the crested tamarin) both this Colombian primate, the cotton-top tamarin, and its Costa Rican and Panamanian relative, the Geoffroy's tamarin. Other authoritative sources categorize the two groups as separate species. Although I have generally used Wolfheim as a primary reference, in this instance I prefer the latter system.

Neyman, 1977, is my principal source on cotton-top habitat destruction in Colombia, and for material on capture of and trade in cotton-top tamarins. Information on breeding problems can be found in Clapp and Tardif, 1985, though much of my information is based upon conversations with Dr. Nora Johnson and Dr. Ronald Hunt of the New England Regional Primate Center. Information on cotton-tops and ulcerative colitis can be found in Knox, 1984. More particular information on Lynne Ausman's research appears in Pease, 1987, though much of my information is based upon conversations with Dr. Ausman. An estimate of the number of cotton-tops in captivity can be found in Tardif, 1985; and Tardif and Colley, 1987; my estimate of more than 1,500 is based upon recent conversations with Dr. Tardif.

The historical background on live primates serving humankind is largely based on chapters 4 and 8 of Morris and Morris, 1966. I have taken several quotations directly from that source, including those of Meunier (pp. 240, 241). Other quotations taken from Morris and Morris are of the article in the *Daily Mail* on the London Zoo's first orangutan, p. 100; and of L. Harrison Matthews in his 1963 speech, p. 251. Information on NASA's Spacechimp project is from "Ham Dies," *IPPL Newsletter* 10 (Apr. 1983): 12; "Monkeys Go into Space," *IPPL Newsletter* 12 (Aug. 1985): 3; and "Space Chimp," *IPPL Newsletter* 8 (Jan. 1981), p. 11. The quotation is from "Monkeys Go into Space." Information and comments on the beach photography business in Spain and the Canary Islands are based on Van Hoorn, 1980. The story of Mickey Antalek's chimpanzees is based on "Circus Chimps," *IPPL Newsletter* 11 (Dec. 1984): 5–7; and "Circus Chimps," *IPPL Newsletter* 12 (Apr. 1985): 10. Information on United States zoos offering their surplus primates to laboratories comes from "Zoo-Lab Links," *IPPL Newsletter* 12 (Aug. 1985): 4. Information on the San Diego Zoo gibbon is from "San Diego," *IPPL Newsletter* 5 (Apr. 1978): 17. Information, including the quotations, on the safari parks in England offering their surplus primates to laboratories is from "International Primatological," *IPPL Newsletter* 5 (Dec. 1978): 15. The story of polio research is largely based on Dowling, 1977, pp. 202–221; the data on the importation of monkeys and their use for polio research are mostly from Eudey and Mack, 1984.

Data on international trade in primates generally are from Kavanagh, 1984; and Mack and Mittermeier, 1984A and B. Regarding the rough estimates for the world trade at midcentury, Mack and Mittermeier, referring to Kavanagh, 1984, write of "well over 200,000 primates traded"; Kavanagh, however, arrives at his rough estimate of trade volume for that period by doubling the figure of 200,000 for United States imports. More general information on monkeys in laboratories is from Caldecott and Kavanagh, 1984; much of the statistical information on terminal and painful research, including the quoted "pain or distress," is from Mack and Mittermeier. Examples of terminal and painful research: the first *item*, including the quotation, is mostly based on Pratt, 1985; also, "Update," *IPPL Newsletter* 11 (Dec. 1984): 11. The second *item*, including quotations, is mostly from Barnes, 1982; also, "Taub's Reaction," *IPPL Newsletter* 9 (Jan. 1982): 12 (which includes Taub's quoted remarks); and "Taub Gets Grant," *IPPL Newsletter* 10 (Aug. 1983): 7. The third *item*, including most of the quotations, is from "Monkey Depression," *IPPL Newsletter* 8 (May 1981): 8, 9; comments and quotations attributed to Suomi, though, are from "Suomi Denounces," *IPPL Newsletter* 8 (Sept. 1981): 13. The other *items*, including quotations, in the order in which they are used, are from "India Bans," *IPPL Newsletter* 5 (Apr. 1978): 2–4; "Baboons," *IPPL Newsletter* 5 (Apr. 1978): 9; "Controversy," *IPPL Newsletter* 5 (Aug. 1978): 12; "Chimpanzees Killed," *IPPL Newsletter* 9 (Nov. 1982): 6; "Primates Used," *IPPL Newsletter* 8 (Sept. 1981): 12; "United States," *IPPL Newsletter* 8 (Sept. 1981): 12; "Primates Used," *IPPL Newsletter* 9 (May 1982): 9, 10; "Monkeys Placed," *IPPL Newsletter* 6 (Dec. 1979): 6; "India Bans," *IPPL Newsletter* 5 (Apr. 1978), 2–4 (the Monroe quotation); "The Radiation," *IPPL Newsletter* 5 (Apr. 1978): 4–6; and "Value of Nuclear," *IPPL Newsletter* 7 (June 1980): 6, 7 (including the quotation).

The data on remaining rhesus populations in India are cited in "Indian Rhesus," *IPPL Newsletter* 9 (Nov. 1982): 7. Material on capturing gorillas can be found in IUCN, 1982A; Cousins, 1978C; Denis, 1963; Fossey, 1983; Merfield, 1954; Merfield and Miller, 1956; and Schaller, 1964. Material on the capture and trade of orangutans comes from Harrisson, 1963; Rijksen and Rijksen-Graatsma, 1975; and Wolfheim, 1983, pp. 723–726; also Caras, 1966; Jones, 1982; and MacKinnon, 1977. Information on Gajo capture techniques is from Rijksen and Rijksen-Graatsma, 1975; the story of the eleven baby orangs is from Caras, 1966, pp. 49, 50. Soini's account of the capture of primates in Peruvian Amazonia, including the quotation, is from Soini, 1972; Tsalickis's account of capture in Colombian Amazonia, including quotations, is from Tsalickis, 1972. The statistics on mortality are based on Tsalickis, and on Thorington, 1972; but the general anecdotes on major shipping losses are from "320 Vervets and Baboons," *IPPL Newsletter* 8 (Sept. 1981): 8; "A Tale," *IPPL Newsletter* 7 (Oct. 1980): 3; "480 Monkeys," *IPPL Newsletter* 6 (Dec. 1979): 5; and "Two Hundred," *IPPL Newsletter* 6 (Aug. 1979): 11. The "most thorough survey of shipping mortality" is cited in Mack and Eudey, 1984; the estimates on shipping mortality by Charles Dasorno are cited in "480 Monkeys," *IPPL Newsletter* 6 (Dec. 1979): 5.

The story of India's trade in rhesus macaques leading to the 1978 ban, including quotations, is based on "India Bans," *IPPL Newsletter* 12 (Apr. 1978): 2–4; and "The Rhesus Monkey Certification," *IPPL Newsletter* 5 (Oct. 1978): 9, 10; the quoted remarks of a scientist from Lederle are from "State Department," *IPPL Newsletter* 6 (Dec. 1979): 7–9. The story of MOL Enterprises and Bangladesh is largely based on Long, 1982A, 1982B, 1982C; "Bangladesh Monkey," *IPPL Newsletter* 9 (Nov. 1982): 10; also "Supreme Court," *IPPL Newsletter* 12 (Apr. 1985): 13. Kenneth Green's survey and words are from "Protection Proposed," *IPPL Newsletter* 5 (Aug. 1978): 11.

Statistics on international trade are based on Kavanagh, 1984 (table 19). More specific details on the Peruvian trade are based on Soini, 1972. Some of the general information (including details on Leticia's free port status) is from Inskipp and Wells, 1979, pp. 23–26; other general information on smuggling is from Kavanagh and Bennett, 1984. The more anecdotal material on gibbon smuggling into Singapore is from "The Story," *IPPL Newsletter* 10 (Dec. 1983): 3–14; on gibbon smuggling into Japan, from "Primate Smuggling," *IPPL Newsletter* 11 (Dec. 1984): 11; and on the Bangkok Wildlife Company, " 'Khun Khampheng,' " *IPPL Newsletter* 5 (Dec. 1978): 8. Material on the importation of gibbons by the University of California at Davis Oncology Laboratory, including quotations, is from "Origin Of," *IPPL Newsletter* 7 (Mar. 1980): 2, 3.

Information on international controls over the wildlife trade, particularly on CITES, is largely from Inskipp and Wells, 1979; and Kavanagh and Bennett, 1984. For information on the trade in chimpanzees, see McGiffin, 1985; "Sierra Leone," *IPPL Newsletter* 10 (Aug. 1983): 6, 7; also Teleki, 1980; and Eckholm, 1985. Some of the details on Belgium's role as a laundering nation are from "Wildlife Trade in Belgium," 1981; but the story of George Munro's animal business in Belgium is based on "Pygmy Chimpanzees," *IPPL Newsletter* 10 (Aug. 1983): 3–5. The story of Bolivia's role as a South American outlet for smuggled animals includes much from "Bolivia to Re-enter," *IPPL Newsletter* 12 (Aug. 1985): 3; the quotations are all from that source. Most of the discussion of the problems in implementing CITES is based on Inskipp and Wells; and Kavanagh and Bennett.

12 / ZOOS

The information on douc langurs comes mainly from Lippold, 1977; also Kavanagh, 1983A, pp. 138–141; Van Peenen, Light, and Duncan, 1971; and Wolfheim, 1983, pp. 631–635. For further information on douc langur behavior, see Gochfeld, 1974; and Kavanagh, 1972; also Lippold, p. 524. Dr. Dao Van Tien's range estimates are described by Constable, 1982. On "Western experts," see Wolfheim, 1983, p. 769; and Southwick, Siddiqui, and Siddiqui, 1970, p. 1052.

Information on the hunting of douc langurs can be found in Lippold, 1977; and Wolfheim, 1983, pp. 631–635. Information on hunting since the war can be found in Constable, 1982. Myers, 1980A, provides important particular information on deforestation in Cambodia (p. 76) and Laos (pp. 76–79), and on the effects of shifting cultivation in Vietnam (pp. 114–116). Other accounts of douc habitat loss are based on Wolfheim, 1983, p. 631, and on the account of habitat in the Mt. Sontra region in Lippold, 1977. Figures on postwar Vietnam are largely from Karnow, 1983, especially p. 31. Figures on American bombing and its effects are from Clifton and Moreau, 1985; Orians and Pfeiffer, 1970; and Pfeiffer, 1973. The quotation on remaining bomb craters is from Clifton and Moreau, p. 59. Information on the use of defoliants and their effects is from Clifton and Moreau; Lippold; Orians and Pfeiffer; and Tschirley, 1969. The statistics on the amount of land sprayed "before the decade was over" are from Tschirley; the figure of 20 million gallons is cited in "Vietnam's Forests," *IPPL Newsletter* 12 (Dec. 1985): 7. The quotations attributed to Tschirley are from Tschirley, pp. 783, 784, and 786. The "1982 report" is Constable's; the quotation from that report can be found on p. 252. The quotation attributed to "observers" in 1985 is from Clifton and Moreau, p. 59.

Brief comments on Cue Phuong National Park in Vietnam are based on Constable, 1982; "enormous" is his word, p. 250. My information on the breeding of douc langurs in North America is almost wholly based upon conversations with Dr. Diane Brockman and Gale Foley of the San Diego Zoo, and with Dr. Lois Lippold of San Diego State University. Further information on douc langurs in the Cologne Zoo can be found in Hick, 1975.

General information on zoos and captive breeding: primary sources are Campbell, 1980; Conway, 1980; Ehrlich, 1980, pp. 248–263; Kleiman, 1980; and several chapters in Martin, ed., 1975. The opening portion of this section is based upon Doherty and Smith, 1984; Durrell, 1975; and Martin, 1975; statistics on rare animals bred in zoos by 1976 are from Ralls and Ballou, 1983. Information on the successful captive breeding of the Arabian oryx, Père David's deer, and Przewalski's horse, can be found in several places. My sources are as follows. For the Arabian oryx, Caras, 1966, pp. 1–5; Fisher, Simon, and Vincent, 1969, pp. 148–151; Homan, 1975; Simon and Geroudet, 1970, pp. 161–166. On Père David's deer, Diole, 1974, pp. 59, 60; Fisher, Simon, and Vincent, pp. 132, 133; and Kohl, 1982. For Przewalski's horse, see Caras, pp. 15–20; Fisher, Simon, and Vincent, pp. 161–166; and Volf, 1975.

Information on the problems of captive breeding comes from Conway, 1980; Kear, 1975; Kleiman, 1980; Martin, 1975A and B; and Maynes, 1975. The story of the London Zoo's Monkey Hill exhibit has been told many times in many places; my sources are Eimerl and DeVore, 1965, p. 40; and Sparks, 1982, pp. 233–236. Foose, 1983, was my primary source of information on the Species Survival Plan, as well as Foose, n.d. Material on the Siberian tiger is from Foose, 1983; and Simon and Geroudet, 1970, pp. 114–127. Information on the problems and successes of reintroduction is from Campbell, 1983; Conway, 1980; Diole, 1974, pp. 311, 312; Erickson, 1975; Fyfe, 1975; Kear, 1975; and Wayre, 1975. Background material on genetics and genetic problems of small populations is largely based on Rothwell, 1979; Senner, 1980; Soulé, 1983; and Templeton and Read, 1983.

The early background on captive breeding of lemurs is from Grzimek and others, 1968, p. 289. Information on the lesser mouse lemurs at Rotterdam is from Glatston, 1981. The report on captive breeding of ruffed lemur subspecies is based on a personal communication with Dr. Diane Brockman of the San Diego Zoo. Information on the Duke University Primate Center is from Anderson and Brown, 1984, pp. 42–52; "The Duke University Primate Center," n.d.; and conversations with Dr. Linda Taylor. Information on captive breeding of indris is from Simon and Geroudet, p. 190. Information on captive breeding of aye-ayes is largely based on Grzimek and others, p. 297.

My main source for the section on captive breeding of the lion-tailed macaque is Gledhill, 1985; other sources include Gledhill, 1984; Green and Minkowski, 1977, p. 300; Foose and Conway, 1982; and a personal communication with Laurence Gledhill of the Woodland Park Zoo, Seattle. See also Terry, 1985.

On the orangutan in captivity, my main sources are Brambell, 1975; Grzimek and others, 1968, pp. 503–520; Harrisson, 1963; Jones, 1982; and Perry, 1976; also MacKinnon, 1971. The Harrisson quotation is from Harrisson, p. 138. The second quotation is of David Martin, quoted in Perry, 1976, p. 263. The discussion of the 1974 population is based upon material in Perry. The story of the birth and hand rearing of an orang in the Toronto Zoo is from Cole and others, 1979. The discussion of current problems with the captive population is based on a conversation with Marvin Jones of the San Diego Zoo.

Some of the early history of gorillas in captivity comes from Schaller, 1964. Statistical information on captive gorillas is from IUCN, 1982A. General suggestions about the needs of captive gorillas are based on comments by Fossey, 1983. The story of Mouila at the Apenheul Sanctuary is based on Mager, 1981. The story of the Lincoln Park Zoo Project is taken from Rosenthal, 1983. Conversations with Dr. Lester Fisher of the Lincoln Park Zoo provided most of the information about the AAZPA captive breeding project.

Additional information on the genetic issues of captive breeding: on inbreeding, see Ralls and Ballou, 1983; also Soulé, 1980. On details of Mendelian genetics and inbreeding, see Chambers, 1983; Rothwell, 1979; Mendel, 1965 (1926); Olby, 1966; and Ronan, 1982. On genetic variability, see Carson, 1983; Foose, 1983; and Franklin, 1980.

13 / WORDS

The brief survey of the ways people distinguish themselves from animals is largely based on Thomas, 1984, an excellent history. The list of early definitions of "the human animal" is taken from that source, pp. 31, 32; the quotation of Price is from p. 32. The quotations of Aristotle, Marco Polo, and Battell are from Yerkes and Yerkes, 1929, pp. 3, 4, 8, and 10. Information on Tyson is from Bourne, 1974, pp. 281, 282. The chromosomal comparison of chimpanzees and humans is described in Premack and Premack, 1983, p. 2. The comparison of ape and human brains is based upon Bourne, 1974, pp. 437–450. Information about prehuman skulls is from Rensberger, 1984. Thomas, p. 125, pointed out the significance of God's comment to Adam (*Paradise Lost*, viii, line 374); I first became aware of Locke's "Beasts abstract not" in Sagan, 1977. The quotations of Bingley and Lambe are both from Thomas, pp. 35 and 137. See also Gallup, 1970; and Griffin, 1981, 1984.

The Darwin quotation is from Darwin, 1859, 1871, p. 446. The story of Lewis Henry Morgan is from Morgan, 1868; Resek, 1960; and Sparks, 1982, pp. 114–118. The Morgan quotations are from Morgan, pp. viii–ix, 263, 256, and 249. The account of Clever Hans is based upon Sparks, pp. 144–146; and Chevalier-Skolnikoff, 1981, pp. 61–66. The background material on ape language studies is based upon Gardner and Gardner, 1969; Patterson and Linden, 1981; Premack and Premack, 1972; Crail, 1983; and Sagan, 1977. The report on Patterson's work with Koko and critical objections to the work are based mostly on Patterson and Linden, 1981; and on Chevalier-Skolnikoff, 1981; Cohn, 1984; Crail, 1983; Gorney, 1985; Patterson, 1984; and Patterson, Cornwall, and Share, 1984. Test comparisons of human and chimpanzee strength are reported in Bourne, 1974. Quotations or exchanges with Koko are from Patterson and Linden: problems with "shell" and "rock," p. 80; problem with "drink," p. 77; Michael's description of a bird, p. 173; Koko's joke with red lint, pp. 80, 81; Michael's lying, p. 174; Koko's lying, pp. 182, 183; "Time bye good bye," p. 91; what Koko has in common with humans, pp. 187, 188; about death, pp. 190, 191. Other quotations and exchanges with Koko: how to insult is from Hiller, 1984; the exchange with Wilkinson is from Wilkinson, 1984; Koko with the mirror, from Patterson; "Fine animal gorilla," from Patterson, Cornwall, and Share. The quoted remarks of critics are from Gorney, 1985, p. B4; the description of Koko crying at the news of All Ball's death is from the same source. Chomsky is quoted in Crail, p. 70; I've not been able to find the original quotation.

Works Cited

Aken, Kron Mide, and Michael Kavanagh. 1983. "Species Conservation Priorities in the Tropical Forests of Sarawak, Malaysia." Unpublished manuscript.

Alvim, Paulo de. 1977. "The Balance between Conservation and Utilization in the Humid Tropics with Special Reference to Amazonian Basin." In *Extinction Is Forever,* ed. Ghillean T. Prance and Thomas S. Elias, pp. 347–352. New York: New York Botanical Garden.

Anderson, Norman D., and Walter R. Brown. 1984. *Lemurs.* New York: Dodd, Mead.

Arditti, Joseph. 1966. "Orchids." *Scientific American* 214 (Jan.): 70–78.

Asibey, Emmanuel O. A. 1972. "Ghana's Progress." *Oryx* 11 (Sept.): 470–475.

———. 1974. "Wildlife as a Source of Protein in Africa South of the Sahara." *Biological Conservation* 6 (Jan.): 32–39.

———. 1978. "Primate Conservation in Ghana." In *Recent Advances in Primatology: Conservation,* vol. 2, ed. D. J. Chivers and W. Lane-Petter, pp. 55–74. New York: Academic Press.

Bachrach, Max. 1953. *Fur: A Practical Treatise,* 3d ed. New York: Prentice-Hall.

Barnes, Donald. 1982. "Seventeen Monkeys Seized in Raid on Laboratory." *International Primate Protection League Newsletter* 9 (Jan.): 11, 12.

Barrett, Suzanne W. 1980. "Conservation in Amazonia." *Biological Conservation* 18: 209–235.

Bates, Henry Walter. 1863. *The Naturalist on the River Amazons, A Record of Adventures, Habits of Animals, Sketches of Brazilian and Indian Life, and Aspects of Nature Under the Equator, During Eleven Years of Travel.* 2 vols. London: John Murray.

Battiata, Mary. 1986A. "The Gorillas' Friend and Her Grisly Death." *Washington Post,* Jan. 24, pp. D1, D8.

———. 1986B. "Murder and the Private War." *Washington Post,* Jan. 25, pp. C1, C6, C7.

Begley, Sharon, and Bob Cohn. 1986. "The Silent Summer." *Newsweek,* June 23, pp. 64–66.

Begley, Sharon, Gerald C. Lubenow, and Mark Miller. 1986. "Silent Spring Revisited?" *Newsweek,* July 14, pp. 72, 73.

Bomford, Elizabeth. 1981. "On the Road to Nosy Mangabe." *International Wildlife* 2 (Jan.–Feb.): 20–24.

Booth, A. H. 1956. "The Distribution of Primates in the Gold Coast." *Journal of the West African Science Association* 2 (Aug.): 122–133.

———. 1958. "The Zoogeography of West African Primates: A Review." *Bulletin de l'Institut Français D'Afrique Noire* 20A (Apr.): 587–622.

Borner, Markus. 1976. "Sumatra's Orang-utans." *Oryx* 13 (Feb.): 290–293.

Bourne, Geoffrey H. 1974. *Primate Odyssey*. New York: G. P. Putnam's Sons.

Brambell, M. R. 1975. "Breeding Orang-Utans." In *Breeding Endangered Species in Captivity*, ed. R. D. Martin, pp. 235–243. New York: Academic Press.

Brown, Lester R. 1984. "Conserving Soils." In *State of the World 1984*, ed. Lester R. Brown and others, pp. 53–73. New York: W. W. Norton.

Butynski, Thomas M. 1985. "Primates and their Conservation in the Impenetrable (Bwindi) Forest, Uganda." *Primate Conservation* 6 (Jul.): 68–72.

Caldecott, Julian Oliver, and Michael Kavanagh. 1984. "Use of Primates and Captive Breeding Programs Outside the United States." In *The International Primate Trade,* vol. 1, ed. David Mack and Russell A. Mittermeier, pp. 137–152. Washington, D.C.: Traffic (U.S.A.).

Camacho, Jorge Hernandez, and Thomas R. Defler. 1985. "Some Aspects of the Conservation of Non-human Primates in Colombia." *Primate Conservation* 6: (July): 42–50.

Campbell, Sheldon. 1980. "Is Reintroduction a Realistic Goal?" In *Conservation Biology: An Evolutionary-Ecological Perspective*, ed. Michael E. Soulé and Bruce A. Wilcox, pp. 263–269. Sunderland, Mass.: Sinauer Associates.

Caras, Roger A. 1966. *Last Chance on Earth: A Requiem for Wildlife*. New York: Chilton Books.

Carson, Hampton L. 1983. "The Genetics of the Founder Effect." In *Genetics and Conservation: A Reference for Managing Wild Animals and Plant Populations,* ed. Christine M. Schonewald-Cox and others, pp. 189–200. Menlo Park, Calif.: Benjamin/Cummings.

Caufield, Catherine. 1985. *In the Rainforest*. New York: Alfred A. Knopf.

Cerquera, Jairo Ramirez. 1985. "S.O.S. for the Cotton-top Tamarin (*Saguinus oedipus*)." *Primate Conservation* 6 (July): 17–19.

Chambers, Steven M. 1983. "Genetic Principles for Managers." In *Genetics and Conservation: A Reference for Managing Wild Animal and Plant Populations,* ed. Christine M. Schonewald-Cox and others, pp. 15–46. Menlo Park, Calif.: Benjamin-Cummings.

Chen, Allan. 1987. "Unraveling Another Mayan Mystery." *Discover* 8 (June): 40–46.

Chevalier-Skolnikoff, Suzanne. 1981. "The Clever Hans Phenomenon, Cuing, and Ape Signing: A Piagetian Analysis of Methods for Instructing Animals." In *The Clever Hans Phenomenon: Communication with Horses, Whales, Apes, and People,* ed. Thomas A. Sebok and Robert Rosenthal, pp. 60–93. New York: New York Academy of Sciences.

Chivers, David. 1977. "The Lesser Apes." In *Primate Conservation,* ed. Prince Rainier III and Geoffrey Bourne, pp. 539–598. New York: Academic Press.

Clapp, Neal K., and Suzette D. Tardif. 1985. "Marmoset Husbandry and Nutrition." *Digestive Diseases and Sciences* 30 (Dec.): 17S–23S.

Clifton, Tony, and Ron Moreau. 1985. "A Wounded Land." *Newsweek,* Apr. 15, pp. 58–61.

Coe, Malcolm. 1980. "African Wildlife Resources." In *Conservation Biology: An*

Evolutionary-Ecological Perspective, ed. Michael E. Soulé and Bruce A. Wilcox, pp. 273–302. Sunderland, Mass.: Sinauer Associates.

Cohn, Ron. 1984. "Background and History of the Gorilla Language Research Project." *Gorilla* 7 (June): 8.

Coimbra-Filho, Adelmar F., and Russell A. Mittermeier. 1977. "Conservation of the Brazilian Lion Tamarins (*Leontopithecus rosalia*)." In *Primate Conservation,* ed. Prince Rainier III and Geoffrey H. Bourne, pp. 59–91. New York: Academic Press.

———. 1983. "Update on the Rio de Janeiro Primate Center." *IUCN/SSC Primate Specialist Group Newsletter* 3 (Mar.): 36, 37.

Cole, Marilyn, and others. 1979. "Notes on the Early Hand-rearing of an Orang-utan *Pongo pygmaeus* and Its Subsequent Reintroduction to the Mother." In *1979 International Zoo Yearbook,* vol. 19, ed. P. J. S. Olney, pp. 263, 264. London: Zoological Society of London.

Committee on Selected Biological Problems in the Humid Tropics. 1982. *Ecological Aspects of Development in the Humid Tropics.* Washington, D.C.: National Academy Press.

Constable, Isabel D., and others. 1985. "Sightings of Aye-Ayes and Red Ruffed Lemurs on Nosy Mangabe and the Masoala Peninsula." *Primate Conservation* 5 (Jan.): 59–62.

Constable, J. D. 1982. "Visit to Vietnam." *Oryx* 16 (Feb.): 249–254.

Conway, William G. 1980. "An Overview of Captive Propagation." In *Conservation Biology: An Evolutionary-Ecological Perspective,* ed. Michael E. Soulé and Bruce A. Wilcox, pp. 199–208. Sunderland, Mass.: Sinauer Associates.

Cousins, Don. 1978A. "Gorillas — A Survey." *Oryx* 14 (June): 254–258.

———. 1978B. "Gorillas — A Survey." *Oryx* 14 (Nov.): 374–376.

———. 1978C. "Man's Exploitation of the Gorilla." *Biological Conservation* 13 (June): 287–297.

Crail, Ted. 1983. *Apetalk & Whalespeak: The Quest for Interspecies Communication.* Chicago: Contemporary Books.

Curry-Lindahl, Kai. 1964. "The Congo National Parks since Independence." *Oryx* 7 (Aug.): 233–239.

Da Fonseca, Gustavo A. B. 1985. "Observations on the Ecology of the Muriqui (*Brachyteles arachnoides* E. Geoffroy 1806): Implications for Its Conservation." *Primate Conservation* 5 (Jan.): 48–52.

Dandelot, P. 1971. "Order Primates." In *The Mammals of Africa: An Identification Manual,* ed. J. Meester and H. W. Setzer. Washington, D.C.: Smithsonian Institution Press.

Darwin, Charles. 1859, 1871. *The Origin of Species and The Descent of Man.* New York: Random House (Modern Library).

De Assumpçao, C. Torres. 1983. "Ecological and Behavioural Information on *Brachyteles arachnoides.*" *Primates* 24 (Oct.): 584–593.

Denis, Armand. 1963. *On Safari: The Story of My Life.* New York: E. P. Dutton.

Diamond, Jared M. 1975. "The Island Dilemma: Lessons of Modern Biogeographic Studies for the Design of Natural Reserves." *Biological Conservation* 7: 129–146.

———. 1980. "Patchy Distributions of Tropical Birds." In *Conservation Biology: An Evolutionary-Ecological Perspective,* ed. Michael E. Soulé and Bruce A. Wilcox, pp. 57–74. Sunderland, Mass.: Sinauer Associates.

Dickinson, J. C., III. 1972. "Alternatives to Monoculture in the Humid Tropics of Latin America." *Professional Geographer* 24 (Aug.): 217–222.

Dickinson, Robert E. 1981. "Effects of Tropical Deforestation on Climate." In *Blowing in the Wind: Deforestation and Long-range Implications,* ed. Vinson H. Sutlive, Nathan Altshuler, and Mario D. Zamora, pp. 411–442. Williamsburg, Va.: Department of Anthropology, College of William and Mary.

Dietz, Lou Ann. 1984. "Golden Lion Tamarins Reintroduced to Wild!" *Focus* 6 (July/Aug.): 6.

Diole, Philippe. 1974. *The Errant Ark: Man's Relationship with Animals.* New York: G. P. Putnam's Sons.

Doherty, Shawn, and Vern E. Smith. 1984. "The Scandal of Atlanta's Zoo." *Newsweek,* June 18, p. 41.

Dosso, Henri, Jean Lois Guillaumet, and Malcolm Hadley. 1981. "The Tai Project: Land Use Problems in a Tropical Rain Forest." *Ambio* 10: 120–125.

Dowling, Harry F. 1977. *Fighting Infection: Conquests of the Twentieth Century.* Cambridge, Mass.: Harvard University Press.

Du Chaillu, Paul B. 1861. *Explorations & Adventures in Equatorial Africa: With Accounts of the Manners and Customs of the People, and of the Chace of the Gorilla, Crocodile, Leopard, Elephant, Hippopotamus, and Other Animals.* London: John Murray.

"The Duke University Primate Center." n.d. Publication of the center.

Durrell, Gerald. 1975. "Foreword." In *Breeding Endangered Species in Captivity,* ed. R. D. Martin, pp. vii–xii. New York: Academic Press.

Eckholm, Erik. 1985. "Will There Be Enough Chimps for Research?" *New York Times,* Nov. 19, pp. C1, C3.

Ehrlich, Paul R. 1980. "The Strategy of Conservation, 1980–2000." In *Conservation Biology: An Evolutionary-Ecological Perspective,* ed. Michael E. Soulé and Bruce A. Wilcox, pp. 329–344. Sunderland, Mass.: Sinauer Associates.

Ehrlich, Paul, and Anne Ehrlich. 1981. *Extinction.* New York: Random House.

Eimerl, Sarel, and Irven DeVore. 1965. *The Primates.* New York: Time (Life Nature Library).

Erickson, R. C. 1975. "Captive Breeding of Whooping Cranes at the Patuxent Wildlife Research Center." In *Breeding Endangered Species in Captivity,* ed. R. D. Martin, pp. 99–114. New York: Academic Press.

Eudey, Ardith, and David Mack. 1984. "Use of Primates and Captive Breeding Programs in the United States." In *The International Primate Trade,* vol. 1, ed. David Mack and Russell A. Mittermeier, pp. 153–180. Washington, D.C.: Traffic (U.S.A.).

Fairlie, Henry. 1979. "Conservation — The False Issue." *New Republic,* Oct. 20, pp. 17–20.

Fisher, James, Noel Simon, and Jack Vincent. 1969. *The Red Book: Wildlife in Danger.* London: Collins.

Flavin, Christopher. 1988. "The Heat Is On." *World Watch* 1 (Dec.): 10–20.

Foose, Thomas J. 1983. "The Relevance of Captive Populations to the Conservation of Biotic Diversity." In *Genetics and Conservation: A Reference for Managing Wild Animal and Plant Populations,* ed. Christine M. Schonewald-Cox and others, pp. 374–401. Menlo Park, Calif.: Benjamin-Cummings.

———. n.d. "The Species Survival Plan (SSP) of the AAZPA." Unpublished manuscript.
Foose, Thomas J., and William G. Conway. 1982. "Models for Population Management of Lion-Tailed Macaque Resources in Captivity (a Working Paper)." Unpublished manuscript.
Fossey, Dian. 1983. *Gorillas in the Mist.* Boston: Houghton Mifflin.
Franklin, Ian Robert. 1980. "Evolutionary Change in Small Populations." In *Conservation Biology: An Evolutionary-Ecological Perspective,* ed. Michael E. Soulé and Bruce A. Wilcox, pp. 135–149. Sunderland, Mass.: Sinauer Associates.
Fyfe, R. 1975. "Breeding Peregrine and Prairie Falcons in Captivity." In *Breeding Endangered Species in Captivity,* ed. R. D. Martin, pp. 133–141. New York: Academic Press.
Gallup, Gordon G., Jr. 1970. "Chimpanzees: Self-Recognition." *Science* 167 (Jan.–Mar.): 86, 87.
Gardner, R. Allen, and Beatrice T. Gardner. 1969. "Teaching Sign Language to a Chimpanzee." *Science* 165 (July–Sept.): 664–672.
Glatston, Angela R. 1981. "The Husbandry, Breeding and Hand-Rearing of the Lesser Mouse Lemur *Microcebus murinus* at Rotterdam Zoo." In *1981 International Zoo Yearbook,* vol. 21, ed. P. J. S. Olney, pp. 131–137. London: Zoological Society of London.
Gledhill, Laurence G. 1984. "SSP Coordinator's Report." *Lion-Tales: Lion-Tailed Macaque Newsletter* 1 (Winter): 1, 2.
———. 1985. "The History and Status of the Lion-tailed Macaque in North America." *Primate Conservation* 5 (Jan.): 38, 39.
Gochfeld, Michael. 1974. "Douc Langurs." *Nature* 247 (Jan.): 167.
Goodall, Alan G. 1978. "On Habitat and Home Range in Eastern Gorillas in Relation to Conservation." In *Recent Advances in Primatology: Conservation,* vol. 2, ed. D. J. Chivers and W. Lane-Petter, pp. 81–83. New York: Academic Press.
Goodall, Alan G., and Colin P. Groves. 1977. "The Conservation of Eastern Gorillas." In *Primate Conservation,* ed. Prince Rainier III and Geoffrey Bourne, pp. 599–637. New York: Academic Press.
Goodwin, Harry A., and others. 1978. *Red Data Book,* vol. 1. Morges, Switzerland: International Union for the Conservation of Nature.
Gorney, Cynthia. 1985. "When the Gorilla Speaks." *Washington Post,* Jan. 31, pp. B1, B4.
Green, Steven, and Karen Minkowski. 1977. "The Lion-Tailed Monkey and Its South Indian Rain Forest Habitat." In *Primate Conservation,* ed. Prince Rainier III and Geoffrey Bourne, pp. 289–337. New York: Academic Press.
Greenland, D. J. 1975. "Bringing the Green Revolution to the Shifting Cultivator." *Science* 190 (Nov.): 841–844.
Griffin, Donald R. 1981. *The Question of Animal Awareness.* New York: Rockefeller University Press.
———. 1984. *Animal Thinking.* Cambridge, Mass.: Harvard University Press.
Groom, A. F. G. 1973. "Squeezing Out the Mountain Gorilla." *Oryx* 12 (Oct.): 207–215.
Groombridge, Brian. 1984. "A New Locality for the Lion-tailed Macaque." *Oryx* 18 (July): 144–147.

Grzimek, Bernhard, and others, ed. 1968. *Grzimek's Animal Life Encyclopedia,* vol. 10. New York: Van Nostrand Reinhold.

Hall, Elizabeth, and Max Hall. 1964. *About the Exhibits,* 3d ed. Cambridge, Mass.: Museum of Comparative Zoology, Harvard University.

Happold, David. 1972. "Mammals in Nigeria." *Oryx* 11 (Sept.): 469.

Harcourt, A. H. 1977. "Virunga Gorillas — The Case against Translocations." *Oryx* 13 (Feb.): 469–472.

Harcourt, A. H., and A. F. G. Groom. 1972. "Gorilla Census." *Oryx* 11 (May): 355–363.

Harding, Robert S. O. 1984. "Primates of the Kilimi Area, Northwest Sierra Leone." *Folia Primatologica* 42: 96–114.

Harrisson, Barbara. 1963. *Orang-utan.* Garden City, N.Y.: Doubleday.

Hatton, John, Nick Smart, and Roy Thomson. 1984. "In Urgent Need of Protection — Habitat for Woolly Spider Monkey." *Oryx* 18 (Jan.): 24–29.

Henderson-Sellers, A. 1977. "The Effect of Land Clearance and Agricultural Practices upon Climate." In *Blowing in the Wind: Deforestation and Long-range Implications,* ed. Vinson H. Sutlive, Nathan Altshuler, and Mario D. Zamora, pp. 443–485. Williamsburg, Va.: Department of Anthropology, College of William and Mary.

Herrera, Rafael, and others. 1981. "How Human Activities Disturb the Nutrient Cycles of a Tropical Rainforest in Amazonia." *Ambio* 10: 109–114.

Heseltine, Nigel. 1971. *Madagascar.* New York: Praeger Publishers.

Hick, U. 1975. "Breeding and Maintenance of Douc Langurs at Cologne Zoo." In *Breeding Endangered Species in Captivity,* ed. R. D. Martin, pp. 223–233. New York: Academic Press.

Hiller, Barbara. 1984. "Conversations with Koko." *Gorilla* 8 (Dec.): 8, 9.

Homan, W. G. 1975. "Breeding the International Herd of Arabian Oryx at Phoenix Zoo." In *Breeding Endangered Species in Captivity,* ed. R. D. Martin, pp. 285–292. New York: Academic Press.

Hornaday, William T. 1885. *Two Years in the Jungle.* New York: Charles Scribner's Sons.

Humphreys, John. 1982. "Plants That Bring Health — or Death." *New Scientist,* Feb. 25, pp. 513–516.

Hutton, Angus F. 1949. "Notes on the Snakes and Mammals of the High Wavy Mountains, Madura District, South India." *Journal of the Bombay Natural History Society* 48 (Dec.): 691–694.

Inskipp, Tim, and Sue Wells. 1979. *International Trade in Wildlife.* London: Earthscan.

International Primate Protection League Newsletter 1–12. 1977–1985. Anonymous or unattributed articles from this newsletter are not additionally listed in this bibliography. Such articles are listed in the Notes under a short version of the article's title, followed by *IPPL Newsletter,* and the appropriate issue and page reference.

IUCN. 1982A. *A Conservation Strategy for the Great Apes.* IUCN Conservation Monitoring Centre / Global Environment Monitoring System / IUCN-SSC Primate Specialist Group / World Wildlife Fund / United Nations Environmental Programme.

———. 1982B. *A Conservation Strategy for Threatened Central and South American*

Primates. IUCN Conservation Monitoring Centre / Global Environment Monitoring System / IUCN-SSC Primate Specialist Group / World Wildlife Fund / United Nations Environmental Programme.

"Ivory Coast Poaching Ring." 1980. *TRAFFIC Bulletin* 2: 88.

Jeffrey, Sonia M. 1970. "Ghana's Forest Wildlife in Danger. *Oryx* 10 (May): 240–243.

———. 1975A. "Ghana's New Forest National Park." *Oryx* 13 (Apr.): 34–36.

———. 1975B. "Notes on Mammals from the High Forest of Western Ghana (Excluding Insectivora)." *Bulletin de l'Institut Fondamental d'Afrique Noire* 30: 950–973.

Johns, Andrew. 1985. "Current Status of the Southern Bearded Saki (*Chiropotes satanas satanas*)." *Primate Conservation* 5 (Jan.): 28.

Johns, Andrew, and J. M. Ayres. 1987. "Southern Bearded Sakis beyond the Brink." *Oryx* 21 (July): 164–167.

Johnsingh, A. J. T. 1984. "A Threat to Kalakadu Wildlife Sanctuary." *Hornbill (Bombay Natural History Society Newsletter)* 3: 17–19.

Jolly, Alison. 1987. "Madagascar: A World Apart." *National Geographic* 171 (Feb.): 149–183.

Jones, Marvin L. 1982. "The Orang-utan in Captivity." In *The Orang utan: Its Biology and Conservation,* ed. Leobert E. M. De Boer, pp. 17–37. The Hague: Dr. W. Junk.

Jones, T. S. 1970. "Notes on the Commoner Sierra Leone Mammals." *Journal of the Nigerian Field Society* 35 (Oct.): 4–18.

Karanth, K. Ullas. 1985. "Ecological Status of the Lion-tailed Macaque and Its Rainforest Habitats in Karnataka, India." *Primate Conservation* 6 (July): 73–84.

Karnow, Stanley. 1983. *Vietnam: A History*. New York: Viking Press.

Karr, James R. 1973. "Ecological and Behavioural Notes on the Liontailed Macaque (*Macaca silenus*) in South India." *Journal of the Bombay Natural History Society* 70 (Apr.): 191, 192.

Kartawinata, Kuswata, and others. 1981. "The Impact of Man on a Tropical Forest in Indonesia." *Ambio* 10: 115–119.

Kavanagh, Michael. 1972. "Food-Sharing Behavior within a Group of Douc Monkeys (*Pygathrix nemaeus nemaeus*)." *Nature* 239 (Oct.): 406, 407.

———. 1983A. *A Complete Guide to Monkeys, Apes and Other Primates*. New York: Viking Press.

———. 1983B. "The Role of Sarawak's National Parks and Wildlife Sanctuaries." *Planter, Kuala Lumpur* 59: 507–512.

———. 1984. "A Review of the International Primate Trade." In *The International Primate Trade,* vol. 1, ed. David Mack and Russell A. Mittermeier, pp. 49–89. Washington, D.C.: Traffic (U.S.A.).

Kavanagh, Michael, and Elizabeth Bennett. 1984. "A Synopsis of Legislation and the Primate Trade in Habitat and User Countries." In *The International Primate Trade,* vol. 1, ed. David Mack and Russell A. Mittermeier, pp. 19–48. Washington, D.C.: Traffic (U.S.A.).

Kear, J. 1975. "Breeding of Endangered Wildfowl as an Aid to Their Survival," and "Returning the Hawaiian Goose to the Wild." In *Breeding Endangered Species in Captivity,* ed. R. D. Martin, pp. 49–60, 115–123. New York: Academic Press.

Kevles, Bettyann. 1976. *Watching the Wild Apes: The Primate Studies of Goodall, Fossey, and Galdikas*. New York: E. P. Dutton.

King, F. Wayne. 1984. "Preservation of Genetic Diversity." In *Sustaining Tomorrow: A Strategy for World Conservation and Development*, ed. Francis R. Thibodeau and Hermann H. Field, pp. 41–55. Hanover, N.H.: University Press of New England.

Kleiman, Devra G. 1977. "Characteristics of Reproduction and Sociosexual Interactions in Pairs of Lion Tamarins (*Leontopithecus rosalia*) during the Reproductive Cycle." In *The Biology and Conservation of the Callitrichidae*, ed. by Devra G. Kleiman, pp. 181–190. Washington, D.C.: Smithsonian Institution Press.

———. 1980. "The Sociobiology of Captive Propagation." In *Conservation Biology: An Evolutionary-Ecological Perspective*, ed. Michael E. Soulé and Bruce A. Wilcox, pp. 243–261. Sunderland, Mass.: Sinauer Associates.

———. 1984. "The National Zoo's Role in an International Primate Conservation Program." *IUCN/SSC Primate Specialist Group Newsletter* 4 (Mar.): 45, 46.

Kleiman, Devra G., Jonathan D. Ballou, and Ronald F. Evans. 1982. "An Analysis of Recent Reproductive Trends in Captive Golden Lion Tamarins (*Leontopithecus r. rosalia*) with Comments on Their Future Demographic Management." In *1982 International Zoo Yearbook*, vol. 22, ed. P. J. S. Olney, pp. 94–101. London: Zoological Society of London.

Kleiman, Devra G., and Marvin Jones. 1977. "The Current Status of *Leontopithecus rosalia* in Captivity with Comments on Breeding Success at the National Zoological Park." In *The Biology and Conservation of the Callitrichidae*, ed. Devra G. Kleiman, pp. 215–218. Washington, D.C.: Smithsonian Institution Press.

Knox, Richard A. 1984. "Ulcerative Colitis Clue Found." *Boston Globe*, Dec. 17, pp. 17, 22.

Kohl, Larry. 1982. "Père David's Deer Saved from Extinction." *National Geographic* 162 (Oct.): 478–485.

Kreig, Margaret B. 1964. *Green Medicine: The Search for Plants That Heal*. New York: Rand McNally.

Krishnan, M. 1971. "An Ecological Survey of the Larger Mammals of Peninsular India." *Journal of the Bombay Natural History Society* 68 (Dec.): 503–555.

Kummer, Hans. 1971. *Primate Societies: Group Techniques of Ecological Adaptation*. Chicago: Aldine Atherton.

Lessem, Don. 1985. "Wildlife: How Big an 'Ark'?" *Boston Globe*, Apr. 29, pp. 37, 40.

———. 1986. "Big Dams: A Flood of Concerns." *Boston Globe*, Mar. 17, pp. 45, 46.

Lindburg, Don. 1980. "Status and Captive Reproduction of the Lion-tailed Macaque." In *1980 International Zoo Yearbook*, vol. 20, ed. P. J. S. Olney, pp. 60–64. London: Zoological Society of London.

Lippold, Lois K. 1977. "The Douc Langur: A Time for Conservation." In *Primate Conservation*, ed. Prince Rainier III and Geoffrey H. Bourne, 513–538. New York: Academic Press.

Liu, Edwin H., and Mary Jo W. Godt. 1983. "The Differentiation of Populations over Short Distances." In *Genetics and Conservation: A Reference for*

Managing Wild Animal and Plant Populations, ed. Christine M. Schonewald-Cox and others, pp. 78–95. Menlo Park, Calif.: Benjamin/Cummings.
Long, James. 1982A. "Monkey Dealer Agreeable to Trade Ban Mediation." *Oregon Journal,* Mar. 17.
———. 1982B. "Polio Vaccine Maker Denies Supply Problem." *Oregon Journal,* Mar. 22.
———. 1982C. "Sen. Packwood Comes to Aid of Monkey Business." *Oregon Journal,* Mar. 16.
Lovejoy, Thomas E. 1980. "Discontinuous Wilderness: Minimum Areas for Conservation." *Parks* 5: 13–15.
———. 1982. "The Tropical Forest — Greatest Expression of Life on Earth." In *Primates and the Tropical Forest,* ed. Russell A. Mittermeier and Mark J. Plotkin, pp. 45–48. Pasadena, Calif.: California Institute of Technology.
Lowes, R. H. G. 1970. "Destruction in Sierra Leone." *Oryx* 10 (Sept.): 279–310.
Lusigi, Walter J. 1981. "New Approaches to Wildlife Conservation in Kenya." *Ambio* 10: 87–92.
MacArthur, Robert H. 1972. *Geographical Ecology: Patterns in the Distribution of Species.* New York: Harper & Row.
MacArthur, Robert H., and Edward O. Wilson. 1967. *The Theory of Island Biogeography.* Princeton, N.J.: Princeton University Press.
Mack, David, and Ardith Eudey. 1984. "A Review of the U.S. Primate Trade." In *The International Primate Trade,* vol. 1, ed. David Mack and Russell A. Mittermeier, pp. 91–136. Washington, D.C.: Traffic (U.S.A.).
Mack, David, and Russell A. Mittermeier. 1984A. "The International Primate Trade: Summary, Update and Conclusions." In *The International Primate Trade,* vol. 1, ed. David Mack and Russell A. Mittermeier, pp. 181–185. Washington, D.C.: Traffic (U.S.A.).
———. 1984B. "Introduction." In *The International Primate Trade,* vol. 1, ed. David Mack and Russell A. Mittermeier, pp. 15–17. Washington, D.C.: Traffic (U.S.A.).
MacKinnon, John. 1971. "The orang-utan in Sabah Today." *Oryx* 11 (May): 141–191.
———. 1973. "Orang-utans in Sumatra." *Oryx* 12 (May): 234–242.
———. 1977. "The Future of Orang-utans." *New Scientist* (June): 697–699.
Mager, Wim. 1981. "Stimulating Maternal Behavior in the Lowland Gorilla *Gorilla g. gorilla* at Apeldoorn." In *1981 International Zoo Yearbook,* vol. 21, ed. P. J. S. Olney, pp. 138–143. London: Zoological Society of London.
"Malaria." 1986. *Harvard Medical School Health Letter* 11 (Jan.): 5–7.
Mallinson, Jeremy. 1984A. "Golden-Headed Lion Tamarin Contraband a Major Conservation Problem." *IUCN/SSC Primate Specialist Group Newsletter* 4 (Mar.): 23–25.
———. 1984B. "Lion Tamarins' Survival Hangs in Balance." *Oryx* 18 (Apr.): 72–78.
Marler, Peter. 1969. "*Colobus guereza*: Territoriality and Group Composition." *Science* 163 (Jan.–Mar.): 93–95.
Martin, R. D. 1975A. "General Principles for Breeding Small Mammals in Captivity." In *Breeding Endangered Species in Captivity,* ed. R. D. Martin, pp. 143–166. New York: Academic Press.

———. ed. 1975B. *Breeding Endangered Species in Captivity*. New York: Academic Press.

Maynes, G. M. 1975. "Breeding the Parma Wallaby in Captivity." In *Breeding Endangered Species in Captivity,* ed. R. D. Martin, pp. 167–170. New York: Academic Press.

McGiffin, Heather. 1985. "History of Primate Conservation in Sierra Leone," and "A New National Park in Sierra Leone." *International Primate Protection League Newsletter* 12 (Dec.): 2–3, 3–5.

Mendel, Gregor. 1965 (1926). *Experiments in Plant Hybridization,* ed. J. H. Bennett. London: Oliver & Boyd.

Merfield, Fred G. 1954. "The Gorilla of the French Cameroons." *Zoo Life* 9 (Autumn): 84–94.

Merfield, Fred G., and Harry Miller. 1956. *Gorilla Hunter*. New York: Farrar, Straus and Cudahy.

Miller, Gerrit S., Jr. 1903. "Seventy New Malayan Mammals." *Smithsonian Miscellaneous Collections* 45 (Nov.), no. 1420.

Milton, Katherine, and Carlos de Lucca. 1984. "Population Estimate for *Brachyteles* at Fazenda Barreiro Rico, Sao Paulo State, Brazil." *IUCN/SSC Primate Specialist Group Newsletter* 4 (Mar.): 27, 28.

Mittermeier, Russell A. 1973. "Colobus Monkeys and the Tourist Trade." *Oryx* 12 (May): 113–117.

———. 1982. "The World's Endangered Primates: An Introduction and a Case Study — The Monkeys of Brazil's Atlantic Forests." In *Primates and the Tropical Rainforest,* ed. Russell A. Mittermeier and Mark J. Plotkin, pp. 11–22. Pasadena, Calif.: California Institute of Technology.

———. 1987. "Monkey in Peril." *National Geographic* 171 (Mar.): 387–395.

Mittermeier, Russell A., and Adelmar F. Coimbra-Filho. 1977. "Primate Conservation in Brazilian Amazonia." In *Primate Conservation,* ed. Prince Rainier III and Geoffrey H. Bourne, pp. 117–166. New York: Academic Press.

Mittermeier, Russell A., and John F. Oates. 1985. "Primate Diversity: The World's Top Countries." *Primate Conservation* 5 (Jan.): 41–48.

Mittermeier, Russell A., Celio Valle, and Adelmar F. Coimbra-Filho. 1985. "Update on the Muriqui." *Primate Conservation* 5 (Jan.): 28–30.

Mittermeier, Russell A., and others. 1982. "Conservation of Primates in the Atlantic Forest Region of Eastern Brazil." In *1982 International Zoo Yearbook,* vol. 22, ed. P. J. S. Olney, pp. 2–17. London: Zoological Society of London.

Mohnot, S. M. 1978. "The Conservation of Non-Human Primates in India." In *Recent Advances in Primatology,* vol. 2, ed. D. J. Chivers and W. Lane-Petter, pp. 47–53. New York: Academic Press.

Morgan, Lewis H. 1868. *The American Beaver and His Works*. Philadelphia: J. B. Lippincott.

Morris, Ramona, and Desmond Morris. 1966. *Men and Apes*. New York: McGraw-Hill.

Myers, Norman. 1979. *The Sinking Ark: A New Look at the Problem of Disappearing Species*. New York: Pergamon Press.

———. 1980A. *Conversion of Tropical Moist Forests*. Washington, D.C.: National Academy of Sciences.

———. 1980B. "The Problem of Disappearing Species: What Can Be Done?" *Ambio* 9: 229–235.

———. 1981. "The Hamburger Connection: How Central America's Forests Become North America's Hamburgers." *Ambio* 10: 3–8.

———. 1983. *A Wealth of Wild Species: Storehouse for Human Welfare*. Boulder, Colo.: Westview Press.

———. 1984A. "Genetic Resources in Jeopardy." *Ambio* 13: 171–174.

———. 1984B. *The Primary Source: Tropical Forests and Our Future*. New York: W. W. Norton.

———. 1985. "Tropical Deforestation and Species Extinctions." *Futures* 17 (Oct.): 451–463.

Napier, J. R., and P. H. Napier, eds. 1967. *A Handbook of Living Primates*. New York: Academic Press.

Neyman, Patricia A. 1977. "Aspects of the Ecology and Social Organization of Free-ranging Cotton-Top Tamarins (*Saguinus oedipus*) and the Conservation Status of the Species." In *The Biology and Conservation of the Callitrichidae,* ed. Devra G. Kleiman, pp. 39–71. Washington, D.C.: Smithsonian Institution Press.

"1988 World Population Data Sheet." 1988. Washington, D.C.: Population Reference Bureau.

Oates, John F. 1977. "The Guereza and Man." In *Primate Conservation,* ed. Prince Rainier III and Geoffrey Bourne, pp. 419–467. New York: Academic Press.

———. 1982. "In Search of Rare Forest Primates in Nigeria." *Oryx* 16 (Oct.): 431–436.

Olby, Robert C. 1966. *Origins of Mendelism*. London: Constable.

Oldfield, Margery L. 1981. "Tropical Deforestation and Genetic Resources Conservation." In *Blowing in the Wind: Deforestation and Long-range Implications,* ed. Vinson H. Sutlive, Nathan Altshuler, and Mario D. Zamora, pp. 277–346. Williamsburg, Va.: Department of Anthropology, College of William and Mary.

Olney, P. J. S., ed. 1984. *1983 International Zoo Yearbook,* vol. 23. London: Zoological Society of London.

Oren, David C. 1987. "Grande Carajás, International Financing Agencies, and Biological Diversity in Southeastern Brazilian Amazonia." *Conservation Biology* 1 (Oct.): 222–227.

Orians, Gordon H., and E. W. Pfeiffer. 1970. "Ecological Effects of the War in Vietnam." *Science* 168 (Apr.–June): 544–554.

Padua, Maria Tereza Jorge, Alceo Magnanini, and Russell A. Mittermeier. 1974. "Brazil's National Parks." *Oryx* 12 (June): 452–464.

Patterson, Francine. 1984. "Self-Recognition by *Gorilla gorilla gorilla*." *Gorilla* 7 (June): 2, 3.

Patterson, Francine, Claudia Cornwall, and Elizabeth Share. 1984. "Interspecies Communication and Conservation." *Gorilla* 8 (Dec.): 2–4.

Patterson, Francine, and Eugene Linden. 1981. *The Education of Koko*. New York: Holt, Rinehart and Winston.

Pease, Theresa. 1987. "Nutrition Professor Studies Colon Cancer." *Tufts Journal* 9 (Oct.): 4.

Perry, John. 1976. "Orang-utans in Captivity." *Oryx* 13 (Feb.): 262–264.

Peterson, Cass. 1987. "U.S., Soviets Urged to Join in Saving Rain Forests." *Washington Post*, Apr. 24, p. A21.

Petter, J.-J., and A. Peyrieras. 1974. "A Study of Population Density and

Home Ranges of *Indri indri* in Madagascar." In *Prosimian Biology,* ed. R. D. Martin, G. A. Doyle, and A. C. Walker, pp. 39–48. Pittsburgh: University of Pittsburgh Press.

Pfeiffer, E. W. 1973. "Post-War Vietnam." *Environment* 15 (Nov.): 29–33.

Pirie, N. W. 1967. "Orthodox and Unorthodox Methods of Meeting World Food Needs." *Scientific American* 216 (Feb.): 27–35.

Pollock, J. I. 1975. "Field Observations on *Indri indri:* A Preliminary Report." *Lemur Biology,* ed. Ian Tattersall and Robert W. Sussman, pp. 287–311. New York: Plenum Press.

Postel, Sandra. 1984. "Protecting Forests." In *State of the World 1984,* ed. Lester R. Brown and others, pp. 74–94. New York: W. W. Norton.

Pratt, Dallas. 1985. "Necessary Fuss." *International Primate Protection League Newsletter* 12 (Aug.): 5–9.

Premack, Ann James, and David Premack. 1972. "Teaching Language to an Ape." *Scientific American* 227 (Oct.): 92–99.

Premack, David, and Ann James Premack. 1983. *The Mind of an Ape.* New York: W. W. Norton.

Ralls, Katherine, and Jonathan Ballou. 1983. "Extinction: Lessons from Zoos." In *Genetics and Conservation: A Reference for Managing Wild Animal and Plant Populations,* ed. Christine M. Schonewald-Cox and others, pp. 164–184. Menlo Park, Calif.: Benjamin-Cummings.

Rensberger, Boyce. 1984. "Bones of Our Ancestors." *Science* 84 (Apr.): 28–39.

Resek, Carl. 1960. *Lewis Henry Morgan: American Scholar.* Chicago: University of Chicago Press.

Richard, Alison F. 1982. "The World's Endangered Species: A Case Study on the Lemur Fauna of Madagascar." In *Primates and the Tropical Rainforest,* ed. Russell A. Mittermeier and Mark J. Plotkin, pp. 23–30. Pasadena, Calif.: California Institute of Technology.

Richard, Alison F., and Robert W. Sussman. 1974. "Future of the Malagasy Lemurs: Conservation or Extinction?" In *Lemur Biology,* ed. Ian Tattersall and Robert W. Sussman, pp. 335–350. Pittsburgh: University of Pittsburgh Press.

Rijksen, Herman D. 1982. "How to Save the Mysterious 'Man of the Rain Forest'?" In *The Orang utan: Its Biology and Conservation,* ed. Leobert E. M. De Boer. The Hague: Dr. W. Junk.

Rijksen, Herman D., and Ans G. Rijksen-Graatsma. 1975. "Orang Utan Rescue Work in North Sumatra." *Oryx* 13 (Apr.): 63–73.

Robinson, P. T. 1971. "Wildlife Trends in Liberia and Sierra Leone." *Oryx* 11 (Sept.): 117–121.

Rodman, Peter S. 1977. "Feeding Behavior of Orang-utans of the Kutai Nature Reserve, East Kalimantan." In *Primate Ecology: Studies of Feeding and Ranging Behavior in Lemurs, Monkeys and Apes,* ed. T. H. Clutton-Brock, pp. 383–413. New York: Academic Press.

Ronan, Colin A. 1982. *Science: Its History and Development among the World's Cultures.* New York: Facts on File.

Rosenthal, Mark. 1983. "Lowland Gorillas at Lincoln Park Zoo." *IUCN/SSC Primate Specialist Group Newsletter* 3 (Mar.): 38, 39.

Rothwell, Norman V. 1979. *Understanding Genetics,* 2d ed. New York: Oxford University Press.

Rudran, R., and J. F. Eisenberg. 1982. "Conservation and Status of Wild

Primates in Venezuela." In *1982 International Zoo Yearbook,* vol. 22, ed. P. J. S. Olney, pp. 52–59. London: Zoological Society of London.

Rylands, Anthony B. 1985. "Conservation Areas Protecting Primates in Brazilian Amazonia." *Primate Conservation* 5 (Jan.): 24–27.

Rylands, Anthony B., and Russell A. Mittermeier. 1982. "Conservation of Primates in Brazilian Amazonia." In *1982 International Zoo Yearbook,* vol. 22, ed. P. J. S. Olney, pp. 17–37. London: Zoological Society of London.

———. 1983. "Parks, Reserves and Primate Conservation in Brazilian Amazonia." *Oryx* 17 (Apr.): 78–87.

Sabater Pi, Jorge, and Colin Groves. 1972. "The Importance of Higher Primates in the Diet of the Fang of Rio Muni." *Man* 7 (June): 239–243.

Sagan, Carl. 1977. *The Dragons of Eden: Speculations on the Evolution of Human Intelligence.* New York: Random House.

Sanchez, Pedro A. 1981. "Soils of the Humid Tropics." In *Blowing in the Wind: Deforestation and Long-range Implications,* ed. Vinson H. Sutlive, Nathan Altshuler, and Mario D. Zamora, pp. 347–410. Williamsburg, Va.: Department of Anthropology, College of William and Mary.

Schaller, George B. 1963. *The Mountain Gorilla.* Chicago: University of Chicago Press.

———. 1964. *The Year of the Gorilla.* Chicago: University of Chicago Press.

Schenkel, Rudolf, and Lotte Schenkel-Hulliger. 1967. "On the Sociology of Free-Ranging Colobus (*Colobus guereza caudatus* Thomas 1885)." In *Progress in Primatology,* ed. D. Starck, R. Schneider, and H. J. Kuhn, pp. 185–194. Stuttgart: Gustav Fischer Verlag.

Senner, John W. 1980. "Inbreeding Depression and the Survival of Zoo Populations." In *Conservation Biology: An Evolutionary-Ecological Perspective,* ed. Michael E. Soulé and Bruce A. Wilcox, pp. 209–224. Sunderland, Mass.: Sinauer Associates.

Simon, Noel, and Paul Geroudet. 1970. *Last Survivors: The Natural History of Animals in Danger of Extinction.* New York: World Publishing.

Smith, Nigel J. H. 1976. "Spotted Cats and the Amazon Skin Trade." *Oryx* 13 (July): 362–371.

Soini, Pekka. 1972. "The Capture and Commerce of Live Monkeys in the Amazonian Region of Peru." In *1972 International Zoo Yearbook,* vol. 12, ed. Joseph Lucas and Nicole Duplaix-Hall, pp. 26–36. London: Zoological Society of London.

———. 1982. "Primate Conservation in Peruvian Amazonia." In *1982 International Zoo Yearbook,* vol. 22, ed. P. J. S. Olney, pp. 37–47. London: Zoological Society of London.

Soulé, Michael E. 1980. "Thresholds for Survival: Maintaining Fitness and Evolutionary Potential." In *Conservation Biology: An Evolutionary-Ecological Perspective,* ed. Michael E. Soulé and Bruce A. Wilcox, pp. 151–169. Sunderland, Mass.: Sinauer Associates.

———. 1983. "What Do We Really Know about Extinction?" In *Genetics and Conservation: A Reference for Managing Wild Animal and Plant Populations,* ed. Christine M. Schonewald-Cox and others, pp. 111–124. Menlo Park, Calif.: Benjamin-Cummings.

Soulé, Michael E., Bruce A. Wilcox, and Claire Holtby. 1979. "Benign Neglect: A Model of Faunal Collapse in the Game Reserves of East Africa." *Biological Conservation* 15: 259–272.

Southwick, Charles H., M. Rafiq Siddiqui, and M. Farooq Siddiqui. 1970. "Primate Populations and Biomedical Research." *Science* 170 (Nov.): 1051–1054.

Sparks, John. 1982. *The Discovery of Animal Behavior*. Boston: Little, Brown.

Struhsaker, Thomas T. 1969. "Correlates of Ecology and Social Organization among African Cercopithecines." *Folia Primatologica* 11: 80–118.

———. 1972. "Rain-forest Conservation in Africa." *Primates* 13 (Mar.): 103–109.

Sugiyama, Yukimaru. 1968. "The Ecology of the Lion-tailed Macaque (*Macaca silenus* [Linnaeus]) — A Pilot Study." *Journal of the Bombay Natural History Society* 65 (Aug.): 283–293.

Sussman, Robert W., Alison F. Richard, and Gilbert Ravelojaona. 1985. "Madagascar: Current Projects and Problems in Conservation." *Primate Conservation* 5 (Jan.): 53–59.

Tardif, Suzette D. 1985. "Status of the Cotton-top Tamarin (*Saguinus oedipus*) in Captivity." *Primate Conservation* 6 (July): 38, 39.

Tardif, Suzette D., and Rob Colley. 1987. *International Cotton-top Tamarin Studbook*. Oak Ridge, Tenn.: Oak Ridge Associated Universities.

Tattersall, Ian. 1982. *The Primates of Madagascar*. New York: Columbia University Press.

Taylor, Norman. 1945. *Cinchona in Java*. New York: Greenberg.

Teleki, Geza. 1980. "Report from Sierra Leone." *TRAFFIC Bulletin* 2 (Sept./Oct.): 78.

Templeton, Alan R., and Bruce Read. 1983. "The Elimination of Inbreeding Depression in a Captive Herd of Speke's Gazelle." In *Genetics and Conservation: Reference for Managing Wild Animal and Plant Populations*, ed. Christine M. Schonewald-Cox and others, pp. 241–261. Menlo Park, Calif.: Benjamin-Cummings.

Tenaza, Richard R. 1975. "Territory and Monogamy among Kloss' Gibbons (*Hylobates klossii*) in Siberut Island, Indonesia." *Folia Primatologica* 24: 60–80.

———. 1987. "The Status of Primates and Their Habitats in the Pagai Islands, Indonesia." *Primate Conservation* 8 (Sept.): 104–110.

Tenaza, Richard R., and W. J. Hamilton III. 1971. "Preliminary Observations of the Mentawai Islands Gibbon, *Hylobates klossii*." Folia Primatologica 15: 201–211.

Tenaza, Richard R., and Arthur Mitchell. 1985. "Summary of Primate Conservation Problems in the Mentawai Islands, Indonesia." *Primate Conservation* 6 (July): 36, 37.

Terry, Joanne. 1985. "Infant Hand-rearing and Attempted Reintroduction." *Lion-Tales: Lion-tailed Macaque Newsletter* 2 (Spring): 2, 4.

Thayer, James Steel. 1983. "Nature, Culture and the Supernatural among the Susu." *American Ethnologist* 10 (Feb.): 116–132.

Thomas, Keith. 1984. *Man and the Natural World: A History of the Modern Sensibility*. New York: Pantheon Books.

Thorington, Richard W., Jr. 1972. "Importation, Breeding, and Mortality of New World Primates in the United States." In *1972 International Zoo Yearbook*, vol. 12, ed. Joseph Lucas and Nicole Duplaix-Hall, pp. 16–23. London: Zoological Society of London.

Thornback, Jane, and Martin Jenkins, eds. 1982. *The IUCN Mammal Red Data Book*, Part I. Gland, Switzerland: IUCN.

Index